External Field and Radiation Stimulated Breast Cancer Nanotheranostics

About the Series

Series in Physics and Engineering in Medicine and Biology will allow IPEM to enhance its mission to 'advance physics and engineering applied to medicine and biology for the public good.'

Focusing on key areas including, but not limited to:
- clinical engineering
- diagnostic radiology
- informatics and computing
- magnetic resonance imaging
- nuclear medicine
- physiological measurement
- radiation protection
- radiotherapy
- rehabilitation engineering
- ultrasound and non-ionising radiation.

A number of IPEM–IOP titles are published as part of the EUTEMPE Network Series for Medical Physics Experts.

External Field and Radiation Stimulated Breast Cancer Nanotheranostics

Edited by
Nanasaheb D Thorat and Joanna Bauer
Department of Biomedical Engineering, Wroclaw University of Science and Technology, Poland

IOP Publishing, Bristol, UK

ISBN   978-0-7503-2416-8 (ebook)
ISBN   978-0-7503-2414-4 (print)
ISBN   978-0-7503-2415-1 (mobi)

DOI   10.1088/2053-2563/ab2907

Version: 20190901

IOP Expanding Physics
ISSN 2053-2563 (online)
ISSN 2054-7315 (print)

British Library Cataloguing-in-Publication Data: A catalogue record for this book is available from the British Library.

Published by IOP Publishing, wholly owned by The Institute of Physics, London

IOP Publishing, Temple Circus, Temple Way, Bristol, BS1 6HG, UK

US Office: IOP Publishing, Inc., 190 North Independence Mall West, Suite 601, Philadelphia, PA 19106, USA

Dedicated to
Professor Syed Tofail University of Limerick, Ireland
and Professor S H Pawar, India

Contents

Preface

Around the world, breast cancer remains one of the most significant causes of morbidity and mortality. Every year, the numbers of newly diagnosed cases of breast tumors are growing. Nanomedicine approaches hold tremendous potential for early diagnosis and therapy for breast cancer. After 30 years of remarkable efforts, nanomedicine has achieved the ability to target chemotherapeutic cargoes to breast tumor sites and the capacity of crossing the anatomical barrier as well as accessing the breast after systemic administration. The nanomedicine approach is based on a new generation of 'hybrid' nanoparticles that have been designed as novel targeted delivery devices. Additionally, external physical controls such as light, magnetic field, sound waves and x-ray radiation have the ability to stimulate these hybrid nanoparticles for their actions on-demand. This externally or physically stimulated nanomedicine approach is advancing breast cancer theranostics (therapy + diagnostics).

Almost a decade after biomedical use of externally stimulated cancer nanomedicine it is interesting to see how much progress in this field has really advanced breast cancer theranostics. There has been great development in materials science, bioengineering, nanodevices and even in neuro-oncology. But how close are we to realizing the long-held dreams of this novel breast cancer theranostics approach? Among the various types of hybrid nanostructures, organic–inorganic, organic–organic, smart liposomes occupy an almost starring role in nanomedicine. Hybrid nanostructures have astonishing physical properties in terms of responsiveness to magnetic field, light, ultrasound etc, and their stability against chemical and physical degradation have led to researchers devising ever more ingenious biomedical applications. But will they have a role in breast cancer theranostics?

This book seeks to provide an up-to-date literature overview and information on the progress of externally stimulated breast cancer nanomedicine. The nanomedicine approach in breast cancer theranostics is centered around the use of nanovehicles for drug delivery. Novel hybrid nanostructures in targeted drug delivery and diagnostics continue to fascinate multidisciplinary nanomedicine laboratories. However, there are concerns about whether externally stimulated nanomedicine can ever be used in the clinical stage, considering the well-known doubts about possible nano-toxicity and control over physical field parameters. In the following chapters we summarize the externally stimulated nanomedicine modalities, the types of hybrid nanostructures and their properties. We discuss interactions of physical responsive systems to nanostructures and applications to drug targeting, diagnosis and breast cancer therapy. Finally, smart stimulus-responsive and breast tumor targeted approaches of nanomedicine are addressed.

Most importantly, the emphasis of the book is as much on in its collection of authors, and the need for convergence of multidisciplinary knowledge and skills in this emerging field. The authors contributing different chapters to the book are representatives of a few generations of scientists in this field. Additionally, the book offers first-hand accounts of the use of external field stimulated nanotheranostics tools in the area of cancer. It provides comprehensive information related to current developments in light, magnetic, ultrasound and radiation techniques used in cancer therapies and the basic principles used in connection to nanotechnology. Overall the book disseminates the contributors' respective experiences and expertise with both fundamental concepts and practical utilizations of external field stimulated materials to the wider scientific community, oncologists, and medical device manufacturing industrialists.

This book collects the understanding of external filed stimulation-response of a wide range of nanomaterials that has been synthesized over the past three decades and how this understanding has potentially led to cancer theranostics. It brings together contributors from pioneering scientists in the field with a view to promoting interest from readers to find and exploit cancer nanotheranostics. It also provides published research outlines and suggests directions in which future research could go.

The highly interdisciplinary subject matter of this book goes far beyond the areas of expertise of the editors. We would like to express our deep gratitude to all authors for their contributions to the book, compliance with the guidelines and their patience with the onerous editorial process. We would like to thank Professor Syed Tofail, Professor Tewfik Soulimane, Professor Christophe Silien, and team members of MOSAIC group of University Limerick, Ireland for their valuable suggestions. The publisher and senior Commissioning Editor Jessica Fricchione, Editorial Assistants Poppy Emerson and Daniel Heatley must be praised for their active work and support. Finally, we would like to thank friends and family members, Ms Neeta, and our children Paulina and Ninad for their sacrifices and patience during the preparation of this book.

Acknowledgements

It is a moment of gratification and pleasure to look back with a sense of contentment at the long-traveled journey, to be able to recapture some of the fine moments, to be able to thank the funding sources, that made this project possible for the editors.

The project leading to this work has received funding from the European Union's Horizon 2020 research and innovation program under the Marie Sklodowska-Curie grant agreement 'NANOCARGO' 751903. Initial financial support from the Irish Research Council under the Government of Ireland Postdoctoral Fellowship-2015 GOIPD/2015/320 is acknowledged. This book is based on the scientific dissemination activity of COST Action CA 17140 'Cancer Nanomedicine from the Bench to the Bedside' supported by COST (European Cooperation in Science and Technology). Socrates-Erasmus Bilateral Exchange between the Wroclaw University of Science and Technology, Poland and University of Limerick, Ireland (in the years 2017–2019) are also duly acknowledged. Support from Polish Ministry of Science and Education is duly acknowledged.

Editor biographies

Dr Nanasaheb D Thorat

Dr Nanasaheb Thorat is a Marie Curie Fellow (IF) at the Department of Biomedical Engineering at the Wroclaw University of Science and Technology, Poland. Dr Thorat did his Master's degree in Physics in the year 2008 in the Department of Physics, Shivaji University Kolhapur, India. In 2010, he joined the Center for Interdisciplinary Research, D Y Patil University, Kolhapur, India for doctoral studies under the guidance of Professor S H Pawar. He received his PhD in Physics with Gold Medal and was awarded the 'Certificate of Research Excellence' by D Y Patil University, Kolhapur for excellent contributions in research for the year 2013–2014. After completion of his doctoral studies in 2014, he joined the Samsung Biomedical Research Institute, Sungkyunkwan University, Suwon South Korea as a senior researcher. In 2015, he was awarded an Irish Research Council Fellowship at the University of Limerick, Ireland with Professor Tofail Syed (www.mosaicteam.eu). In 2018 Dr Thorat received the European Commission's H2020 Marie Skłodowska-Curie Fellowship (IF) in Poland/Switzerland. He is a recipient of various other prestigious fellowships including, Japanese Society for the Promotion of Science (JSPS) Fellowship in Japan 2017, Government of Ireland IRC fellowship in Ireland 2015, Government of Israel PBC Outstanding Fellowship in Israel 2015, Science and Engineering Research Board, Government of India Overseas Fellowship 2016. He has also received a second Marie Curie Fellowship at the University of Oxford to start in 2020.

Dr Thorat's contributions to the area of nanobiotechnology and theranostics have been recognized internationally through invited and keynote lectures in prestigious and exclusive scientific symposia and fora. Dr Thorat was also involved in an innovation partnership research project that involved three world leading pharmaceutical companies Jansen, GSK, *Pfizer*. His work involved upskilling to spectroscopic techniques such as x-ray photoelectron spectroscopy and Raman spectroscopy to probe the drug/material interface.

Dr Thorat is currently serving as a Management Committee member of the European Commission's Horizon 2020 Cooperation in Science and Technology (COST) Programme. Dr Thorat is an active member of various scientific societies such as European Materials Research Society (EMRS), International Magnetic Society, European Nanomedicine Society and many others.

Dr Thorat has interests in outreach and volunteering activities in various non-scientific platforms where he elaborates the societal impact of his and others' laboratory-based research to the wider community. He is thus *au fait* in communicating with members of both the scientific and non-scientific communities, from Lindau Nobel Laureate meetings to American Chemical Society and to European Materials Research Society symposiums to popularization of analytical science to school children in the Royal Society of Chemistry (RSC) Spectrosocopy in a Suitcase program.

Dr Thorat has a strong record of collaborative research in science. In addition to his current research at Wroclaw University of Technology, Poland and University of Zurich, Switzerland, Dr Thorat has been deeply engaged in collaborative work with eminent scientists from the UK, Japan, USA, Germany, Korea, Ireland, India, and Australia. Of particular note is his collaboration with Professor Helen Townley, University of Oxford, Professor Yususke Yamauchi, (Editor: *APL Materials*) NIMS, Japan, and has published papers on cancer nanotheranostics and chemoresistance in cancer. He has also had an excellent collaboration and interactions with over 30 junior and senior colleagues in India, South Korea, Japan, Taiwan and Ireland with whom he has co-authored his publications.

Dr Thorat has published 45 peer reviewed journal research papers and articles, three books and six book chapters; H index: 25, total citations: 1400. He has presented five keynote speeches, five invited talks and five oral presentations at prestigious scientific peer-conferences, received international acclaim and awards for research contributions, generated research funds in excess of > €550 000, supervised students/junior researchers and actively participated in outreach and scientific dissemination for the service of the wider community.

Professor Joanna Bauer

Dr Joanna Bauer is an Assistant Professor at the Department of Biomedical Engineering at the Wroclaw University of Science and Technology, Poland. She received her PhD in Applied Physics with Honors, as well as an MSc Eng. in Biomedical Engineering with Honors and an MSc Eng. in Management and Marketing from the same university. She is an experienced academic researcher specializing in multidisciplinary biomedical engineering projects related to early diagnosis, personalized and preventive medicine, nano-biomaterials functionalization and characterization, infrared imaging, as well as pattern recognition and image analysis. Dr Bauer is an author and co-author of more than 90 publications, 60 conference presentations, including over 20 plenary and invited lectures. She has over 17 years' experience in carrying out national and international scientific projects within programmes such as State Committee for Scientific Research (KBN), Deutscher Akademischer Austauschdienst (DAAD) European Social Funds (ESF), FP6, FP7 and H2020. Dr Bauer has also been a Research Fellow and Visiting Scientist in a number of scientific institutions worldwide such as Charite Medizin University of Berlin, Germany; Laser-und Medizin-Technologie GmbH, Berlin, Germany; Ural Federal University, Russia; University Politehnica of Bucharest, Romania; University of Limerick, Ireland; Comenius University, Slovakia, University of Aveiro, Portugal. Both her publication and research profile show her ability in driving innovation and outstanding scientific breakthroughs in the highly challenging field of bioengineering. This was confirmed by many industrial/scientific-excellence awards including for example Siemens Research Prize for Outstanding Achievements in Technology and Science, Appreciation Certificate from the European Association for Predictive, Preventive and Personalized Medicine, Development Award in Technical Sciences for Young Distinguished Scientists from the Polish Minister of Science and Education and Distinction of the Polish Society for Biomedical Engineering.

Contributors

Dr Sachin Otari

Dr Sachin V Otari received his Master's degree in 2009 from the Department of Microbiology, Shivaji University Kolhapur, India. In 2010, he joined Center for Interdisciplinary Research, D Y Patil University, Kolhapur for doctoral studies under the guidance of Professor S H Pawar and Professor S J Ghosh to study the interaction of nanoparticles with biological materials for possible application in biomedicine. During his PhD program he was awarded the 'Certificate of Research Excellence' by D Y Patil University, Kolhapur for the excellent contribution in research area for the year 2013–2014 and 2014–2015. After completion of his doctoral studies in the year 2015, he joined the Department of Chemical engineering, Konkuk University, Seoul South Korea with a Brain Pool Fellowship as a research assistant professsor. During the three years of the fellowship, he worked on the synthesis of various forms of nanoparticles for the immobilization of peptides, enzymes, whole cells for industrial application. Since 2018, he has been working as a post -doctoral fellow in the Department of Chemical Engineering, Konkuk University, Seoul South Korea under BK-21 projects, working on the modification of enzymes with nanoparticles in the production of value added products. Dr Otari has published more than 30 research articles in high impact international journals such as *Chemical Engineering Journal, Energy, Scientific Reports, Journal of Hazardous Materials* etc. He has also contributed three book chapters published by reputed publishers such as the Elsevier book entitled *Hybrid Nanostructures for Cancer Theranostics*.

Dr Rakesh Patil

Dr Rakesh M Patil (MSc, PhD), is an active researcher. He completed his post graduation (MSc) in Biochemistry from Shivaji University, Kolhapur and PhD in Biochemistry from D Y Patil University, Kolhapur, India. He qualified in the Graduate Aptitude Test in Engineering (GATE) and CSIR-National Eligibility Test (NET) in life sciences in 2010 and worked as a Junior Research Fellow (JRF) on a DST, India funded project. Dr Patil has published 18 peer reviewed journal articles and one book chapter (total citations ~650, H index:16). He is presently working in the Regional Forensic Science Laboratory, Kolhapur as a Gazetted Officer and earlier he worked on DNA Fingerprinting in the Directorate of Forensic Science Laboratories, Kalina, Santacruz (E), Mumbai. His current scientific interests include DNA fingerprinting, fabrication of iron oxide nanoparticles and their core shells for biomedical applications. In particular, his research is aimed at theranostic applications of these nanoparticles and their core shells.

Dr Madhuri P Anuje

Dr Madhuri Anuje completed her Masters degree in Medical Physics from D Y Patil University, Kolhapur, India and one-year internship in Medical physics from KLEs Belgaum Cancer Hospital Belgaum, Karnataka, India. She is a certified Radiological Safety Officer level III (RSO-III) Conducted by Bhabha Atomic Research Centre (BARC), Mumbai, India. She has two and half years of teaching experience in Medical Physics at D Y Patil University, Kolhapur, India. Currently she is working as Medical Physicist/Radiological Safety Officer at BSDT's Integrated Cancer Treatment and Research Centre, Wagholi, Pune, Maharashtra, India. She is registered for PhD in D Y Patil University, Kolhapur, India and is working on the topic 'Role of Nanoparticles as a Radiotherapy Sensitizer'. Her current research work involves improving radio-therapy using nanoparticles as a sensitizer.

Dr Abdul K Parchur

Abdul K Parchur received MSc and PhD degrees in physics from Dr Hari Singh Gour University, India, in 2010. He is currently a Researcher with the Department of Radiology, The Medical College of Wisconsin, Milwaukee, WI, USA. His research involves the development of image-guided theranostic nanoparticles for preclinical therapeutic interventions. He has authored or co-authored over 20 peer reviewed publications and three book chapters that have been cited more than 940 times with an H-factor of 15. His research interests span the field of nanomedicine with a special emphasis on plasmonic nanomaterials, cancer theranostics, near infrared imaging, magnetic resonance imaging, and computed tomography imaging.

Dr Vijaykumar Jadhav

Dr Vijaykumar Jadhav is a Research Scientist in the Material Science and Engineering Department at GTIIT in 2019. He was a recipient of a Government of Ireland Postdoctoral Fellowship in 2016 to develop advanced materials for 3D Li-ion batteries. He was awarded Young Scientist through Lindau Nobel Laureate Meeting, Lindau, Germany. He is an active reviewer for ACS, Elsevier, RSC, and Springer journals. He has published 32 peer reviewed journal articles and one book chapter. He so far has delivered three Invited Talks and two oral presentations at prestigious scientific peer-conferences with international acclaim and awards, generated research funds in excess of >€ 350 000, supervised students/junior researchers, and actively participated in outreach and scientific dissemination for the service of the wider community. He has had several collaborations with eminent scientists across the globe including Australia, Saudi

Arabia, Korea, China, Ireland, Poland, and India. He also has had excellent collaboration and interactions with over 25 junior and senior colleagues in India, South Korea, China, and Ireland with whom he has co-authored his publication. As an asset Dr Jadhav successfully qualified in the National eligibility test (NET) with 364th rank, conducted by CSIR, in Physics, New Delhi, India which helped him to get PhD admission which is also one of the essential eligibility criteria for Assistant Professor Position in various national universities in India. He has demonstrated outstanding ability as an independent researcher, great teamwork and international network in six countries on three continents (Asia, Europe, and Australia). Embracing several countries and cultures is important in someone, looking to dedicate a life to exploration in science and technology.

Dr Nitesh Kumar

Dr Nitesh Kumar completed his Masters in Pharmaceutical Sciences specializing in the field of Pharmacology in 2008. He was awarded his doctoral degree in 2013. He is currently working as Assistant Professor-Selection Grade in the Department of Pharmacology, Manipal College of Pharmaceutical Sciences, MAHE, Manipal, India. As well as being an active academician, he is a sincere researcher, which is reflected by his 51 research publications in a variety of national and international journals of repute with an H-index of 14 as per Scopus. His areas of interest include development of nanoformulation (liposomes), pharmacokinetic study of drugs and formulation, screening molecules for their anticancer activity, neurological profile and hepatoprotective action of drugs and their associated formulations. He has received a total of four awards for best paper/ poster at various conferences. He has completed research projects worth more than 1 crore rupees from various funding agencies including government and company sponsored at various capacities such as principal investigator/co-investigator/key investigators. He is a dedicated team player as can be seen in his publications where the number of authors are greater than five. Currently, he is the guide for three research scholars and co-guide for nine research scholars pursuing their doctoral works.

Ahmaduddin Khan

Ahmaduddin Khan received an MSc in Nanoscience and Nanotechnology from Babasaheb Bhimrao Ambedkar University, India in 2015. Currently he is pursuing a PhD from Vellore Institute of Technology (VIT) in the field of magnetic nanoparticles mediated drug delivery for cancer therapy. His research area includes the synthesis of functionalised magnetic nanoparticles, drug delivery, hyperthermia, *in vitro* as well as *in vivo* studies.

Dr Niroj Kumar Sahu

Niroj Kumar Sahu obtained his Master of Science (Physics) from Sambalpur University, Master of Technology (Ceramic Engineering) from NIT, Rourkela and PhD (Materials Engineering) from IIT Bombay, India. He is currently working as an Associate Professor in the Centre for Nanotechnology Research, VIT, Vellore, India. His research area includes Physics and Chemistry of nanomaterials, development of functionalised magnetic nanomaterials for various biomedical applications, magnetic hyperthermia, thermo-chemotherapy, and nanomaterials for energy storage.

Dr Rohini D Kitture

Dr Rohini D Kitture obtained her Masters in Physics from Shivaji University, Kolhapur and PhD in Physics from University of Pune, India. After pursuing her doctorate, she joined a flagship project funded by Defence Research and Development Organization (DRDO), Government of India, at Defence Institute of Advanced Technology (DIAT), Pune, India. During her tenure of more than four years in DIAT, she has submitted a drug delivery prototype to Defence Bioengineering and Electromedical Laboratory (DEBEL), India for scaling up, filed a patent on antimicrobial formulation. She has published 30+ highly cited research articles in various international journals like *Applied Physics Letters*, *IEEE Sensors*, *Journal of Alloys and Compounds, International Journal of Nanomedicine*. Besides international journals, she has also contributed to some international books, published by Springer, Elsevier etc. She has addressed various International Conferences, including the Frontier Forum Meeting on Nanoscience 2017 at Dubrovnik, Croatia, ICNPM 2015 at Kottayam, Kerala, First Nano-Bio-Med 2011, Trieste, Italy. She has also played an active role as organizing committee member/secretariat of some reputed international conferences, including the Third Nano-Bio-Med 2015, at IIT Bombay, under the auspices of International Center for Theoretical Physics, Italy and IIT Bombay.

She is a recipient of International Fellowship-Visiting Scientist by Karlsruhe House of Young Scientists (KHYS), Karlsruhe Institute of Technology, Karlsruhe, Germany. Recently she has joined Springer Nature Technology and Publishing Solutions, as a Senior Scientific Writer, contributing to the Nanomaterials database, highly beneficial for researchers across the globe. Her research interests include, but are not limited to, functionalized nanomaterials for various healthcare applications like diagnostics, therapeutics, sensing, metamaterials and optical fibers for various selective sensing applications.

Dr Arpita Pandey Tiwari

Dr Arpita Pandey Tiwari has completed her research in the area of Nanobiotechnology from D Y Patil University Kolhapur. She completed her post graduation from Punjabi University, Patiala in the subject of Human Genetics. Her research interests include fabrication of nanomaterials and their use in various biomedical applications. She has more than nine years of research experience and four years of teaching experience. Currently she is working as Assistant Professor in Department of Stem cell and Regenerative medicine, D Y Patil University, Kolhapur.

Dr Sonali S Rohiwal

Dr Sonali S Rohiwal received her Master's degree in 2010 from Department of Biochemistry, New Arts Commerce and Science College, Ahmednagar, India. In 2011, she joined Center for Interdisciplinary Research, D Y Patil University, Kolhapur for doctoral studies under the guidance of Professor S H Pawar and to study synthesis and characterization of bovine serum albumin nanoparticles and their conjugates with dextran for wound healing potential. After completion of her doctoral studies in 2015, she joined Indian Institute of Technology, Kharagpur, West Bengal as senior research fellow with department of biotechnology fellowship. Since 2016, she has been working as postdoctoral fellow at Laboratory of Cell Regeneration and Plasticity, Institute of Animal Physiology and Genetics, The Czech Academy of Sciences, vvi. Libechov, Czech Republic, working on nanofiber membranes for retinal disease and nano-particles for genome editing applications. Dr Sonali has published 13 research articles in high impact international journals. She also contributed to one book chapter. She has generated a research fund of €54 000, supervised students/junior researchers and actively participated to surpass in the scientific community.

IOP Publishing

External Field and Radiation Stimulated Breast Cancer
Nanotheranostics

Nanasaheb D Thorat and Joanna Bauer

Chapter 1

Introduction to external field stimulation modalities

Nanasaheb D Thorat, Rakesh M Patil, Sachin Otari and Joanna Bauer

A number of external field stimulators relying on physical energy forces are currently being used to activate and enhance the targeted drug delivery platform to specific locations, such as cancerous tumors, within the targeted organs. The widely explored physical stimulators are: light, magnetic fields, ultrasound, electric fields, radio-frequency, x-ray, temperature gradients and mechanical forces. Physical stimulation strategies are primarily focused on advancing the remote controlling rate and duration of drug delivery, while targeting explicit locations of the body (such as tumors) for therapy, to deliver the chemotherapeutic cargoes into the tumors. This non-conventional drug delivery approach has been coupled with nanotechnology to develop 'smart drug carriers' in recent years. Physically stimulated cancer therapeutics have great potential as an advanced targeted and personalized medicine approach that can reduce the present limitations and unwanted side effects. Furthermore, stimuli responsive nanomedicine has offered an alternative to conventional cancer therapeutics platforms on account of its spatiotemporally controllable properties. This chapter aims to cover a number of such physical triggering modalities and their basic underlying physics. Several advanced techniques such as phototriggering, phototherapy magnetic guiding, thermotherapy, sonodynamic therapy or sonoporation, will be covered in the chapter. Special emphasis will be placed on nanotechnology advanced physical stimulation strategies owing to the experience of the authors' laboratory in this area.

1.1 Introduction

Over the past decade, extensive research on cancer has been ongoing, unfortunately only a few reports have led to changes in patient outcome. Cancer is indeed characterized by an extremely poor prognosis and has only improved a little despite surgery, chemotherapy and radiotherapy. Development of the next generations of cancer therapy modalities has thus been a crucial requirement of oncology. Among the many different novel cancer therapeutic strategies that have been explored in the last two decades, therapies that are responsive to external physical stimuli coupled with nanoparticles (NPs) that are capable of responding to light, magnetic fields, ultrasound, radio-frequency, or x-ray, have attracted substantial interest from many researchers across the world [1–3]. The external field triggered nanotheranostics (therapy + diagnostics) approach can allow precise control of the drug release, timing, intensity, duration of drug delivery and treatment monitoring. Such regulatory mechanisms are important to allow the patient, team of clinicians or even an algorithm to modulate the treatment response to match the actual needs of the patient, however, it requires a preplanned treatment procedure to be followed [4].

Extraordinarily, external field stimulated-responsive breast cancer therapies have facilitated therapeutic efficacies and reduced side effects in numerous pre-clinical animal studies. This therapeutic approach is found to be advantageous over conventional chemotherapy and radiotherapy, though their clinical solicitations are still mostly at primary stages. (1) Several external field stimulated (EFS) therapies can spatially regulate the therapeutic outcome only in the diseased region (e.g. breast tumor) by focusing physical stimuli specifically on the tumor without triggering loss of normal tissues. (2) Nearly all EFS therapies are upgrading the effect of therapeutic drugs in the tumor region to improve the therapeutic efficacy. (3) Certain types of EFS such as magnetic field and ultrasound may be able to control drug administration and release in the tumor, greatly improving the therapeutic outcome and reducing their systemic toxicity. (4) Many EFS-responsive therapies not only directly kill cancer cells, but also trigger or enhance adjunct cancer therapies to achieve the desired synergistic therapeutic effects via different mechanisms.

The concept of the EFS-responsive nanomedicine approach is satisfying General George S Patton's statement 'A good battle plan that you act on today can be better than a perfect one tomorrow,' by facilitating early cancer diagnosis along with a perfect therapeutic outcome [5]. It is crucial to detect cancer as early as possible in its pre-malignant/malignant form, and recently EFS-responsive nanomedicine has been developed to address this issue. A potential advantage of EFS systems is that they do not demand a specific ligand or tissue property to facilitate targeting: triggered therapeutic release will occur wherever the energy source is directed. In this chapter, we summarize the updated state-of-the-art research on different types of EFS modalities developed for remotely-triggered cancer theranostics (figure 1.1). The clinical relevance of each modality for breast cancer theranostics will then be covered in individual chapters in the book. Furthermore, the chapter will also address the concept of hybrid nanostructures that are coupled with EFS for cancer theranostics. Even though different EFS-responsive hybrid nanostructures might have been discussed in some review articles [3, 6, 7], herein, we will only summarize each EFS modality coupled with various types of hybrid nanostructures, which are indeed designed for remotely-triggered breast cancer theranostics. Moreover, special insights will be, as far as we know for the first time, dedicated to the underlying physics and physical mechanisms of the remotely-triggered release from those delicately-designed hybrid nanostructures.

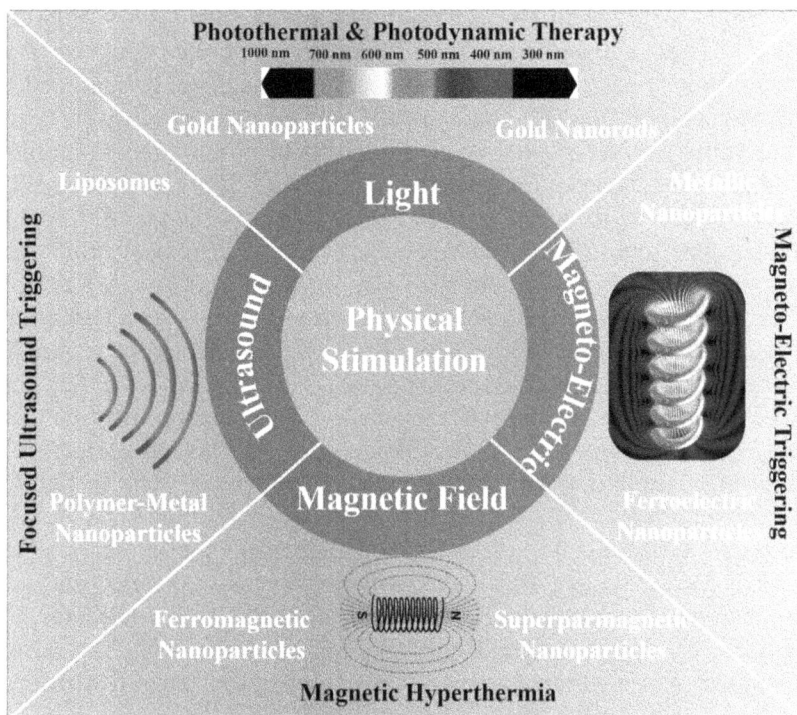

Figure 1.1. External field modalities coupled with nanomedicine to treat cancer.

1.2 External field stimulation modalities

Several physical energy sources have been used to trigger hybrid nanostructures. In recent times various research groups have invented novel ways to utilize nanoscale hybrid materials that are capable of responding to external physical stimulus such as (i) light (ii) magnetic fields (iii) ultrasound (iv) radiofrequencies (v) x-rays. Novel hybrid nanostructures are highly sensitive to external stimulus and can perform 'on demand' cancer theranostics [8]. The term 'theranostics' symbolizes a new class of treatment options that are capable of providing both therapy and diagnostics to patients [9]. Currently, light, magnetic field and ultrasound are widely being studied as external stimulation modalities. In recent years, nanomedicine combined with radio-frequency, x-ray and electric field have also been proposed as novel physical stimulation strategies.

As of today, comprehensive EFS modalities are engineered and designated that are proficient as remotely controlled cancer theranostics modalities. These multi-modalities are responding to more than one stimulus all together in both internal and external means. EFS relies on smart hybrid nanostructures and facilitates an appropriate way for incorporated chemotherapeutic cargo activation in the tumor. Furthermore, after local internalization EFS-responsive hybrid nanostructures on demand release can be achieved with enhancing chemotherapeutic action [10]. All the EFS modalities including magnetic fields, ultrasound, light and temperature gradients can act as enhancers for chemotherapeutic actions of hybrid nano-structures [11]. Every single form of physical energy has been associated with novel EFS techniques in cancer treatment; for example, a magnetic field is used to treat cancer by magnetoporation [12] and magnetic field drug targeting in the tumor [13], electric current/potential for electroporation [14] and iontophoresis [15] in cancer, ultrasound for sonodynamic therapy [16] and sonoporation [17], pulsed light for optoporation and drug release inside the tumor [18], and temperature for thermo-poration and hyperthermia therapy of the tumor [19]. EFS has several advantages over the conventional cancer nanomedicine approach such as: (i) EFS activated on demand or remotely controlled chemotherapeutic cargoes and enhances the triggered drug delivery to tumor, (ii) EFS coupled with magnetic nanoparticles (MNPs) induce hyperthermia ablation of cancerous tissue, coupled with metallic gold NPs can induce photo-hyperthermia, (iii) many hybrid nanostructures can enhance radiotherapy in a clinical setting under EFS, and (iv) EFS such as light, magnetic and ultrasound stimulation can permeate the blood–brain barrier (BBB) using hybrid nanostructures.

As far as EFS theranostics are concerned, nanomedicine or nanotheranostics, however, may have a unique advantage. Unlike conventional cancer therapies, for example chemotherapy which applies to the whole body, EFS nanotheranostics are often localized or have a targeted approach on the tumor [8]. Not only tumor therapy, but also real time monitoring of chemotherapeutic drugs distributed in the tumor or whole body in the course of treatment is vital for maximal therapeutics outcome. Ongoing chemotherapy clinical protocols have a very narrow monitoring window before and after the treatment. On the other hand, physically stimulated nanotheranostics either light, magnetic field, ultrasound, or others allows the real monitoring window of distribution of therapeutic agents. EFS or radiation stimulated nanotheranostics is opening new windows for real time monitoring when it is combined with imaging techniques. Furthermore, EFS nanotheranostics in combination with radio imaging modalities such as magnetic resonance imaging (MRI) or computed tomography (CT) is accommodating the monitoring of drug doses and therapeutic responses.

1.3 External field stimulation modalities for cancer theranostics

1.3.1 Magnetically stimulated cancer theranostics

Magnetic fields (MF) are non-invasive in nature and have excellent human tissue penetration. Owing to this characteristic of MF, it has been used in clinics for whole body MRI. In recent years, MF has been used to activate drug delivery into the tumor region through field dependent thermal and non-thermal effects. High frequency (>10 kHz) alternating magnetic fields (AMF) responsive MNPs can generate heat by utilizing different physical mechanisms. Magnetic field stimulation of MNPs to specific regions, especially tumors within the body, has interested researchers for over 30 years. The report on magnetic guidance of MNPs to a particular location of the body was published by Freeman *et al* in the 1960s, and the evidence of MNP accumulation under an additional MF have been witnessed [20].

In the following years, many MNPs were demonstrated for magnetic tumor targeting. The first clinical trial on 14 cancer patients of chemotherapeutic drug conjugated MNPs and magnetic targeting was performed by Lübbe *et al* in 1996 [21]. For the last two decades, the exchange of MF amplitude or energy into a localized non-invasive spatial heating has been investigated to develop magnetic hyperthermia against cancerous tumor. In the meantime, the external applied magnetic field (AMF)-generated magneto-thermal effect was exploited as a relevant external stimulus to trigger a chemotherapeutic cargo release from magnetic nano-carriers. The combination of MFH and MF stimulated drug release was envisioned as a complementary and encouraging therapeutic modality 'magneto-chemother-apy' [22–24]. Now, the basic physical principles of radio-frequency enabled AMF heating, namely MFH and tumor imaging modality MRI are quite well developed. The field strength and frequency that have been used in these systems have diverged significantly. Furthermore, the penetration depth of the MF decreases as the frequency increases. Generally, to avoid non-specific heating of surrounding tissues at the time of MH or MRI, the product of the field frequency and intensity of the AMF should not exceed 4.85×10^8 A m^{-1} s^{-1} (figure 1.2). Magnetically guided drug delivery can be activated mechanically by low frequency or static fields [4].

The next generation EFS modalities based on magnetic guiding also utilizes the mechanical forces generated via MF termed as 'magneto-mechanical' effect. Magneto-mechanics is currently being used as a powerful manipulation tool of

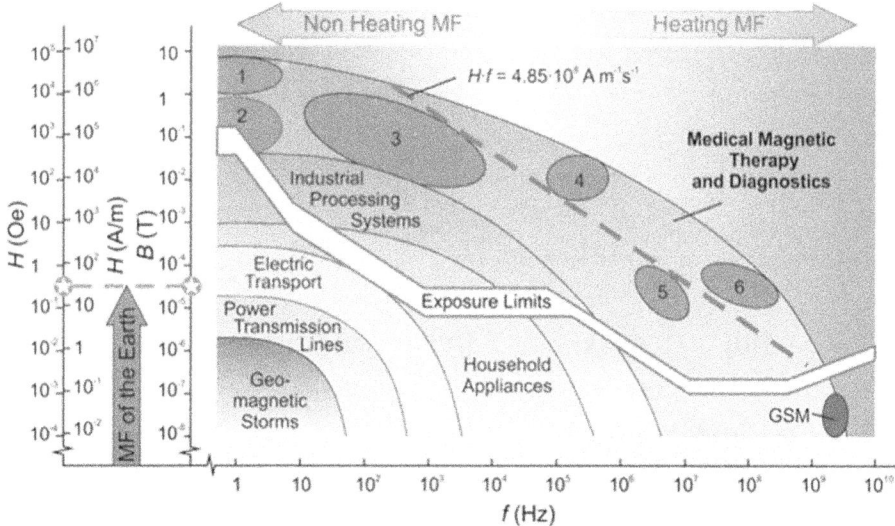

Figure 1.2. The map of natural and technogenic magnetic fields. The broken red lines represent the exposure limits for the fields according to the International Commission on Non-Ionizing Radiation Protection guidelines (ICNIRP Guidelines 1998). The upper line shows the limit for occupational exposure, and the lower line is for general public exposure. The dashed line represents the condition Hf = 4.85×10^8 A m^{-1} Hz^{-1}. 1 MRI (steady MF); 2 magnetostatic therapy; 3 pulse magnetic therapy; 4 radio-frequency magnetic hyperthermia (RFMH); 5 high frequencies inductive heating; 6 MRI (RFMF). Reproduced with permission from reference [25]. Copyright © Springer Science+Business Media Dordrecht 2017.

MNPs inside variable tumor micro environments (figure 1.3). Furthermore, a combination of dynamic AMF or static MF arrangements with MNPs allows conversion of electromagnetic to mechanical energy [26]. This novel EFS approach renders MNPs, useful in cancer research to damage malignant cell membranes and incur cell damage under exerting magneto-mechanical stresses of functionalized MNPs specifically bound to cell surfaces.

The next generation cancer theranostics approach used for interventional oncology allows the diagnosis and treatment of cancer using targeted minimally invasive procedures performed under image guidance. It employs mainly MRI, CT, x-ray and ultrasound imaging to monitor the targeted and precise cancer treatment of solid tumors located in various organs such as the breast, lung, brain and colon in the human body. Recent advances in EFS allow minimally invasive MRI guided detection, characterization, targeting and therapy of deeply situated solid tumors. Thus, the new generation image guided cancer theranostics involves (a) preprocedure planning (pinpointing tumor volume); (b) intraprocedural targeting (magnetic guiding using MNPs); (c) intraprocedural tumor monitoring during the treatment; (d) intraprocedural control; and (e) postprocedure treatment assessment [27].

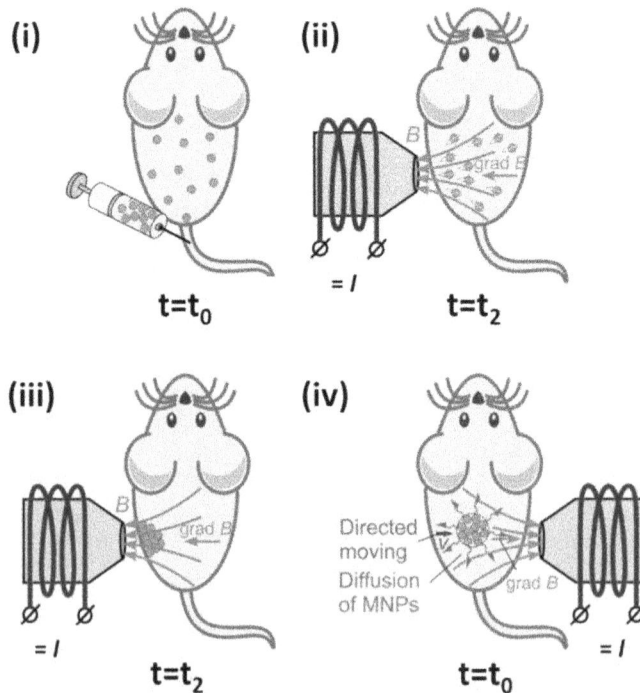

Figure 1.3. (a)–(d) The scheme for non-invasive formation of an f-MNPs 'cloud' in the internal areas of an organism via two alternatively switched gradient magnetic fields on both sides of the object. Reproduced with permission from reference [25]. Copyright © Springer Science+Business Media Dordrecht 2017.

1.3.2 Light stimulated cancer theranostics

Light is a fundamental part of the life of every living organism on Earth, for example many plants and animals employ specific types of optical waveguides to facilitate photosynthesis and visual acuity using sunlight [28]. Photosynthesis processes initiated in plants and small organisms is a good example of external field stimulation to convert sunlight energy to chemical energy. The modern practice of light in clinical medicine began in the early nineteenth century, with prompt developments in the understanding of both the physical nature of light and fundamental light–matter interactions (figure 1.4). Light activated cancer theranostics harness the versatility of light in manipulating photoactive molecules, hybrid nanostructures and polymers. New generation light activated cancer therapies used in nanomedicine require a properly selected light source. Today, photodynamic

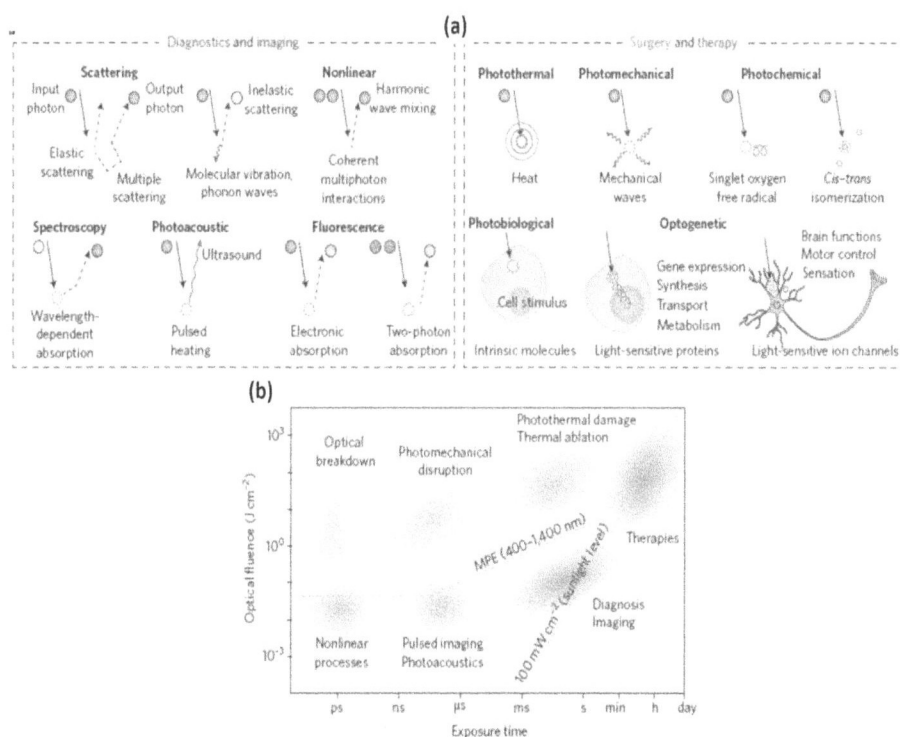

Figure 1.4. Light–tissue interactions. (a) Representative optical mechanisms used in diagnosis and imaging (left), and surgery and therapy (right). Circular objects denote incoming and outgoing photons, with their trajectories indicated by solid and dotted arrows. The colors of the circles represent the spectrum of the light; dotted circles indicate the absorption of input photons. For the case of therapy, specific effects of light on tissue and cells are indicated. (b) Optical techniques mapped according to their optical fluence and exposure time (either total illumination time for continuous-wave light, or pulse duration for pulses). Colors represent medical areas: green for diagnostics, pink for surgery and blue for therapy. MPE, maximum permissible exposure. Reproduced with permission from reference [29]. Copyright © 2017 Macmillan Publishers Limited, part of Springer Nature.

therapy (PDT) and photothermal therapy (PTT) has a projected modality in oncology, and is well established in dermatology, ophthalmology, dentistry, cosmetics. The combined global markets for drugs and devices used in light activated clinical therapies was estimated to be US$630 million in 2014 [29].

Conventional PDT employs a range of illumination sources, including halogen, arcing lamps, lasers and light-emitting diodes to activate a photosensitizer (PS). However, the tissue penetration depth, absorption and scattering of visible light within tissues, narrows down the PDT outcome to superficial tumors. Utilizing light in the near-infrared (NIR) range (700–1100 nm) minimizes the light attenuation in solid tumors, and when combined with NIR-sensitive PSs such as hybrid nano-structures, can maximize the depth of treatment. Even under these optimal conditions, the transmission into tissue is limited to only 5–10 mm [30]. In addition to the advantageous properties of different nanostructures for PDT, largely hybrid nanostructures discussed in chapters 2 and 5, exhibit intrinsic photosensitizing properties. Most hybrid nanostructures can generate a long-lived triplet energy state by absorbing blue–violet light and promote the production of reactive oxygen species (ROS). Compared with small-organic molecule PS, the hybrid nanostruc-tures have much better photostability and less photo-degradation *in vivo*. These hybrid nanostructures serve as both type I and type II PDT agents since they are producing large amounts of ROS in an oxygen-dependent and oxygen-independent mode.

Since the early 1990s, hyperthermic therapy, for example photo mediated thermal therapy (PTT) using externally applied light to heat tumors, has been suggested as an adjunct to chemotherapy and radiotherapy. This light activated adjunct therapeutics has an enhanced treatment response for various tumors including breast cancer [31]. In more recent years, the use of various nano thermal agents to selectively localize heat at localized anatomical sites has shown considerable potential. Strongly light responsive hybrid nanostructures, such as gold nanorods, can induce thermal ablation of tumor cells. Gold nanorods absorb NIR light through the surface plasmon resonances effect and have gone through various clinical trials for breast, lung and head and neck tumors [32]. In recent times, different types of hybrid nanostructures have been designed to combine PDT and PTT together. The combination remarkably improved breast cancer response to treatment at pre-clinical stage [33].

Light stimulation has been used widely to control the release kinetics of chemo-therapeutic drugs in a spatio-temporal manner at irradiated sites. This external remote light stimulus can minimize the initial burst of drug at non-target sites before light exposure [34]. Owing to sufficiently deep penetration of NIR light, drug release rates can be accurately controlled by NIR light intensity and exposure duration. Recently novel organic–inorganic hybrid nanostructures have been developed for high therapeutic efficacy for cancer treatment [35]. Importantly, the hybrid nano-structure based light activated drug delivery system is completely harmless and degradable *in vivo* and is a unique concept for precision cancer therapy to release cancer drugs. The other benefit of light activation is the precise control on premature release of drugs before reaching the tumor. To prevent such premature drug release light-switchable nanoassemblies are coated with cells. Cancer chemotherapeutics require sustained drug delivery to tolerate side effects, light-switchable spatio-temporal release could enable the pulsatile delivery of chemotherapeutics without side effects. Light triggered spatio-temporal release of drugs, proteins and other biomolecules using light-switchable systems implanted beneath the tumor might be an area that warrants future development.

1.3.3 Ultrasound stimulated cancer theranostics

Owing to its non-ionizing nature, the fact that it is non-invasive and its capacity to extend to further depths inside the body than most extracorporeal physical stimuli, ultrasound has established significant modality as a means of promoting chemo-therapeutics delivery in cancer theranostics. Ultrasonic energy has the ability to penetrate deep into tissues and deposit energy, thereby inducing mechanical, thermal, and cavitation effects in the tumor tissues and destroying the tumor vasculature, at the same time triggering coagulative necrosis of cancer tissues (figure 1.5) [36]. Ultrasound stimulated cancer theranostics relies on remote activation of small gaseous pockets by ultrasound and the resulting acoustic

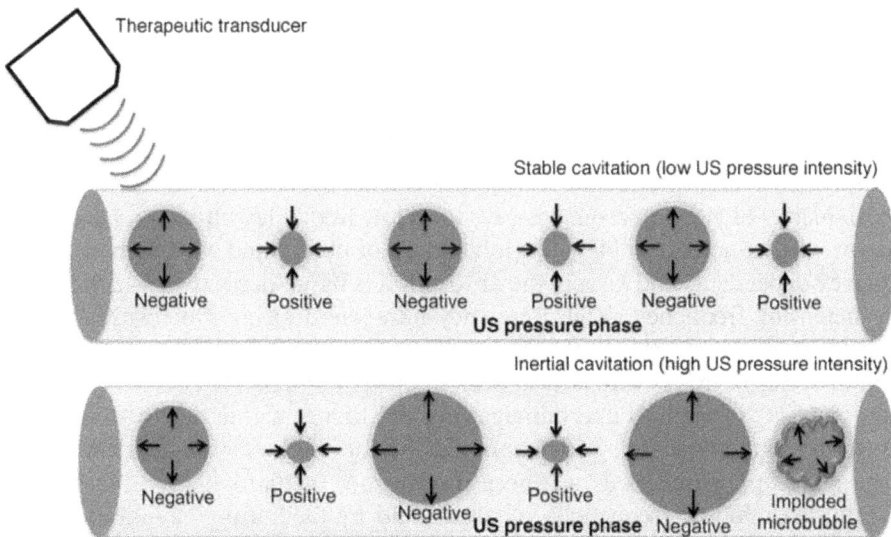

Figure 1.5. Schematic drawing of the principles of stable and inertial cavitation. The type of cavitation strongly depends on pressure intensity. When relatively low pressure intensities are applied, the negative and positive pressure phases of the ultrasound (US) waves cause respective growth and shrinkage of microbubbles, which can repeat stably for many cycles. Such stable oscillation of microbubbles which depends on their resonance frequency, is known as stable cavitation. In contrast, when relatively high pressure intensities are applied, microbubbles violently grow to a much larger size followed by energetic collapse, a phenomenon known as inertial cavitation. Reproduced with permission from reference [38] under the terms of the Creative Commons Attribution Non-Commercial License. Copyright © 2017 Korean Society of Ultrasound in Medicine (KSUM).

cavitation has been used to enhance the extravasation, penetration, and distribution of anticancer agents into solid tumors [37].

A principal advantage of ultrasound, as an external stimulus, is its aptitude to propagate through all deeply situated biological tissues enabling non-invasive control of drug release and tissue responses. Owing to tailorable tissue penetration depth, using ultrasound, one can focus the ultrasound energy into the region of interest such as a tumor to stimulate drug-loaded nanostructures. Ultrasound stimulation generates two physical mechanisms, heat and cavitation, to enhance drug release and energize the drug intake of the tumor. Currently, high-intensity focused ultrasound and low frequency ultrasound treatments are used in external field triggered cancer theranostics to achieve cell membrane permeability of drugs [39]. In the intended therapeutic approach, chemotherapeutic drug-loaded ultrasound-sensitive NPs would be intravenously administered, then discharged under ultrasound energy while in the intravascular blood volume of the target tumor region such as breast cancer and then drug diffuse across the tumor tissues. Ultrasound triggering, however, has a number of disadvantages compared to the stimulated cancer theranostic approach: for example microbubbles are physically restricted to the vasculature with poor circulatory stability, typically exhibiting half-lives of a few minutes [40]. This issue is addressed by developing different types of ultrasound responsive hybrid nanostructures over the past 5–10 years as an alternative means of promoting ultrasound triggered cancer theranostics and that is detailed in chapter 6.

On the basis of intensity and frequency, ultrasound stimulation can be classified as low frequency (20–100 kHz) or high frequency (>1 MHz) ultrasound. Low frequency ultrasound is used to transient cavitation-induced drug release from nanoliposomes. High frequency ultrasound is used to focus deeply situated tumors with diameters of a few micrometers, whereas low frequency ultrasound is unable to penetrate the tissue barrier [41]. The interaction of ultrasound with human organs is frequency dependent, and largely the absorption is tissue-dependent which increases with increasing frequency. High frequency ultrasound at high intensities generates heat and damages tissues by thermal energy. For instance, high frequency ultrasound of about 1.7 MHz with a focal peak intensity of $\sim$1500 W cm^{-2} can heat tissue to above 56 °C in seconds and damages normal tissues within a short duration [4]. Ultrasound, either coupled with nanomedicine or used as drug carriers alone, is encouraging opportunities as a cancer theranostics mediator in combination with clinically feasible low frequencies of ultrasound by facilitating cavitation at lower pressures. Although both low and high frequency ultrasound stimulated nano-medicines have revealed desirable rewards, the damaging side effects of intensive ultrasound should also be paid more attention. For the successful cancer theranostic clinical translation of ultrasound, intensity, frequency, duty cycle and time of exposure need to be optimized for better performance [42]. Ultrasound is already being used in clinical imaging, however safety, biocompatibility, and the balance between less damage and sufficient energy when coupled with nanomedicine remain challenges. Despite the existence of extensive research work on the ultrasound stimulated nanotheranostics effect of these individual components, the true potential of the stimulated cancer theranostics lies in the synergistic external field therapeutic modalities that combine active targeting, and triggered therapeutics.

1.3.4 Radiation (x-ray) stimulated cancer theranostics

Radiation therapy (XRT) is one of the most commonly used treatments for many types of cancers including breast cancer. The XRT module is currently used as an acute modern therapeutics approach to curative and adjuvant treatment of cancers. The underlying physics of XRT is to bombard the tumor with ionizing radiation and damage the cancer cell's DNA either by direct ionization or through generation of free radicals by ionization of water or oxygen molecules [43]. This ionization radiation can control the growth of cancerous cells. Using the nanomedicine approach, the dose of delivered ionizing radiation can be amplified, for example high atomic (Z) numbers termed as Z-materials can amplify radiation dose via an enhancement of the photoelectric effect (figure 1.6).

External beam radiotherapy (EBRT) is used in modern clinical oncology to treat superficial and deeply situated tumors. EBRT utilizes x-ray beams produced by orthovoltage units, or linear accelerators that may be spatially oriented, with beams shaped using multileaf collimators in order to maximize the specificity for the target [43]. A range of energy bandwidths is available for different EBRT

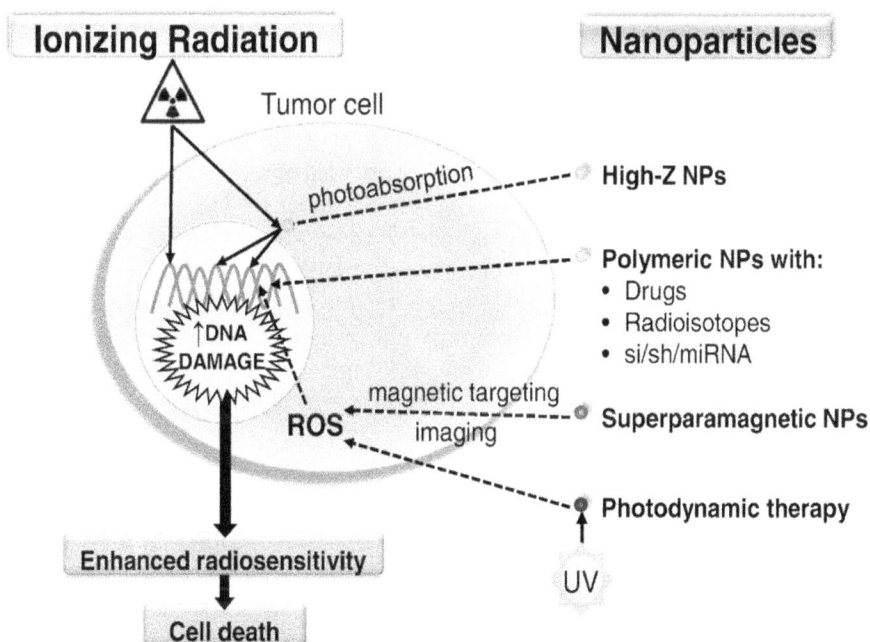

Figure 1.6. Schematic representation of different types of nanoparticles (NPs) and their effects in an irradiated tumor cell. The NPs increase the amount of radiation-induced DNA damage in the cell which results in enhanced cell death. ROS: reactive oxygen species. Reproduced with permission from reference [44]. Copyright © 2015 Elsevier B.V.

targets: 40–100 kV (kilovoltage or 'superficial' x-rays) for skin cancers or other exposed structures; as well as 100–300 kV (orthovoltage) and 4–25MV (megavoltage or 'deep' x-rays) if deeply situated. In recent years, different high Z-NPs have been found to cause a dose enhancement of radiotherapy. High Z-NPs have the predominant photoelectric absorption values that can induce the initiation of Auger emission (release of electrons) or photons. High atomic number materials have a higher cross section and higher photoelectric absorption and this is a function of the atomic number of the atom, approximately determined as Z^4. Thus, an irradiated NP with a higher Z-value has a greater possibility of emitting secondary radiation compared to lower Z value NPs. The emitted secondary electrons induce radiolysis of the surrounding water and production of highly reactive oxygen species (oxidizing free hydroxyl radicals), resulting in DNA damage and cell death by necrosis. Since these radiation sources have low energies, tissue penetration in the range of 10–100 nm can be achieved by proper adjustment. This process leads to localized energy deposition into the targeted site, thus selective tumor destruction is possible. High Z NPs in combination with linear accelerator (Linac) irradiation in the clinical stage set-up can stimulate various biological effects including ROS generation, induction of DNA damage that finally results in tumor destruction [44]. In clinical oncology, the dose enhancement with radiation stimulated nanomedicine approach is ideal for patients receiving radiotherapy for breast, mandibular and head and neck cancers. Nevertheless, using the targeting ability of NPs, active tumor targeting and imaging efficiency of radiotherapy can be improved.

Gold ($Z = 79$) NPs are a widely studied high Z material for NP-induced radiation sensitization. The popularity of gold NPs for radiation sensitization is due to gold's strong photoelectric absorption coefficient and biocompatibility. Recently, various hybrid gold nanostructures have been developed to produce singlet oxygen under an x-ray-trigger. For example, liposomes incorporating gold NPs and photosensitizer verteporfin have been proposed to destroy tumors under 6 MeV x-ray radiation [45]. Importantly, this x-ray-mediated liposome release strategy offers prospects for deep tissue photodynamic therapy, by removing its depth limitation. Thus, radiation stimulated nanoassembly produces synergistic effects in the course of standard radiotherapy combined with chemotherapy delivered via x-ray-triggered liposomes. For further details on radiation triggered cancer nanomedicine, readers are encouraged to read chapters 6 and 7.

1.4 Conclusion

Finally, compared to a non-stimulated cancer nanomedicine approach, external field triggered nanotheranostics, for example light, magnetic, ultrasound and x-rays with the suitable wavelength, field strength and energy, can penetrate the human body, straightforwardly. The higher penetration activates chemotherapeutic cargo release in deeply situated tumor tissues once the external stimulus triggered hybrid nanostructures reach their target. This feature is opening many opportunities for cancer research and clinical medicine. Many single therapies currently suffer from limited penetration depth and efficient triggered release. External stimulation however, beneficial to have access to multimodality treatment, is far from reality but is feasible from a clinical point of view. EFS is giving evidence of better therapeutic effects over the single therapeutic or radiation therapy at reduced toxicity and adverse effects. Furthermore, EFS cancer theranostics options are producing desired therapeutic effects at a significant cost of short- and long-term toxicity. The pre-clinical reality, however, shows great variability in animal studies, which can be further resolved by the selection of physiologically relevant standards for the development of EFS nanomedicines to increase overall patient survival.

Acknowledgement

The project leading to this work has received funding from the European Union's Horizon 2020 research and innovation programme under the Marie Sklodowska-Curie grant agreement 751903. This publication is based on work from COST Action CA 17140 'Cancer Nanomedicine from the Bench to the Bedside' supported by COST (European Cooperation in Science and Technology).

References

[1] El-Sawy H S, Al-Abd A M, Ahmed T A, El-Say K M and Torchilin V P 2018 Stimuli-responsive nano-architecture drug-delivery systems to solid tumor micromilieu: past, present, and future perspectives *ACS Nano* **12** 10636–64

[2] Yao J, Feng J and Chen J 2016 External-stimuli responsive systems for cancer theranostic *Asian J. Pharm. Sci.* **11** 585–95

[3] Ji D-K, Ménard-Moyon C and Bianco A 2019 Physically-triggered nanosystems based on two-dimensional materials for cancer theranostics *Adv. Drug Deliv. Rev.* **138** 211–32

[4] Wang Y and Kohane D S 2017 External triggering and triggered targeting strategies for drug delivery *Nat. Rev. Mater.* **2** 17020

[5] Park S M, Aalipour A, Vermesh O, Yu J H and Gambhir S S 2017 Towards clinically translatable *in vivo* nanodiagnostics *Nat. Rev. Mater.* **2** 17014

[6] Liu J, Detrembleur C, Mornet S, Jérôme C and Duguet E 2015 Design of hybrid nanovehicles for remotely triggered drug release: an overview *J. Mater. Chem.* B **3** 6117–47

[7] Pinel S, Thomas N, Boura C and Barberi-Heyob M 2019 Approaches to physical stimulation of metallic nanoparticles for glioblastoma treatment *Adv. Drug Deliv. Rev.* **138** 344–57

[8] Chen Q, Ke H, Dai Z and Liu Z 2015 Nanoscale theranostics for physical stimulus-responsive cancer therapies *Biomaterials* **73** 214–30

[9] Sneider A, VanDyke D, Paliwal S and Rai P 2017 Remotely triggered nano-theranostics for cancer applications *Nanotheranostics* **1** 1–22

[10] El-Sawy H S, Al-Abd A M, Ahmed T A, El-Say K M and Torchilin V P 2018 Stimuli-responsive nano-architecture drug-delivery systems to solid tumor micromilieu: past, present, and future perspectives *ACS Nano* **12** 10636–64

[11] Lakshmanan S, Gupta G K, Avci P, Chandran R, Sadasivam M, Jorge A E S and Hamblin M R 2014 Physical energy for drug delivery; poration, concentration and activation *Adv. Drug Deliv. Rev.* **71** 98–114

[12] Wo F, Xu R, Shao Y, Zhang Z, Chu M, Shi D and Liu S 2016 A multimodal system with synergistic effects of magneto-mechanical, photothermal, photodynamic and chemo

therapies of cancer in graphene-quantum dot-coated hollow magnetic nanospheres *Theranostics* **6** 485–500

[13] Wu M and Huang S 2017 Magnetic nanoparticles in cancer diagnosis, drug delivery and treatment *Mol. Clin. Oncol.* **7** 738–46

[14] Kim K and Lee W G 2017 Electroporation for nanomedicine: a review *J. Mater. Chem.* B **5** 2726–38

[15] Huber L A, Pereira T A, Ramos D N, Rezende L C D, Emery F S, Sobral L M, Leopoldino A M and Lopez R F V 2015 Topical skin cancer therapy using doxorubicin-loaded cationic lipid nanoparticles and iontophoresis *J. Biomed. Nanotechnol.* **11** 1975–88

[16] Wan G-Y, Liu Y, Chen B-W, Liu Y-Y, Wang Y-S and Zhang N 2016 Recent advances of sonodynamic therapy in cancer treatment *Cancer Biol. Med.* **13** 325–38

[17] Rinaldi L *et al* 2018 Sonoporation by microbubbles as gene therapy approach against liver cancer *Oncotarget* **9** 32182–90

[18] Baumgart J, Humbert L, Boulais É, Lachaine R, Lebrun J-J and Meunier M 2012 Off-resonance plasmonic enhanced femtosecond laser optoporation and transfection of cancer cells *Biomaterials* **33** 2345–50

[19] Thorat N D, Bohara R A, Noor M R, Dhamecha D, Soulimane T and Tofail S A M 2017 Effective cancer theranostics with polymer encapsulated superparamagnetic nanoparticles: combined effects of magnetic hyperthermia and controlled drug release *ACS Biomater. Sci. Eng.* **3** 1332–40

[20] Frei E H 1969 Magnetism and medicine *J. Appl. Phys.* **40** 955–7

[21] Lübbe A S *et al* 1996 Clinical experiences with magnetic drug targeting: a phase I study with 4'-epidoxorubicin in 14 patients with advanced solid tumors *Cancer Res.* **56** 4686–93

[22] Thorat N D, Bohara R A, Malgras V, Tofail S A M, Ahamad T, Alshehri S M, Wu K C-W and Yamauchi Y 2016 Multimodal superparamagnetic nanoparticles with unusually enhanced specific absorption rate for synergetic cancer therapeutics and magnetic resonance imaging *ACS Appl. Mater. Interfaces* **8** 14656–64

[23] Thorat N D, Bohara R A, Noor M R, Dhamecha D, Soulimane T and Tofail S A M 2017 Effective cancer theranostics with polymer encapsulated superparamagnetic nanoparticles: combined effects of magnetic hyperthermia and controlled drug release *ACS Biomater. Sci. Eng.* **3** 1332–40

[24] Thorat N D, Bohara R A, Tofail S A M, Alothman Z A, Shiddiky M J A, A Hossain M S, Yamauchi Y and Wu K C-W 2016 Superparamagnetic gadolinium ferrite nanoparticles with controllable Curie temperature—cancer theranostics for MR-imaging-guided magneto-chemotherapy *Eur. J. Inorg. Chem.* **2016** 4586–97

[25] Golovin Y I, Klyachko N L, Majouga A G, Sokolsky M and Kabanov A V 2017 Theranostic multimodal potential of magnetic nanoparticles actuated by non-heating low frequency magnetic field in the new-generation nanomedicine *J. Nanopart. Res.* **19** 63

[26] Maniotis N, Makridis A, Myrovali E, Theopoulos A, Samaras T and Angelakeris M 2019 Magneto-mechanical action of multimodal field configurations on magnetic nanoparticle environments *J. Magn. Magn. Mater.* **470** 6–11

[27] Kim D-H 2018 Image-guided cancer nanomedicine *J. Imaging* **4** 18

[28] Shabahang S, Kim S and Yun S H 2018 Light-guiding biomaterials for biomedical applications *Adv. Funct. Mater.* **28** 1706635

[29] Yun S H and Kwok S J J 2017 Light in diagnosis, therapy and surgery *Nat. Biomed. Eng.* **1** 0008

[30] Tang R, Habimana-Griffin L M, Lane D D, Egbulefu C and Achilefu S 2017 Nanophotosensitive drugs for light-based cancer therapy: what does the future hold? *Nanomedicine* **12** 1101–5

[31] Chen J, Ning C, Zhou Z, Yu P, Zhu Y, Tan G and Mao C 2019 Nanomaterials as photothermal therapeutic agents *Prog. Mater. Sci.* **99** 1–26

[32] Kang X *et al* 2017 Photothermal therapeutic application of gold nanorods-porphyrin-trastuzumab complexes in HER2-positive breast cancer *Sci. Rep.* **7** 42069

[33] Candido N M, De Melo M T, Franchi L P, Primo F L, Tedesco A C, Rahal P and Calmon M F 2018 Combining photodynamic therapy and chemotherapy: Improving breast cancer treatment with nanotechnology *J. Biomed. Nanotechnol.* **14** 994–1008

[34] Shim G, Ko S, Kim D, Le Q V, Park G T, Lee J, Kwon T, Choi H G, Kim Y B and Oh Y K 2017 Light-switchable systems for remotely controlled drug delivery *J. Control. Release* **267** 67–79

[35] Qiu M *et al* 2018 Novel concept of the smart NIR-light–controlled drug release of black phosphorus nanostructure for cancer therapy *Proc. Natl. Acad. Sci.* **115** 501–6

[36] Chen Y *et al* 2011 Multifunctional mesoporous composite nanocapsules for highly efficient MRI-guided high-intensity focused ultrasound cancer surgery *Angew. Chem. Int. Ed.* **50** 12505–9

[37] Mannaris C, Teo B M, Seth A, Bau L, Coussios C and Stride E 2018 Gas-stabilizing gold nanocones for acoustically mediated drug delivery *Adv. Healthc. Mater.* **7** 1800184

[38] Mullick Chowdhury S, Lee T and Willmann J K 2017 Ultrasound-guided drug delivery in cancer *Ultrasonography* **36** 171–84

[39] Sengupta S, Khatua C, Jana A and Balla V K 2018 Use of ultrasound with magnetic field for enhanced *in vitro* drug delivery in colon cancer treatment *J. Mater. Res.* **33** 625–37

[40] Lee J Y, Crake C, Teo B, Carugo D, de Saint Victor M, Seth A and Stride E 2017 Ultrasound-enhanced siRNA delivery using magnetic nanoparticle-loaded chitosan-deoxycholic acid nanodroplets *Adv. Healthc. Mater.* **6** 1601246

[41] Elkhodiry M A, Momah C C, Suwaidi S R, Gadalla D, Martins A M, Vitor R F and Husseini G A 2016 Synergistic nanomedicine: passive, active, and ultrasound-triggered drug delivery in cancer treatment *J. Nanosci. Nanotechnol.* **16** 1–18

[42] Zhou L, Wang H and Li Y 2018 Stimuli-responsive nanomedicines for overcoming cancer multidrug resistance *Theranostics* **8** 1059–74

[43] Cooper D R, Bekah D and Nadeau J L 2014 Gold nanoparticles and their alternatives for radiation therapy enhancement *Front. Chem.* **2** 86

[44] Bergs J W J, Wacker M G, Hehlgans S, Piiper A, Multhoff G, Rödel C and Rödel F 2015 The role of recent nanotechnology in enhancing the efficacy of radiation therapy *Biochim. Biophys. Acta Rev. Cancer* **1856** 130–43

[45] Deng W, Chen W, Clement S, Guller A, Zhao Z, Engel A and Goldys E M 2018 Controlled gene and drug release from a liposomal delivery platform triggered by X-ray radiation *Nat. Commun.* **9** 2713

Chapter 2

Physically responsive nanostructures in breast cancer theranostics

Sougata Ghosh, Pravin D Patil and Rohini D Kitture

Resulting in almost 9.6 million deaths, cancer is the second leading cause of death globally; it is known to be one of the major health concerns, indicating the urgency to develop precise, early diagnostic methods and effective therapeutics. Recent developments in nanoscience and nanotechnology have offered some encouraging results; yet every technique still faces some challenges, limiting widespread application. An emerging trend which integrates both therapeutics and diagnostics, known as theranostics, has been in focus for various advantageous features over conventional counterparts. In this chapter, we intend to overview the external stimuli (viz light, magnetic, ultrasonic, and ionizing radiation) responsive cancer theranostics techniques, comprising an introduction to the systems stated above, their interesting features, recent progress and future scope for development.

2.1 Introduction

Cancer is known to be one of the major challenges to human health and the second leading cause of death globally. Around 9.6 million deaths due to cancer were reported in 2018 as per the World Health Organization (WHO) fact-sheet [1] (https://www.who.int/news-room/fact-sheets/detail/cancer). It has been reported that early detection of cancer can help in effective treatment with greater probability of survival, at relatively lower expense. However, one of the major challenges in early detection is that no significant symptoms are induced at the early stages.

Extensive study has been carried out in the past few decades to develop precise diagnosis techniques. The development of nanoimaging through computed tomography (CT), ultrasound (US), magnetic resonance imaging (MRI), single-photon emission computed tomography (SPECT), positron emission tomography (PET), and optics are some of the current imaging methods, for early detection of cancer [2].

doi:10.1088/2053-2563/ab2907ch2

Rigorous efforts are ongoing to improve these existing and reliable diagnosis technologies for their widespread application. Moreover, continuous development is going on in the field of cancer therapeutics, to control the mortality rate.

Surgery, among several conventional cancer treatment strategies, is highly invasive. This method cannot typically ensure complete elimination of tumor cells when metastasis has already happened. Poor selectivity, limited therapeutic efficacy, severe toxic side effects including multidrug resistance (MDR), and long-term damage of immune systems are some of the major challenges of the other conventional cancer treatment strategies—chemotherapy and radiotherapy. This defines the need and urgency for the development of a suitable solution. The human body carries out biological processes at the nanoscale, including those essential for life and those that lead to cancer. Therefore, nanotechnology presents researchers with an opportunity to investigate and manipulate macromolecules in real time and during the earliest stages of cancer progression. In this context, multifunctional nano-materials (NM)/nanostructures which integrate precise diagnostics and accurate therapeutics have been in demand, due to the ability to tune the NM/nanostructures for the desired application, offering tremendous opportunities in the domain of cancer theranostics.

Nanoscale devices are 100–1000 times smaller than human cells and similar in size to several biomolecules such as receptors and enzymes. Nanoscale materials (<50 nm) can readily enter most human cells, while those smaller in size (<20 nm) can easily move out of blood vessels as they circulate through the body. The smaller size of nanoscale devices allows them to easily interact with biomolecules located inside and outside the surface of cells, which offers access to several areas of the body and ultimately enhances the potential to diagnose diseases, and deliver therapies in a way previously unimagined. Despite being small in size when compared to cells, nanoparticles are sufficiently large to encapsulate multiple types of small molecule compounds which are a crucial part of the drug delivery system. On the other hand, the relatively large surface area of nanoparticles can accommodate functionalization of several small molecules including ligands, peptides, DNA or RNA strands, aptamers, or antibodies. Apart from being actively targeted to cancer cells while being passively accumulated at the tumor site, nanomaterials can be delivered across traditional biological barriers in the body which makes them highly valuable to cancer research. Nanoscale materials have proven their extraordinary potential to entirely transform cancer therapy with dramatically enhanced efficacy of therapeutic strategies.

The therapy can be initiated/triggered by the internal stimuli like pH, redox etc, or external stimuli like light, magnetic fields, ultrasound etc. In this chapter we will overview the external stimuli triggered nanostructures developed for cancer thera-nostics [3].

2.2 Light-responsive systems

Among various exogenous stimuli mediated therapeutics, light-responsive systems have been preferred owing to their high spatiotemporal precision upon a specific wavelength light irradiation, non-invasiveness and stoichiometric cleavage under light irradiation. Basically, photosensitizer molecules, upon exposure to specific wavelength light irradiation, turn into a highly unstable state by absorbing the energy from light. The energy is then:

(1) utilized by the surrounding oxygen molecules, to generate reactive oxygen species (ROS) damaging the nearby biomolecules, or

(2) converted into heat, thus increasing the local temperature or eventually killing cancer cells, before the molecules return to the ground state.

Based on the mechanism the photosensitizer molecules adopt, the light-responsive system can further be divided into two different types including photodynamic therapy (PDT), photothermal therapy (PTT) etc [4–7].

The first photosensitizer was approved in 1994, by the US Food and Drug Administration (FDA), and since then there has been commendable growth in the light-responsive drug systems for theranostics. For more details about PDT, photosensitizers, readers are encouraged to refer to a detailed report by Abrahamse and Hamblin [8]. Initially, short-wavelength light i.e. ultraviolet (UV) and visible light were the commonly used excitation wavelengths, in theranostics, but these wavelengths did not lie in the so-called biological window. The biological window refers to the spectral range where, due to simultaneous reduction in both absorption and scattering, tissues become partially transparent. Due to the presence and impact of the endogenous light absorbers in living subjects like water, lipids and protein molecules-hemoglobin and oxyhaemoglobin, the UV and visible radiation show limited tissue penetration depth, owing to inherent absorption/scattering. But radiation with wavelengths ranging from 650–900 nm shows minimal absorption, scattering, thus NIR lie in a biological window and are hence preferred for PDT and PTT [4, 9, 10].

The advancement in nanoscience and nanotechnology has offered various promising candidates for real-time imaging and therapeutics as well. As mentioned earlier, the ease of tuning size, shape and other photophysical properties, and surface modification for enhanced results have been the significantly beneficial features of nanomaterials making them the favorite candidates in cancer theranostics.

2.2.1 Photodynamic therapy

In photodynamic therapy (PDT), the incident light energy is absorbed by the photosensitizer molecule, to generate the cytotoxic $^1O_2$ or ROS causing irreversible free radical damage to tissues in the close vicinity, $\sim$20 nm, which induces local tissue apoptosis and necrosis. Besides this, PDT can be helpful in handling the higher dose and more frequent administration of chemotherapy drugs, which needs to be followed to address the MDR effect. The system, showing selectivity towards tumor, with minimal side effects and toxicity towards healthy tissues, faces major drawbacks like lack of stability, solubility etc.

Extensive research in the field of nanoscience has offered promising candidates for PDT, namely metallic nanoparticles like gold nanoparticles (AuNPs), up-conversion nanoparticles (UCNPs), quantum dots (QDs), carbon nanomaterials (CNM) and silica-based nanomaterials etc [4, 8, 11–15].

Recently, Zhang *et al* tried to address the major drawbacks in the AuNPs assisted PDT viz,

  (1) limited the amount of drug loading due to relatively low specific surface area,
  (2) limitations in deep-tissue penetration due to shift of the desirable NIR window to the visible spectral region, which is attributed to frequent clustering and aggregation of the nanoparticles.

To address the issue, Zhang *et al* [16] developed mesoporous silica-coated gold nanorods (Au@SiO$_2$) and studied their theranostic applications. While the mesoporous silica could offer a higher surface area, hence a high drug payload, it could also protect and/or shield the optimized light-transparent window. The functionality of the system was further enhanced by loading chemotherapy drug, doxorubicin hydrochloride (DOX). The schematic illustration of Au@SiO$_2$ as a novel multifunctional theranostic platform for cancer treatment and salient results are shown in figure 2.1.

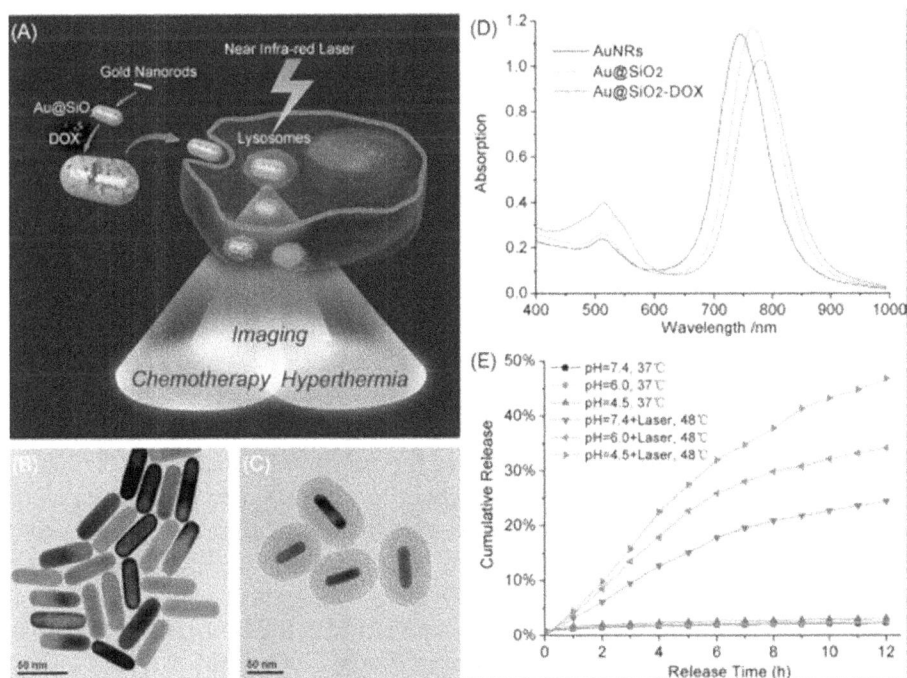

Figure 2.1. (A) Schematic illustration of mesoporous silica-coated gold nanorods (Au@SiO$_2$) as a novel multifunctional theranostic platform for cancer treatment. TEM images of (B) AuNRs and (C) Au@SiO$_2$, (D) extinction spectra of AuNRs, Au@SiO$_2$, and Au@SiO$_2$–DOX, and (E) DOX release profiles from Au@SiO$_2$–DOX with and without NIR laser irradiation at different pHs. Reproduced with permission from [16].

With the laser-triggered release of the DOX drug, the Au@SiO$_2$–DOX system offering chemotherapy under high-power-density laser irradiation, provided hyperthermia therapy, proving it as a promising candidate for multifunctional cancer therapy. Chen *et al* have reviewed some interesting reports on non-metallic nanomaterials in cancer theranostics [3]. This review covers the recent progress in cancer theranostic strategies using mesoporous silica nanoparticles and carbon-based nanomaterials.

Besides inorganic nanomaterial, metal organic frameworks (MOFs) have also been studied for PDT. MOFs when scaled down to the nanoscale, forming nanoscale MOFs (NMOFs), show the advantage of being used as nanocarriers for loading of imaging agents and drugs due to their enhanced permeability and the retention (EPR) effect of cancerous tumors to achieve enhanced tumor accumulation. For example, Liu *et al* [3] have reported nanoscale metal organic frameworks (NMOF) composed by hafnium (Hf^{4+}) and tetrakis (4-carboxyphenyl) porphyrin (TCPP), wherein TCPP works as a photosensitizer. In addition, Hf^{4+} was used as a radio-sensitizer to enhance radiotherapy (RT), due to its strong x-ray attenuation ability. To address the stability issues of the nanoparticles, PEGylation was adopted. The biodistribution of PEGylated Hf–TCPP NMOF (Hf–TCPP NMOF-PEG) nanoparticles into 4T1 tumor bearing Balb/c mice showed encouraging results. Besides stability in physiological solutions, the NMOF nanoparticles after PEGylation, showed reduced *in vitro* cytotoxicity, enhanced blood circulation time, and efficient tumor homing due to the EPR effect after systemic administration. Moreover, the RT and PDT enhanced the overall therapeutic efficacy of the system. The *in vivo* clearance of the NMOF nanoparticles from the body, minimized the long-term toxicity issues, thus indicating that NMOFs are promising candidates for synergistic activity in cancer therapeutics. (Results are shown in figure 2.2.)

Besides these and other such inorganic nanomaterials, organic nanomaterials such as proteins, dendrimers, micelles, liposomes, polysaccharides, hollow polymer microcapsules etc, have been extensively used in the delivery of photosensitizers. Several groups have reviewed the timely developments of such nanomaterials for cancer theranostics [7, 8, 10].

2.2.2 Photothermal therapy (PTT)

While in PDT, the photosensitizers transfer the energy absorbed from the incident radiation to the surrounding oxygen molecules, to generate cytotoxic singlet oxygen or ROS; photothermal therapy (PTT) does not require oxygen to interact with the target cells or tissues but it involves conversion of the absorbed energy into heat to raise the local temperature of tumor thus killing the cancer cells. The radiation belongs to the near-infrared radiation range (from 650–1024 nm), for which tissues are almost transparent to this radiation due to minimal absorption and scattering of the radiation by the body [4–6, 9, 15, 18].

Burke *et al* [19] reported an interesting study on breast cancer stem cells (BCSCs), the tumor initiating stem-like cells. Typically, the 'triple negative' phenotype (estrogen receptor, progesterone receptor, and HER-2 negative) being refractory

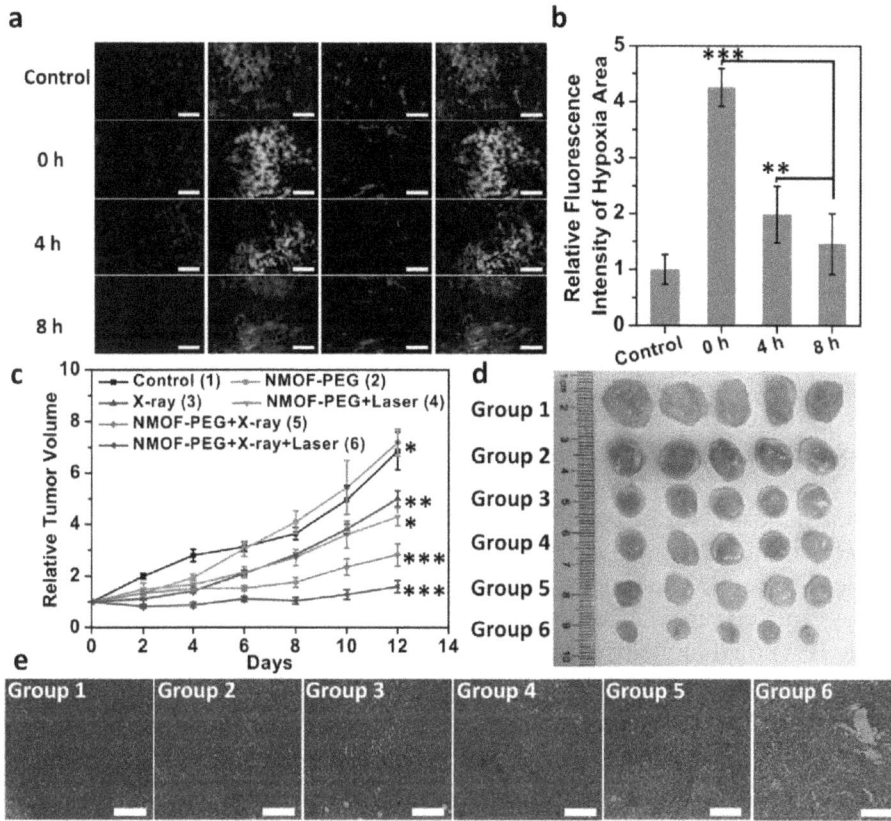

Figure 2.2. *In vivo* combination therapy. (a) Representative immunofluorescence images of tumor slices at different times post RT. The nuclei, blood vessels and hypoxia areas were stained with DAPI (blue), anti-CD31 antibody (red), and anti-pimonidazole antibody (green), respectively. The scale bar: 150 mm. (b) The relative fluorescence intensity of hypoxia are as recorded from more than 10 micrographs for each group. (c) Tumor growth curves of different groups after various treatments indicated: (i) untreated control, (ii) i.v. injection with NMOF-PEG, (iii) x-ray irradiation, (iv) i.v. injection with NMOF-PEG þ Laser, (v) i.v. injection with NMOF-PEG þ x-ray, (vi) i.v. injection with NMOF-PEG þ x-ray þ Laser. Error bars were based on SD of five mice per group. P values were calculated by Tukey's post-test (***$p < 0.001$, **$p < 0.01$, or *$p < 0.05$). (d) Photograph of tumors from mice after receiving various treatments collected on day 14. (e) H&E stained images of tumors collected two days after receiving various treatments. The scale bar: 200 mm. Reproduced with permission from [17].

to chemo/radiotherapy are considered to be the most difficult ones to treat. Considering the impact of such BCSCs, the effect of conventional hyperthermia and nanoparticles-based PTT on HMLER[shEcadherin] breast cancer stem cells was studied. It was seen that these cells were resistant to the conventional hyperthermia whereas CNT-based PTT could promote necrotic cell death, hence is effective against BCSCs. These and many such similar reports indicate the advantages of PTT against the conventional therapeutic system.

A variety of nanomaterials ranging from simple metallic nanoparticles and their various other morphologies to organic–inorganic complex have been studied for this purpose. Gold, since it absorbs NIR through surface plasmon resonance (SPR) has been the favored candidate. Various morphologies of gold have been developed and the efficacy for PTT was studied [14, 20, 21].

Among metal nanoparticles, silver and palladium nanoparticles have also been studied for PTT. Khafaji *et al* [14] reviewed the inorganic nanomaterials used in chemo/PTT. The review covered studies on various gold-based, iron-based and metal sulfide-based nanostructures. Thanks to the ability of surface modification of nanomaterials, various complex nanomaterials have also shown excellent results for their potential application in chemo and PTT. Organic compounds, typically NIR dyes, due to their inherent photophysical properties, ease of conjugation with specific molecules of interest for imaging or therapy and their availability for large-scale chemical synthesis, have been interesting candidates for research [7, 9, 12, 14, 18, 22].

Hybrid nanostructures have been in focus in cancer theranostics, to overcome the drawbacks of the existing materials, achieving improved and promising results. For example, Li *et al* [23] developed a light-responsive biodegradable nanomaterials-based system for cancer management, through enhanced ultrasound (US)/photo-acoustic (PA) dual-modality imaging guided photothermal therapy of melanoma. The conventional drug-loaded microbubbles show poor stability and tumor pene-tration, while their recent nano counterparts, although showing improved perform-ance, were non-biodegradable, hence showed a limited future in biomedical application. To address this issue, Li *et al* [23] combined the excellent inherent properties of gold nanorods, mesoporous silica and perfluoropentane (PFP) to design a PFP-loaded nanorattle with a thin mesoporous silica shell and a gold nanorod (GNR) core (GNR@SiO$_2$–PFP). After the basic characterization, an *in vitro* and *in vivo* imaging study concerning therapeutic efficacy of GNR@SiO$_2$–PFP was evaluated on A375 tumor-bearing mice. The treatment with GNR@SiO$_2$–PFP with NIR laser exposure showed significant tumor growth inhibition as compared to the control samples. The results could be visually observed, as shown in figure 2.3. This evaluation was followed by biocompatibility and biosafety assessment via a hemolysis test. The NIR responsive GNR@SiO$_2$–PFP could generate PFP nanobubbles, showing appreciable permeation into the nonmicrovas-cular tissue, offering excellent imaging, followed by promising tumor inhibition upon NIR laser irradiation. The system being biocompatible and biodegradable, offers applications in cancer nanotheranostics through enhanced dual-modality US/PA imaging guided PTT of cancer.

2.3 Magnetically responsive systems

In addition to light-responsive systems, magnetic responsive systems have also been an outstanding topic of research and have proved their effectiveness. Due to deeper tissue penetration and stability, magnetic nanoparticles (MNPs) have been widely

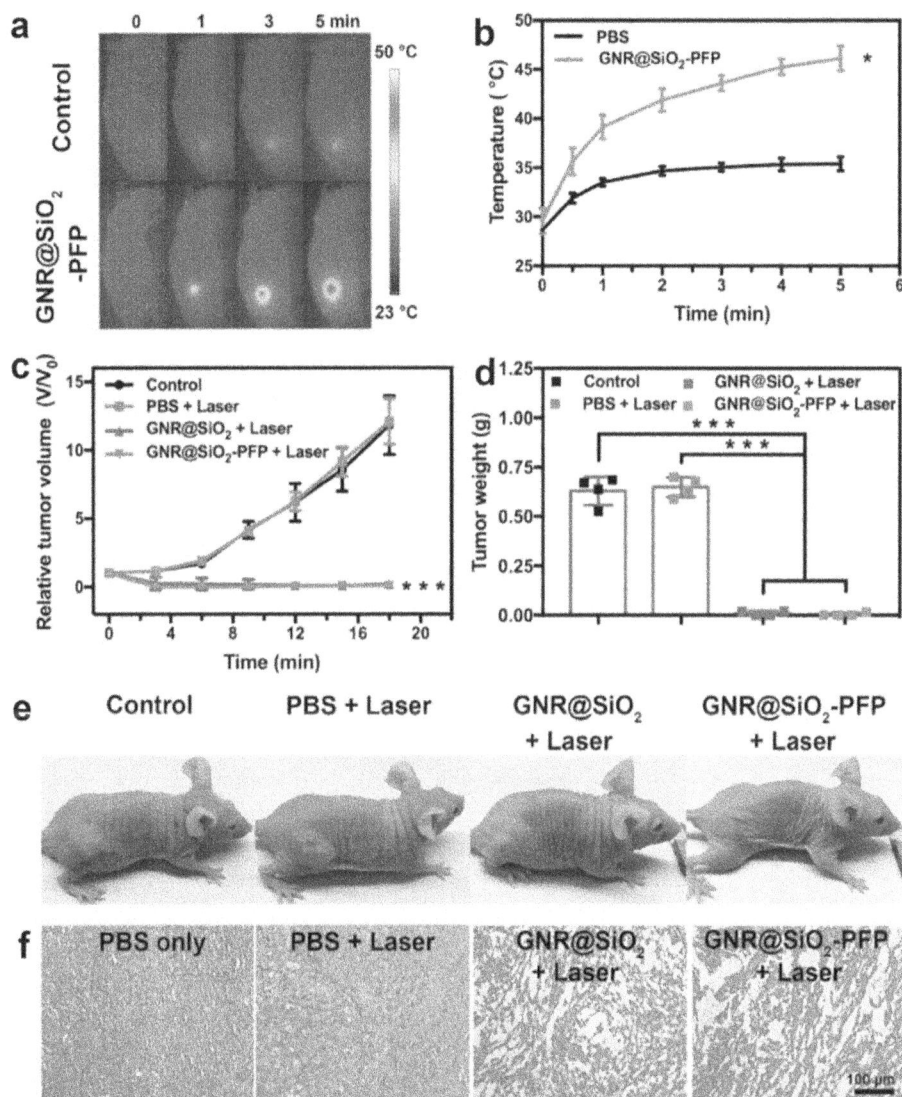

Figure 2.3. *In vivo* therapeutic efficacy of GNR@SiO$_2$–PFP nanorattles in A375 tumor-bearing mice. (a) Representative thermographic images and (b) temperature changes of A375 tumors upon laser irradiation at 24 h post-injection. *$P < 0.05$. (c) Tumor growth curves of different groups after PTT treatment. Tumor volumes have been normalized to the initial sizes. (mean ± SD, $n = 4$, ***$P < 0.001$). (d) Tumor weight of mice on day 18 after different treatments (mean ± SD, $n = 4$, ***$P < 0.001$). (e) Photographs of A375 tumor-bearing mice with various treatments. (f) H&E staining images of tumor sections after different treatments. Reproduced with permission from [23].

studied for their application as **MRI** contrast agents, which overcome the limitations of the conventional organic dyes in combination with nanoplatforms for fluorescent imaging. MRI, characterized by non-invasive, real-time and cost-effective features, has been widely investigated in precisely detecting diseases at an early stage.

Besides imaging, magnetically responsive systems/MNPs see potential applications in the field of cancer therapeutics. This class has several advantages over the light-responsive system, the latter can only be applied if the penetration depth is less than centimeters, and the stimulating radiation is in a 'biological window' etc. In contrast, the intrinsic properties of the MNPs can be tuned and harvested for their use in magnetically targeted and triggered drug delivery and magnetic hyperthermia. A permanent magnetic field (PMF) or alternating magnetic field (AMF) can be used as external stimuli remotely controlling the system. Thus, magnetically responsive systems are strong contenders in the race, for application in cancer theranostics.

2.3.1 Magnetic resonance imaging

MRI, a non-invasive, non-ionizing imaging modality is far superior to the other imaging modalities due to its high penetration depth and spatial resolutions down to 10 µm which can still go to 1 µm in the case of clinical magnets. The possible use of MRI for diagnosis was predicted in the 1970s and was favored as it did not use ionizing radiation yet offered excellent soft tissue contrast [2, 24–27].

Unfortunately, due to the low specificity hence false-positives, the widespread application of MRI in cancer diagnostics is limited [2, 28]. There has been tremendous research in developing magnetic nanomaterials as MRI contrast agents. The agents can be primarily divided into four types depending upon the physical mechanism of generating signals: T1, T2, PARACEST, and hyperpolarization. The initial reports of using contrast agents for MRI were published in the early 1980s. For example, Carr *et al* [29] reported the use of gadolinium–diethylenetriamine penta-acetic acid, also known as Gd–DTPA as MRI agent in human volunteers in 1984. These are the most commonly used T1 contrast agents. In T1 contrast imaging, the water molecules chemically interact with these paramagnetic molecules (Gd^{3+}) giving rise to the T1-weighted signal. The unpaired seven electrons from Gd^{3+} produce a strong magnetic field. This field, in turn, affects the water protons, when they come in very close proximity to the contrast agents, thus modulating their relaxation times along the longitudinal axis (spin–lattice relaxation). The longer relaxation time due to the seven unpaired electrons offers longer signaling time, making Gd^{3+} one of the most conventional MRI contrast agents. Some other forms of Gd^{3+} have also been studied, including gadolinium phosphates ($GdPO_4$), gadolinium oxides (Gd_2O_3) and $GdF_3:CeFn_3$. Mn^{2+} has also been another candidate for T1 contrast imaging, wherein a bright image is formed [27].

With the advancements in nanoscience various other nanomaterials have also been studied for their application as MRI contrast agents. For example, inorganic nanoparticles viz cobalt (Co), Au@Fe, Au@Pt, Pt@Fe, Au@Ag, Au@Fe_2O_3, Au@Co, Au@TiO_2, small and ultrasmall superparamagnetic iron oxide nanoparticles (SPION, USPION) etc [30]. Guerbet AMAG Pharm. Inc. developed Dextran T10 coated SPIONs of the order of 120–180 nm, available with the name Ferumoxides AMI-25 Feridex/Endoremz. Jun *et al* in 2005 reported a biocompatible, water-soluble, magnetic nanocrystal model system wherein the magnetic properties were tuned by varying the nanocrystal size, as shown in figure 2.4 [31].

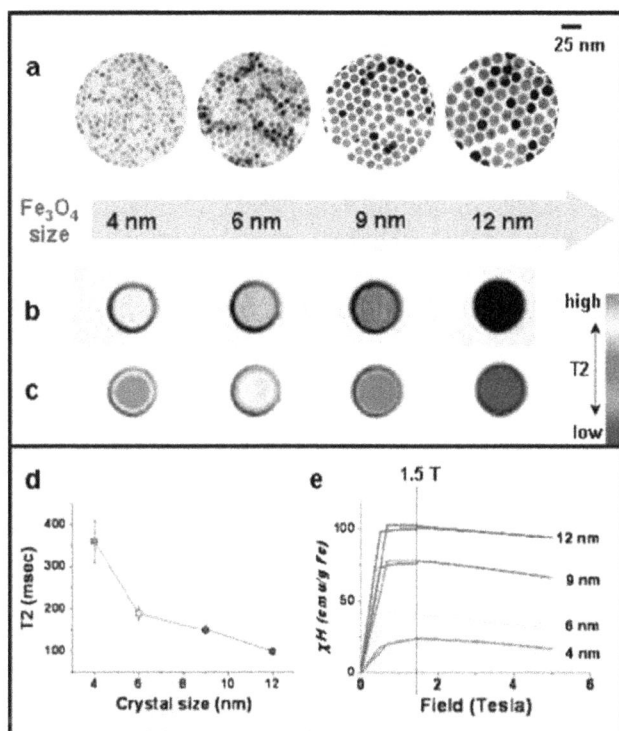

Figure 2.4. Effect of size on magnetism and MR signals of water-soluble Fe_3O_4 nanocrystals. Reprinted with permission from [31].

The nanocrystal exhibited excellent T2 MR images, with potential application in breast cancer diagnosis.

While earlier work emphasized tuning the particle size by varying the capping agent/surfactant, there have been interesting efforts on engineering the nanostructures for the intended applications. For example, Shanavas *et al* reported hybrid nanostructures of magnetic poly(lactide-*co*-glycolide) (PLGA) nanoparticles as the core, with folate–chitosan (fol–cht) conjugate as the shell for cancer theranostics [32]. While the coating could control the undesired burst drug release, the receptor targeted release could also be attained with this hybrid nanostructure.

Besides SPIONs, Gd-based agents have also been studied. A brief review of such other agents is given in table 1 of the article by Weinstein *et al* [33]. These nanomaterials create magnetic field inhomogeneities affecting T2 relaxation (spin–spin relaxation) rate. Interestingly, the physical properties of these nanomaterials can be tuned in order to use them for the imaging of T1 and T2. There are some excellent reviews dedicated to the recent developments in the MR contrast agents [5, 27, 33–36]. Readers are encouraged to refer to these reviews for further details. A detailed discussion on nanostructures for MRI can also be found in [2]. (Excerpt partially reprinted with permission from [2].)

2.3.2 Magnetic hyperthermia and targeted drug delivery system

Magnetic hyperthermia treatment (MHT) relies on the production of heat MNPs under alternating current magnetic field (ACMF), to kill cancerous cells. Heat-generation with the help of MNPs needs to be optimal for the destruction of cancer cells, which requires high magnetization. AMF-to-heat conversion efficiency is governed by their specific absorption rate (SAR). Composition of the MNPs can be tuned to attain high SAR, hence desired temperature on application of AMF. Three heating mechanisms contribute to the SAR namely, hysteresis loss, relaxation loss, and eddy currents. These mechanisms are attributed to the intrinsic and extrinsic characteristics of MNPs; which can be tuned by controlling the shape, particle size, AMF parameters, such as frequency and amplitude etc [37–39].

Considering the advantages of MHT over the conventional therapeutics, various nanoparticles have been studied for MHT. Magnetic metal nanoparticles like iron (Fe), nickel (Ni), and cobalt (Co), due to their high magnetic properties have been extensively studied for MHT; but their poor chemical stability and biocompatibility, high reactivity to oxidation, lead to partial or complete loss of their magnetization. Hence various other alternatives were also developed. Metal oxide MNPs including of iron oxide-Fe_3O_4 nanoparticles (IONPs) were focused for MHT. Various forms of IONPs have been studied for more than a decade and their applications in biomedicine is also well proven. Owing to their biocompatibility, some IONPs have been approved as MRI contrast agents by the FDA. For these biomedical applications, the particles must have combined properties of high magnetic saturation, biocompatibility, and active functional groups at the surface. The surfaces of these particles could be modified through the creation of few atomic layers of organic polymer or inorganic metals such as gold or oxides or weak acids or bases, making them suitable for further functionalization by the attachment of various bioactive molecules. Besides IONPs, nickel oxide and even lanthanum strontium manganite have also been studied for their potential application by tuning their size, shape or by surface modification. Hedayatnasab et al have reviewed the developments in the domain of MHT [37]. The table which portrays the review of various MNPs developed for MHT has been partially reproduced in table 2.1. Readers are encouraged to refer to table 5, of the original article [37].

Surface modification of the MNPs, SPIONs offer significant improvements in the magnetic properties, including difference in saturation magnetization value. Moreover, the limitations in the uncoated nanoparticles, like ready aggregation hence reduction in the magnetization can be ruled out by selecting right coating material, thus enhancing the SAR to achieve the desired rise in temperature for cancer cell destruction.

Various types of coating materials can be selected for this purpose. Figure 2.5 shows the types of coatings for MNPs.

There have been numerous attempts to use the inherent properties of the as synthesized MNPs by integrating the MHT with targeted drug delivery for a smart theranostics approach. Figure 2.6 shows a schematic of such attempts for smart theranostics.

Table 2.1. Magnetic properties and *in vitro* results of varied MNPs applied in MNFHT. Partially reprinted with permission from [37].

Core material	Coating material and details			Magnetic properties		
	Material	Coated particle size (nm)	Coated particle shape	Ms (emu g^{-1})	Hc (Oe)	Mr (emu g^{-1})
$MnFe_2O_4$	CS	18	Cubic	A: 71.45 B: 58.34	0	0
$CoFe_2O_4$	OA	~9.9	Spherical	60.42	0	0
Fe_3O_4	OA	~45	—	—	—	—
$NiFe_2O_4$	PEG	~16	Rod	15	0	0
Fe_3O_4	Phosphate	14	Nearly spherical	A: 63.6	Low	Low
$MnFe_2O_4$	Tetraethyl orthosilicate (TEOS)	14	Spherical	B: ~40	0	0
Fe_3O_4	3-Aminopropyltriethoxy silane, protamine sulfate	10	Spherical	—	—	—
$Zn_{0.9}Fe_{0.1}Fe_2O_4$	—	11	Spherical	12	—	—
Fe_3O_4 hollow spheres	—	~22	Spherical	81	Low	—
$Fe_{63.2}Cr_{11.5}Nb_{0.3}B_{20}$	CS	20–40	—	A: ~42 B: ~34	—	—
Fe_3O_4	—	~13.9	Spherical	III: 81.26	—	—
$Co_xFe_{3-x}O_4$	—	8–8.5	Spherical	X = 0: 72	Low	Low

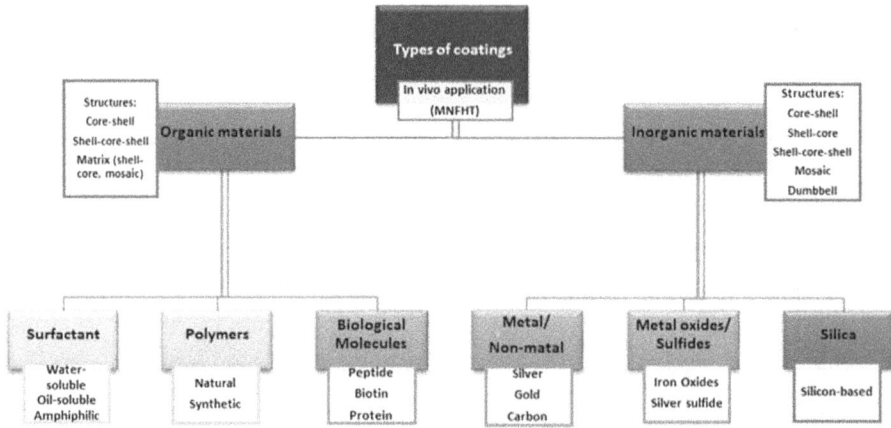

Figure 2.5. Types of coatings for MNPs. Reproduced with permission from [37].

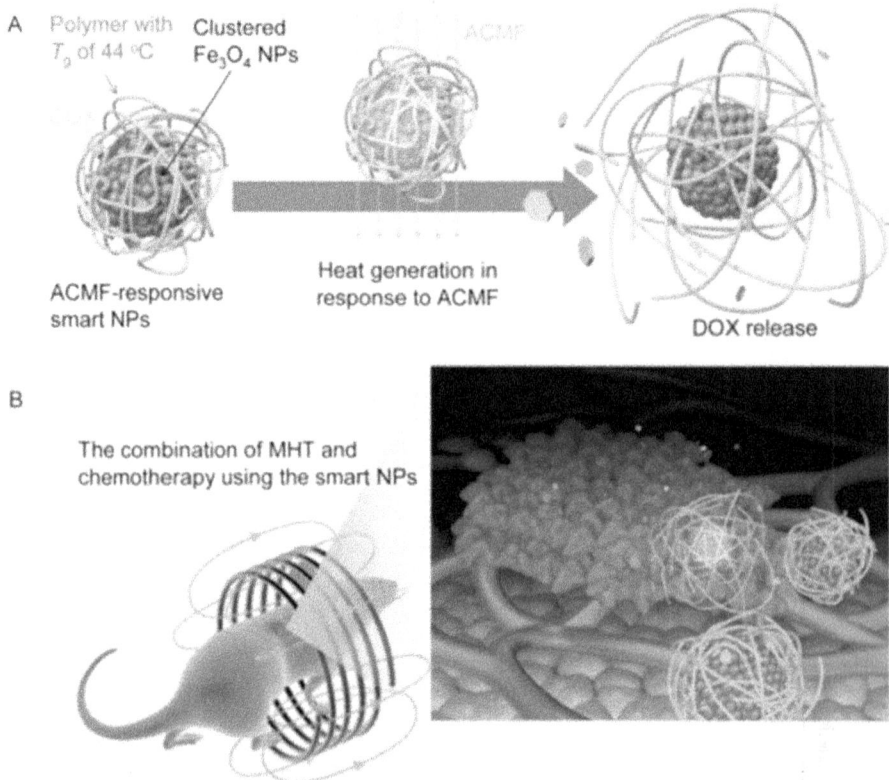

Figure 2.6. (A) Illustration of smart NPs that produce heat in response to AMF and sequentially release DOX. (B) Illustration of cancer treatment with the combination of MHT and chemotherapy using the smart NPs. Reproduced with permission from [40].

Gold nanoparticles (AuNPs) can be used for simultaneous imaging of tumor vasculature and diagnosis leading to enhancement of intratumoral localization of therapeutics. Further, AuNPs are modified with monoclonal antibodies, enabling complete destruction of cancer cells. Cetuximab (C225) conjugated AuNPs shows efficient tumor tissue targeting and RF-induced heating in pancreatic cancer cells. QDs are also considered as potential radiofrequency (RF) responsive nanomaterials for pancreatic cancer cell ablation. Conjugation with specific antibodies helps to target cancer tissue and limit nonspecific distribution. The ErbB family of tyrosine kinase belonging to the epidermal growth factor receptor (EGFR) plays a critical role in regulation of cellular proliferation, survival and differentiation. RF heatable single-walled carbon nanotubes (SWNTs) with unique physical and chemical characteristics can be effectively used to induce thermal cytotoxicity in malignant cells. SWNTs targeted to cancer cells lead to non-invasive RF field treatments to produce lethal thermal injuries specifically on the malignant cells. Nonspecific uptake of theranostic nanomaterials can be blocked by functionalizing with human epidermal growth factor (EGF) as EGF coated magnetic NPs (MNPs) were internalized within the EGF +ve cancer cell lines, such as MCF-7 (breast cancer cells) and Panc-1, respectively. The surface charge of the cancer cells is a function internalization of the therapeutic nanoparticles [41]. Enhanced internalization of EGF-MNPs is characterized by increased negative surface charge. Even a short duration of RF excitation of 10 min treatment can cripple most of the MCF-7 breast cancer cells with ~less than 3% cell viability after treatment with EGF-MNPs (figure 2.7).

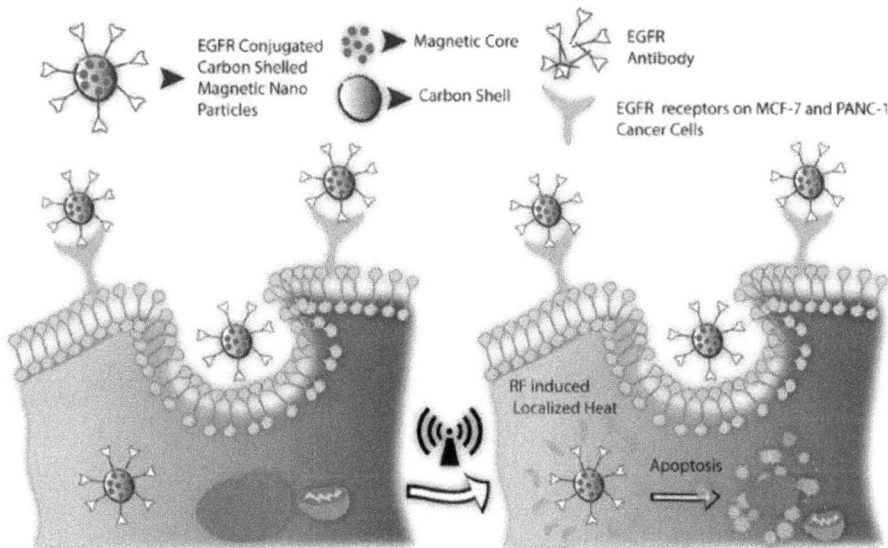

Figure 2.7. Schematic representation for the selective internalization of EGFR modified carbon shelled magnetic nanoparticles in EGFR+ve cancer cells followed by apoptosis on RF excitation. Reproduced with permission from [41].

Kitture *et al* have tried to conjugate the IONPs with curcumin, known as goldmine in medicine, owing to its inherent medicinal values. Citric acid was used as a linker between the IONPs core and curcumin. The linking strategy was carefully developed in order to keep the functional group of curcumin available for its antioxidant and anticancer activities. The magnetic properties of the core IONPs were utilized for magnetic targeting and hyperthermia, whereas additional tumor inhibition and suppression were achieved through curcumin on the surface [38].

2.4 Ultrasonic responsive system

Although there are many available techniques like MRI, CT, PET and optical imaging, ultrasound imaging is advantageous due to its low-cost real-time imaging which is safe as well as portable. Advancement in the fabrication of efficient ultrasound contrast agents (UCA) has led to the improvement of both resolution and sensitivity of clinical ultrasound imaging. Ideally, UCAs are micrometer-sized bubbles with an inner gaseous core usually filled with air, nitrogen, sulfur hexafluoride (SF_6) or perfluorocarbons (PFCs) and a thin shell composed of proteins, surfactants, polymers or lipids [4]. Microbubbles are considered to be the best choice as imaging/diagnostic UCA which can further be modified as drug-loaded vehicles for delivering into blood vessels for therapeutic applications. Unlike traditional soft-shell microbubbles, hard-shell polymeric microbubbles are more stable and can prevent gas diffusion which enables the use of regular air instead of heavy gases for loading purposes. Novel theranostics are fabricated targeting mostly cancer angiogenesis [42]. Polymeric cyanoacrylate microbubbles are synthesized by addition of monomeric butyl-2-cyanoacrylate to an aqueous acidic solution (pH 2.5) containing Triton X-100 (surfactant) under vigorous stirring. Further, these micro-bubbles are linked to vascular endothelial growth factor receptor 2 (VEGFR2) and $\alpha_v\beta_3$ integrin binding ligands which show higher accumulation in squamous cell carcinoma xenografts (HaCaT-ras-A-5RT3) in mice (figure 2.8). Such surface modified microbubbles are highly efficient in multimarker ultrasound imaging. This strategy enables molecular ultrasound mediated detection of increased expression of

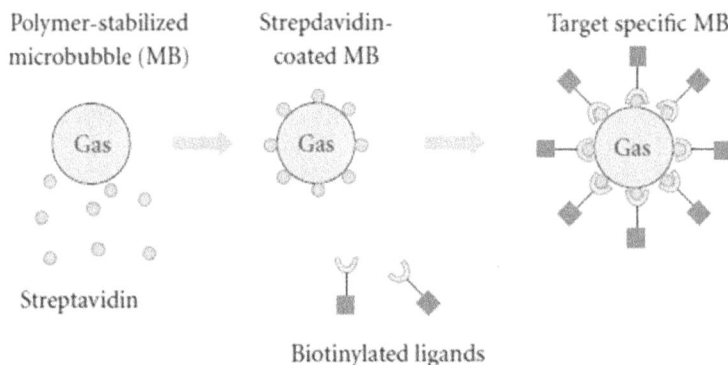

Figure 2.8. Schematic representation of synthetic approach of target-specific microbubbles. Reproduced with permission from [42].

VEGFR2 and $\alpha_v\beta_3$ integrin during tumor growth. Similarly, it can also be used for the evaluation of post-administration therapeutic efficiency of drugs like matrix metalloproteinase inhibitors, in terms of decrease in both marker densities [43].

Rationally constructed UCAs help in US imaging guided therapy. Apart from diagnostic applications, ultrasound therapy is based on both nonthermal effects (cavitation) and thermal effects to achieve tumor-specific drug/gene delivery. On high exposure to ultrasound, UCAs start to cavitate. The generation of micro-streams and shock waves leads to local drug release from UCAs, inducing perforations on cell membranes (sonoporation), thus facilitating intracellular delivery of the released drugs. Additionally, the microjets and shock waves also permeabilizes blood vessels releasing high molecular weight drugs and nanoparticles. This ultrasound triggered drug delivery can efficiently overcome biological barriers, particularly blood–brain barrier for enhancement of the drug response. Moreover, high-intensity focused ultrasound (HIFU) can trigger hyperthermia for direct thermal ablation therapy and enhanced drug delivery [4].

Echogenic drug delivery systems like lipophilic drug, doxorubicin (DOX) loaded perfluoropentane (PFP) nanoemulsion are constructed using a biodegradable block copolymer poly(ethylene glycol)–poly (L-lactide) (PEG–PLLA). PFP nanodroplets vaporize on heating at physiological temperature forming microbubbles with excellent visibility for ultrasound imaging (figure 2.9(a)). When administered using intravenous injection, DOX-PFP nanoemulsions selectively extravasate into the tumor sites characterized by strong ultrasound contrast (figure 2.9(b)). On accumulation in the tumor tissue, small nanobubbles coalesce into larger, highly echogenic microbubbles [44]. Under the influence of ultrasound irradiation DOX is released from the microbubbles (figure 2.9(c)).

Paclitaxel (PTX)-loaded ultrasound triggered nanoemulsions are effective in ovarian, breast, and pancreatic cancers. Biodegradable nanocapsules with perfluorohexane (PFH) as the gas core, and disulfide cross-linked poly (methacrylic acid) (PMAA) with DOX shell has a uniform size of ~300 nm. This DOX-loaded PMMA–PFH nanocapsules can easily penetrate through tumor tissue via enhanced permeability and retention (EPR) effect, offering enhanced US contrast in the tumor site and enabling ultrasound triggered chemotherapy. Ultrasonic triggered nano-bubbles are considered to be promising vehicles for targeted gene delivery owing to sonoporation mediated enhanced extravasation of large macromolecules, such as plasmid DNA, siRNA and drugs to improve delivery into tissue to acquire additional functionalities in both diagnostic imaging and cancer therapy. siRNA loaded octafluoropropane (OFP) nanobubbles (siRNA-NB) are constructed using siRNA-complexed polymeric micelles and gas-cored liposomes (figure 2.10). Exposure to ultrasound facilitates siRNA-NB for siRNA transfection ensuring enhanced gene silencing, leading to significantly improved therapeutic efficacy in a nude mouse glioma model [4, 45]. Ultrasound triggered nanomaterials exhibit deep-tissue penetration and is thus a convenient diagnostic imaging modality. However, often their relatively larger dimension (hundreds of nm or even at the micrometer scale) limits extravasation and reduces the mobility towards target tissues. The overall architecture is also critical as very small bubble size (nanobubbles) largely

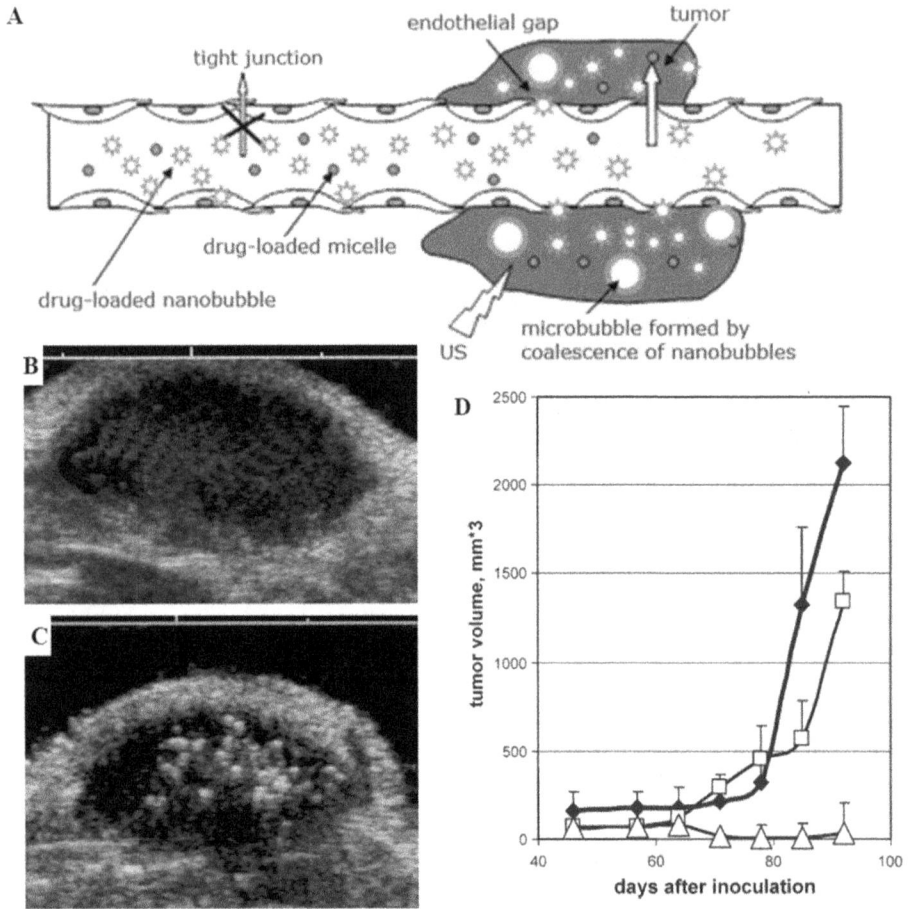

Figure 2.9. Ultrasonic responsive chemotherapy for *in vivo* cancer treatment. (A) Schematic representation of drug targeting through the defective tumor microvasculature using the echogenic drug delivery system. The system comprises polymeric micelles (small circles), nanobubbles (stars), and microbubbles (large circles). (B) US imaging in the tumor before intravenous injection of drug-loaded nanobubbles. (C) Contrast-enhanced US imaging in the tumor after intravenous injection of drug-loaded nanobubbles. (D) Ultrasound triggered therapeutic efficacy on tumor volume (filled diamonds: control; open squares: injection of drug-loaded nanobubbles without US exposure; open triangles: injection of drug-loaded nanobubbles with US exposure at 3 MHz for a 30 s at 2 W cm^{-2}). Reproduced with permission from [44].

compromises their acoustic properties leading to weakened ultrasound signals. Efforts are being made to fabricate 'smarter' UCAs, which although initially start as microbubbles, eventually will be converted to smaller drug-loaded nanoparticles under ultrasonic stimuli for simultaneous diagnosis and therapy in cancer.

2.5 Ionizing radiation triggered system

High-energy radiation such as x-rays and gamma-rays kills cancer cells which is a commonly used strategy in radiation therapy. Heavy metal containing nanohybrids

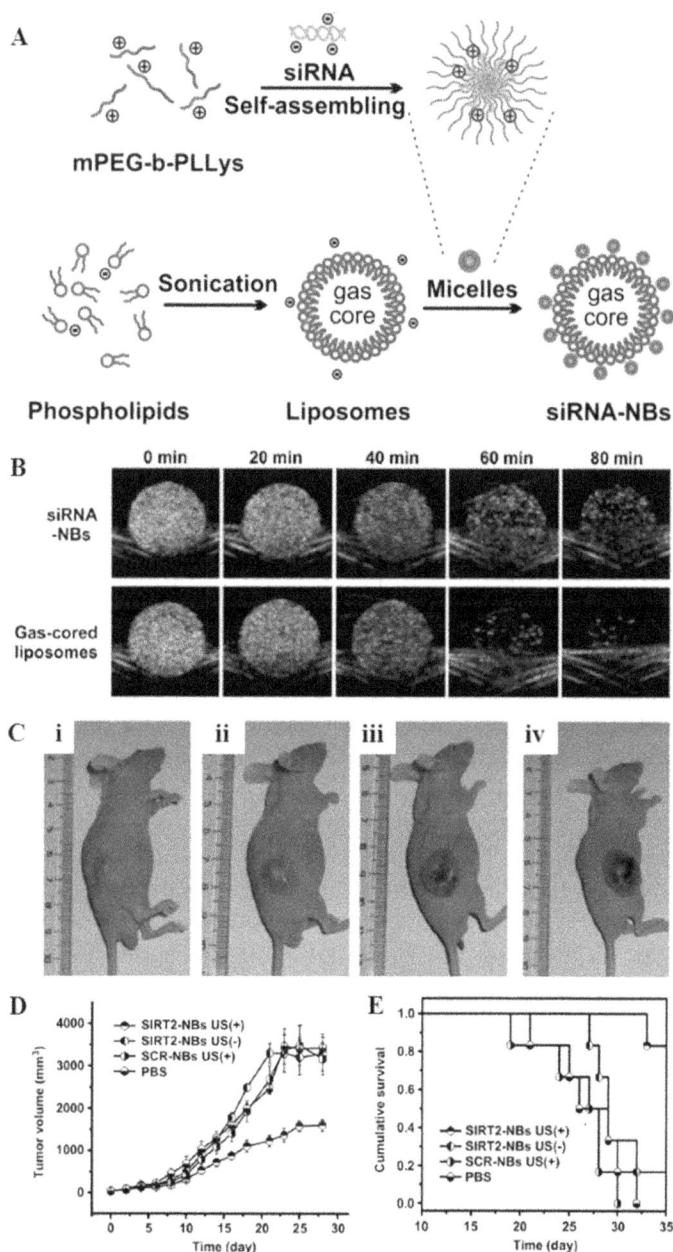

Figure 2.10. Ultrasound responsive siRNA-NBs for *in vivo* cancer treatment. (A) Illustrative preparation of siRNA-NBs complexes using the positively charged siRNA micelles and negatively charged gas-cored liposomes. (B) Stability of the siRNA-NBs and gas-cored liposomes analyzed by contrast-enhanced US imaging *in vitro*. Compared with gas-cored liposomes, the gray-scale intensities of siRNA-NBs complexes declined more slowly, confirming strengthened stability. (C) Photographs of mice at day 28 after different treatments (i) SIRT2-NBs US (+) group, (ii) SIRT2-NBs US (−) group, (iii) SCR-NBs US (+) group and (iv) PBS group; (D) Detection of mean tumor volume using caliper measurement. (E) Cumulative survival of C6 glioma bearing mice after receiving different formulations. Reproduced with permission from [45].

can absorb high-energy ionizing radiation which may enhance the efficacy of radiation therapy [4]. PEG-coated gold nanoparticles of variable sizes are reported to cause a significant decrease in cancer cell survival after gamma radiation. The activity is size dependent where particles with sizes of 12.1 and 27.3 nm show better dispersive distributions in the cells and stronger sensitization effects by both cell apoptosis and necrosis. These particles significantly decrease tumor volume and do not cause spleen and kidney damage. However, an adverse effect and gold accumulation is observed in the liver tissues [46]. Ultrasmall CuS nanoparticles decorated onto the surface of silica-coated rare earth up-conversion nanoparticle leads to integration of photothermal ablation (PTA) with radiotherapy (RT) for enhancement of effective cancer therapy. These multifunctional core/satellite nanotheranostic (CSNT) simultaneously convert near-infrared light into heat for effective thermal ablation as well as induce highly localized radiation damage. Additionally, these particles facilitate excellent up-conversion luminescence/magnetic resonance/computer tomography trimodal imaging [47]. The three-in-one system for magnetic resonance imaging, luminescence imaging, and photodynamic therapy have become very attractive nanomaterials for simultaneous diagnosis and targeted therapy (figure 2.11). In this respect, highly uniform Fe_3O_4/SiO_2 core/shell nanoparticles functionalized by phosphorescent iridium complexes (Ir) are well suited for phosphorescent labeling and simultaneous singlet oxygen generation for apoptosis induction [48].

Similarly, $LaF_3:Tb^{3+}$–meso-tetra (4-carboxyphenyl) porphine (MTCP) nanoparticle conjugates effectively generate singlet oxygen due to x-ray irradiation which makes them efficient photodynamic agents. Folic acid addition helps in the targeting to folate receptors on tumor cells that makes these particles promising deep cancer treatment modality [49]. Integration of a scintillator and a semiconductor as an ionizing radiation-induced PDT agent, achieves synchronous radiotherapy and depth-insensitive PDT with diminished oxygen dependence. In the core–shell Ce^{III}-dopedLiYF$_4$@SiO$_2$@ZnO structure, the down-converted ultraviolet fluorescence from the Ce^{III}-doped LiYF$_4$ nanoscintillator under ionizing irradiation enables the generation of electron–hole pairs in ZnO nanoparticles, giving rise to the formation of biotoxic hydroxyl radicals (figure 2.12). This process enhances antitumor therapeutic efficacy [50].

2.6 Future perspective

The developments in nanoscience have significantly contributed to the development and rapid progress in cancer theranostics in the past few decades. The physically responsive theranostic systems, with incredible therapeutic and detection potential, have been more beneficial, due to their higher treatment specificity accompanied by lower systemic toxicity, against their conventional counterparts like chemotherapy for cancer.

Besides conventional inorganic nanomaterials, polymers, lipids and hybrid nanomaterials/nanostructures have proven to be strong contenders in cancer theranostics, both at research and clinical trial level. By tuning their morphology, surface characteristics and physico-chemical properties, these nanostructures offer

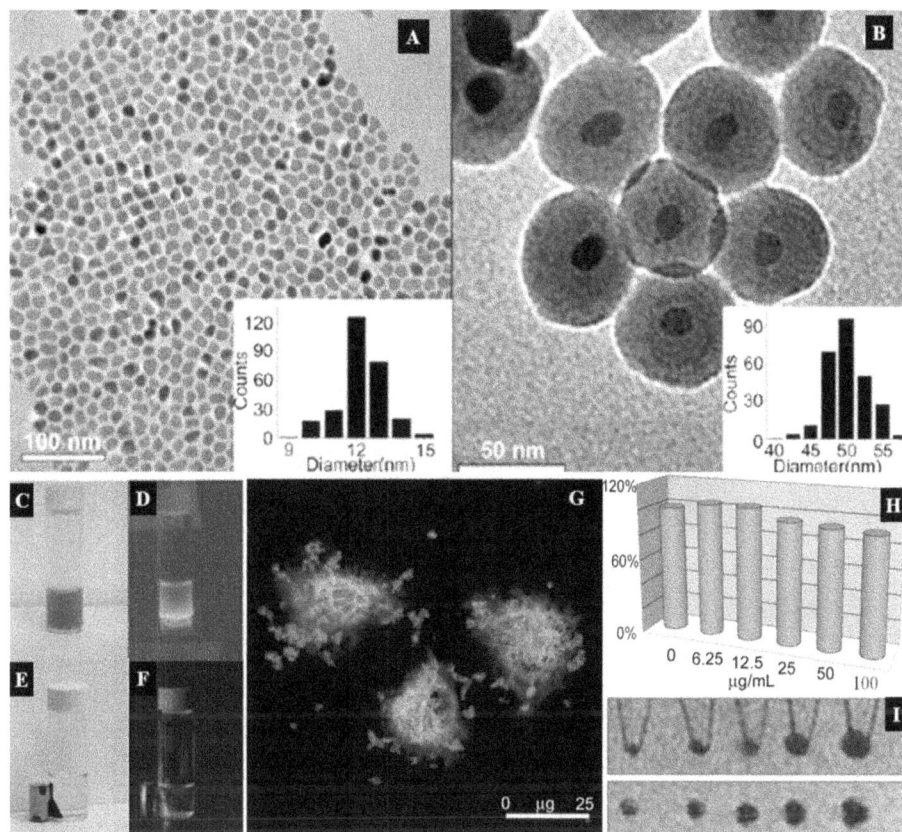

Figure 2.11. TEM images of (A) Fe_3O_4 and (B) $Fe_3O_4/SiO_2(Ir)$ nanoparticles; insets: histograms of particle diameters. (C) The appearance of the $Fe_3O_4/SiO_2(Ir)$ suspension in aqueous solution. (D) The corresponding red emission produced under excitation by a 366 nm UV lamp. (E) The appearance of $Fe_3O_4/SiO_2(Ir)$ nanoparticles upon placement of a magnet at one side of the vial, and (F) under UV excitation. (G) Confocal image of HeLa cells treated with 100 mg mL^{-1} $Fe_3O_4/SiO_2(Ir)$. The DNA, cytoskeleton, and nanoparticles are shown in blue, green, and red, respectively. (H) MTT assay of HeLa cells treated with $Fe_3O_4/SiO_2(Ir)$ in the range 0–100 mg mL^{-1}. (I) Side and bottom views of MRI of the $Fe_3O_4/SiO_2(Ir)$-treated cells. The concentrations, from left to right, were: 6.25, 12.5, 25, 50, and 100 mg mL^{-1}, respectively. Reproduced with permission from [48].

exceptional control over imaging resolution and spatiotemporal drug release that leads to superior *in vitro* and *in vivo* theranostic efficiency. For example, light-responsive systems have demonstrated their efficacy and significant role in theranostics. Several reports suggest using it with the conventional therapies for enhanced efficacy and reliable results. After removal of primary tumors by surgery, the small tumors can be eradicated with the help of image-guided PTT. Such strategies would offer additional tools to prevent further tumor metastasis. Yet, these systems still have further scope for development owing to some major challenges. For example, in the case of light-responsive systems, strong absorption and scattering faced by the UV and visible light demands for radiation in the

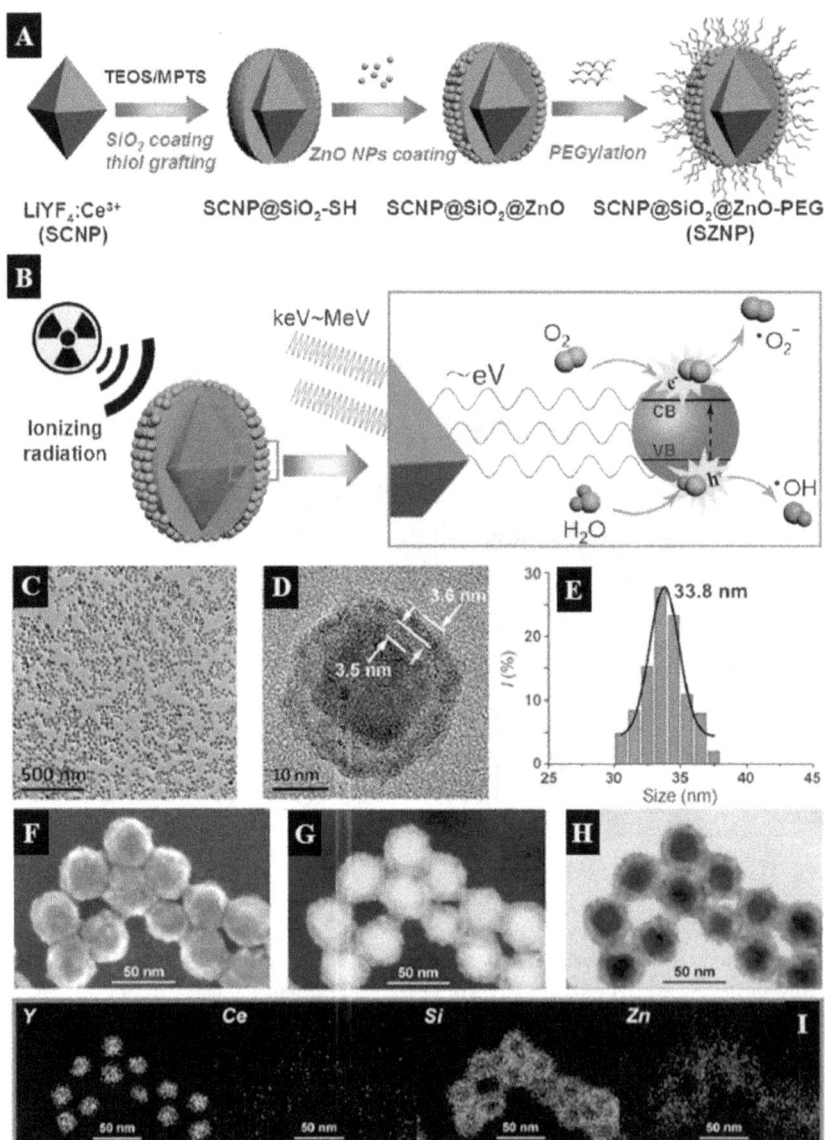

Figure 2.12. (A) Schematic illustration of the synthetic route to monodisperse SZNPs and (B) the mechanism of ionizing radiation-induced photodynamic therapy. The electron–hole (e–h+) pair is formed after exposure to ionizing radiation. TEOS = tetraethyl orthosilicate. MPTS = (3-mercaptopropyl) trimethoxysilane. Structural and compositional characterization of SZNPs. TEM images at (C) low (scale bar = 500 nm) and (D) high (scale bar = 10 nm) magnifications and (E) the corresponding size distribution of SZNPs. STEM image of SZNPs using (F) SEM, (G) dark-field, and (H) bright-field modes. (I) Corresponding element mappings (for Y, Ce, Si, and Zn) of SZNPs. Scale bars in (d)–(g) = 50 nm. Reproduced with permission from [50].

biological window. NIR source has been an excellent alternative source, yet there is huge scope to enhance the photosensitivity of the light-responsive materials. The toxicity and biocompatibility of the nanomaterials is another important aspect which needs to be taken care of. Moreover the power of the laser radiation used should not exceed the threshold of 1 W cm^{-2} [4, 15].

The challenge/limitation of tissue penetration limits has been ruled out in the case of magnetic responsive system for obvious reasons. Besides the FDA approved MRI contrast agents, magnetic tumor targeting and hyperthermia have also entered the clinical trials and proved useful. Recent developments indicated that the coated MNPs perform better than the uncoated ones, in terms of heating efficiency, biocompatibility, dispersion, yet, the coating thickness needs to be optimized in order to avoid any compromise in the Ms and SAR values. Also, the external stimulus, the AMF, plays crucial roles in the output, hence there should be fine tuning in the MNPs composition and the applied field, to achieve the desired results. In addition, the overall treatment cost, which currently seems higher than the conventional systems like CT scans, needs to be brought down.

Recently, several research groups have reported overcoming the disadvantage of the ultrasonic responsive system. The large size of microbubbles as drug carriers, and poor stability and penetration have been addressed by synthesizing hybrid nanostructures, which can offer excellent biocompatibility as well as efficacy. The encouraging *in vivo* results offer wide scope for clinical trials of the system.

While research and development in enhancing the efficacies and overcoming the drawbacks of the above stated systems continue, another significant factor which needs attention is thorough understanding of *in vivo* behaviors of those nano-agents. This includes the biodistribution, pharmacokinetics, degradation, elimination and toxicology profiles, etc. Getting FDA approval for clinical use of these materials after achieving promising results in the preclinical studies is another big task and there is still a long way to go [15].

References

[1] *Latest Global Cancer Data: Cancer Burden Rises to 18.1 Million New Cases and 9.6 Million Cancer Deaths in 2018* 2018 (Geneva, Switzerland: International Agency for Research on Cancer)

[2] Kitture R and Ghosh S 2019 Hybrid nanostructures for *in vivo* imaging *Hybrid Nanostructures for Cancer Theranostics* ed R Ashok Bohara and N Thorat (Amsterdam: Elsevier) ch 10, pp 173–208

[3] Chen Y-C, Huang X-C, Luo Y-L, Chang Y-C, Hsieh Y-Z and Hsu H-Y 2013 Non-metallic nanomaterials in cancer theranostics: a review of silica- and carbon-based drug delivery systems *Sci. Technol. Adv. Mater.* **14** 044407

[4] Chen Q, Ke H, Dai Z and Liu Z 2015 Nanoscale theranostics for physical stimulus-responsive cancer therapies *Biomaterials* **73** 214–30

[5] Ferrari M 2005 Cancer nanotechnology: opportunities and challenges *Nat. Rev. Cancer* **5** 161–71

[6] Ai X, Mu J and Xing B 2016 Recent advances of light-mediated theranostics *Theranostics* **6** 2439–57

[7] Chen H and Zhao Y 2018 Applications of light-responsive systems for cancer theranostics *ACS Appl. Mater. Interfaces* **10** 21021–34

[8] Abrahamse H and Hamblin M R 2016 New photosensitizers for photodynamic therapy *Biochem. J.* **473** 347–64

[9] Estelrich J and Busquets M A 2018 Iron oxide nanoparticles in photothermal therapy *Mol. J. Synth. Chem. Nat. Prod. Chem.* **23** E1567

[10] Awan M A and Tarin S A 2006 Review of photodynamic therapy *Surg. J. R. Coll. Surg. Edinb. Irel.* **4** 231–6

[11] Kim H S and Lee D Y 2018 Near-infrared-responsive cancer photothermal and photo-dynamic therapy using gold nanoparticles *Polymers* **10** E961

[12] Lan G, Ni K and Lin W 2019 Nanoscale metal-organic frameworks for phototherapy of cancer *Coord. Chem. Rev.* **379** 65–81

[13] Youssef Z *et al* 2017 The application of titanium dioxide, zinc oxide, fullerene, and graphene nanoparticles in photodynamic therapy *Cancer Nanotechnol.* **8** 6

[14] Khafaji M, Zamani M, Golizadeh M and Bavi O 2019 Inorganic nanomaterials for chemo/photothermal therapy: a promising horizon on effective cancer treatment *Biophys. Rev.* **11** 335–52

[15] Yao J, Feng J and Chen J 2016 External-stimuli responsive systems for cancer theranostic *Asian J. Pharm. Sci.* **11** 585–95

[16] Zhang Z *et al* 2012 Mesoporous silica-coated gold nanorods as a light-mediated multifunctional theranostic platform for cancer treatment *Adv. Mater.* **24** 1418–23

[17] Liu J *et al* 2016 Nanoscale metal–organic frameworks for combined photodynamic & radiation therapy in cancer treatment *Biomaterials* **97** 1–9

[18] Jaque D *et al* 2014 Nanoparticles for photothermal therapies *Nanoscale* **6** 9494–530

[19] Burke A R *et al* 2012 The resistance of breast cancer stem cells to conventional hyperthermia and their sensitivity to nanoparticle-mediated photothermal therapy *Biomaterials* **33** 2961–70

[20] Huang X, El-Sayed I H, Qian W and El-Sayed M A 2006 Cancer cell imaging and photothermal therapy in the near-infrared region by using gold nanorods *J. Am. Chem. Soc.* **128** 2115–20

[21] Ahmad R, Fu J, He N and Li S 2016 Advanced gold nanomaterials for photothermal therapy of cancer *J. Nanosci. Nanotechnol.* **16** 67–80

[22] Zhao X, Yang C-X, Chen L-G and Yan X-P 2017 Dual-stimuli responsive and reversibly activatable theranostic nanoprobe for precision tumor-targeting and fluorescence-guided photothermal therapy *Nat. Commun.* **8** 14998

[23] Li C *et al* 2018 Light-responsive biodegradable nanorattles for cancer theranostics *Adv. Mater.* **30** 1706150

[24] Geva T 1991 Introduction: magnetic resonance imaging *Pediatr. Cardiol.* **21** 3–4

[25] Condeelis J and Weissleder R 2010 *In vivo* imaging in cancer *Cold Spring Harb. Perspect. Biol.* **2** a003848

[26] Cheon J and Lee J H 2008 Synergistically integrated nanoparticles as multimodal probes for nanobiotechnology *Acc. Chem. Res.* **41** 1630–40

[27] Smith B R and Gambhir S S 2017 Nanomaterials for *in vivo* imaging *Chem. Rev.* **117** 901–86

[28] Key J and Leary J F 2014 Nanoparticles for multimodal *in vivo* imaging in nanomedicine *Int. J. Nanomed.* **9** 711–26

[29] Carr D H *et al* 1984 Gadolinium-DTPA as a contrast agent in MRI *Am. J. Roentgenol.* **143** 215–24

[30] Blasiak B, van Veggel F C J M and Tomanek B 2013 Applications of nanoparticles for MRI cancer diagnosis and therapy *J. Nanomater.* **2013** 148578

[31] Jun Y-W *et al* 2005 Nanoscale size effect of magnetic nanocrystals and their utilisation for cancer diagnosis via magnetic resonance imaging *J. Am. Chem. Soc.* **127** 5732–3

[32] Shanavas A, Sasidharan S, Bahadur D and Srivastava R 2017 Magnetic core–shell hybrid nanoparticles for receptor targeted anti-cancer therapy and magnetic resonance imaging *J. Colloid Interface Sci.* **486** 112–20

[33] Weinstein J S *et al* 2010 Superparamagnetic iron oxide nanoparticles: diagnostic magnetic resonance imaging and potential therapeutic applications in neurooncology and central nervous system inflammatory pathologies, a review *J. Cereb. Blood Flow Metab.* **30** 15–35

[34] Fass L 2008 Imaging and cancer: A review *Mol. Oncol.* **2** 115–52

[35] Erathodiyil N and Ying J Y 2011 Bioimaging applications *Acc. Chem. Res.* **44** 925–35

[36] Pysz M A, Gambhir S S and Willmann J K 2010 Molecular imaging: current status and emerging strategies *Clin. Radiol.* **65** 500–16

[37] Hedayatnasab Z, Abnisa F and Daud W M A W 2017 Review on magnetic nanoparticles for magnetic nanofluid hyperthermia application *Mater. Des.* **123** 174–96

[38] Kitture R *et al* 2012 Fe_3O_4-citrate-curcumin: Promising conjugates for superoxide scavenging, tumor suppression and cancer hyperthermia *J. Appl. Phys.* **111** 064702

[39] Bañobre-López M, Teijeiro A and Rivas J 2013 Magnetic nanoparticle-based hyperthermia for cancer treatment *Rep. Pract. Oncol. Radiother.* **18** 397–400

[40] Hayashi K *et al* 2014 Magnetically responsive smart nanoparticles for cancer treatment with a combination of magnetic hyperthermia and remote-control drug release *Theranostics* **4** 834–44

[41] Rejinold N S, Jayakumar R and Kim Y-C 2015 Radio frequency responsive nano-biomaterials for cancer therapy *J. Control. Release* **204** 85–97

[42] Liu Z, Kiessling F and Gätjens J 2010 Advanced nanomaterials in multimodal imaging: design, functionalization, and biomedical applications *J. Nanomater.* **2010** 894303

[43] Palmowski M *et al* 2008 Molecular profiling of angiogenesis with targeted ultrasound imaging: early assessment of antiangiogenic therapy effects *Mol. Cancer Ther.* **7** 101–9

[44] Rapoport N, Gao Z and Kennedy A 2007 Multifunctional nanoparticles for combining ultrasonic tumor imaging and targeted chemotherapy *J. Natl. Cancer Inst.* **99** 1095–106

[45] Yin T *et al* 2013 Ultrasound-sensitive siRNA-loaded nanobubbles formed by hetero-assembly of polymeric micelles and liposomes and their therapeutic effect in gliomas *Biomaterials* **34** 4532–43

[46] Zhang X-D *et al* 2012 Size-dependent radiosensitization of PEG-coated gold nanoparticles for cancer radiation therapy *Biomaterials* **33** 6408–19

[47] Xiao Q *et al* 2013 A core/satellite multifunctional nanotheranostic for *in vivo* imaging and tumor eradication by radiation/photothermal synergistic therapy *J. Am. Chem. Soc.* **135** 13041–8

[48] Lai C-W *et al* 2008 Iridium-complex-functionalized Fe3O4/SiO2 core/shell nanoparticles: a facile three-in-one system in magnetic resonance imaging, luminescence imaging, and photodynamic therapy *Small* **4** 218–24

[49] Liu Y, Chen W, Wang S and Joly A G 2008 Investigation of water-soluble x-ray luminescence nanoparticles for photodynamic activation *Appl. Phys. Lett.* **92** 043901

[50] Zhang C *et al* 2015 Marriage of scintillator and semiconductor for synchronous radiotherapy and deep photodynamic therapy with diminished oxygen dependence *Angew. Chem. Int. Ed.* **54** 1770–4

Chapter 3

Externally/physically stimulated breast cancer nanomedicine

Sachin V Otari, Joanna Bauer and Nanasaheb D Thorat

The treatment of breast cancer is as challenging as many other cancers due to its high metastasis rate and extremely drug resistant nature. The traditional cancer therapies, i.e. surgery, radiation, and chemotherapy, are essential and effective for limiting the cancer progress rate, but come with significant adverse side effects making the breast cancer patient's life extremely difficult during the course of treatment. The advancement in the current treatment procedure has come with minimal adverse effects, not harming normal healthy cells and providing a better quality of life for the patient. The nanotechnology has provided a major advantage over current therapeutic platforms where the simultaneous diagnosis, treatment and monitoring of responses is practical. Several nanotherapeutic based drugs are currently available in the market for clinical application and several others are in clinical trials. In recent years, remotely triggered cancer therapy based on nano-technology which can be controlled by external or physical stimuli allowing targeted delivery at the tumor site without insertion of a probe, has gained attention and made extensive progress. The external stimuli are photodynamic, photothermal, phototriggered chemotherapeutic release, ultrasound, electro-thermal, magneto-thermal, x-ray, and radiofrequency therapies which control the activity of the nanotherapeutic molecule or release of the cargo at the targeted sites. Various nanostructures based on polymers, inorganic, organic, metallic or non-metallic, magnetic structures have been engineered with imaging molecules, sensitizers or drugs for the externally triggered nanotherapeutic application against breast cancer cells. This chapter discusses briefly the remotely triggered nanotherapeutic platform studied against breast cancer cells and the possible clinical studies. The emergence of this new field along with convectional therapies may have the solution for current healthcare issues but there is desperate need for advancement in the therapeutic platforms due to progress in diseases like breast cancer recurrence.

3.1 Introduction

Breast cancer is the most commonly occurring cancer in women and the second most common cancer overall. There were over two million new cases in 2018 [1]. Though the disease has spread in all races and ethnicity in recent decades, the overall invasive breast cancer incidence rate has been stable or slightly reduced, varying by race and age. The stable or declined rate of breast cancer mortality is due to improved treatment procedures and early detection methods [2]. The use of various available treatments like radiotherapy, chemotherapy, surgery, hormonal therapy etc, has effectively reduced the cancer progress rate [3]. With the developments in the nanotechnology field, extensive research has been focused on the introduction of new and advanced treatment methods based on the nanoparticles, which is an interdisciplinary field focused on design and engineering of the devices based on nanotechnology [4]. Due to the nanoscale size with high surface-to-volume ratio and easy surface modifications, nanoparticles with anticancer drugs can enhance the therapeutic index by modifying pharmacokinetic and pharmacodynamic profiles of drugs. Also, nanomedicine compounds can alter the biodistribution of the drugs and accumulation at the tumor site by enhancing the permeability and retention effect (EPR). Thus nanotechnology based new medical treatment procedure is providing several advantages over the conventional one, including biocompatibility, specificity, passive and active targeting, reduced side effects, engineering multifunctional active molecules [5–12].

A variety of the nanostructure-based polymers, proteins, lipids, organic, and inorganic compounds have been developed for the construction of new cancer therapeutic molecules (figure 3.1) [13, 14]. There are several nanocomposites composed of nanostructures and anticancer drugs on the market which are approved for use as treatment and under clinical trials. The liposomal based doxorubicin, i.e. Doxil™/Caelyx™ was the first anticancer nanomedicine approved in 1995 by the FDA [14–16]. In treatment, Doxil™ showed a nearly 300-fold increased efficiency compared to free doxorubicin. There are several other nanoparticle-based drugs available on the market, e.g. Abraxane® (ABI-007) [Abraxis/Celgene], DaunoXome® [Galen], DepoCyt® [Pacira], Doxil®/Caelyx® [Johnson & Johnson], Genexol-PM® (IG-001) [Samyang Biopharm] for the treatment of breast cancer, pancreatic cancer, non-small-cell lung cancer, Kaposi's sarcoma, neoplastic meningitis, multiple myeloma etc [17–20]. Several other nanodrugs are still under clinical trials [13].

Recently, stimuli responsive nanomaterials for drug release that utilize physical, chemical, or biological stimuli to facilitate drug delivery at targeted sites enhancing the bioaccumulation of the drug in the tumor cells have been developed. This minimizes the systemic exposure of the cancer drugs, reducing the toxicity of the drugs towards healthy cells. These triggers can be internal/endogenous stimuli like pH, ionic strength, shear stress, nucleic acids, peptides, electron transfer reactions in the target tissues [21, 22]. For example, the pH in the exterior part of the tumors is

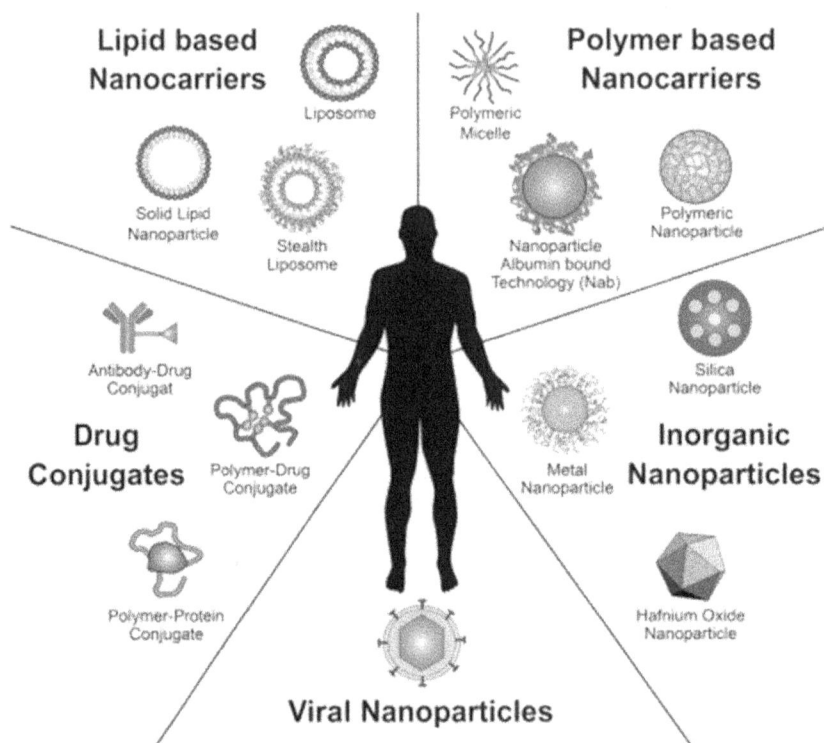

Figure 3.1. Schematic illustration of established nanotherapeutic platforms used for cancer therapy [13], reproduced with permission © ESO.

more acidic than the systemic pH [23]. Using this condition, the pH responsive liposome encapsulated drug can be released in the tumor site increasing the efficiency of the anticancer drugs [24]. In another example, due to over-expression of proteases and glucuronidases on tumor cells, multistage gelatin nanoparticles can be used to penetrate tumor cells after MMP-2 triggered shrinkage overcoming anatomical and physiological barriers [25]. Physical stimuli are usually applied externally to trigger the release of drugs from the nanocarriers. These physical stimuli may involve ultrasound, electromagnetism, light, and temperature for release of the cargo at the desired location targeting specific tissues [26, 27]. For instance, thermoresponsive nanocarriers such as ThermoDox, a phase III nanocarrier, can be used to release the doxorubicin drug on local hyperthermia induction. In addition, local hyperthermia increases the vascular permeability which facilitates anticancer drug delivery deep into the tumor [28]. The conducting polypyrrole nanoparticles were demonstrated for the drug release on an external electric field [29].

This chapter briefly discusses the various externally controlled nanotheranostics for breast cancer applications with their development and clinical potential. These remotely or physically stimulated theranostic moieties consist of engineered nanoparticles with anticancer molecules and external triggers like magnetic field, x-ray, radiofrequency, ultrasonic waves and light. These externally stimuli-based nanomedicines have shown their potential application *in vitro* breast cancer treatment, whereas the use of these moieties for clinical purposes need to overcome the challenges by extensive research and development. The following chapter gives a brief summary of the treatments which are traditionally- and nanotechnology-based for breast cancer treatment and briefly discusses the various physically stimuli-based theranostic breast cancer examples focusing on possible *in vivo* application.

3.2 External/physical nanomedicine for breast cancers

The term 'theranostic' was first coined by Funkhouser, it represents the use of the treatment procedure for therapy as well as for diagnostic purpose [30]. The ultimate purpose of this treatment procedure is to monitor the progress of the cancer cells, target the delivery of the anticancer drugs and analyze the reaction of the drugs towards cancer cells. The use of nanomaterials for the theranostic application has led to a rise in new emerging therapeutics, i.e. 'nanotheranotics' therapy, where nanomaterials applied for the diagnosis and therapeutic purpose also act as the drug or carrier for the drug at targeted site or enhance the efficacy of the anticancer drugs. Due to the small size and large surface area, these nanoparticles can not only be modified based on the desired functionalization, allowing the modified drug to circulate in the blood and deliver at the target site, but also allow high loading of the imaging and therapeutic drugs on the surface, lowering the side effects of the drugs to the healthy and normal cells over standard chemotherapeutic drugs. The intrinsic optical properties of the nanoparticles allow them to be used as active molecules for diagnostic purposes. The properties of the nanoparticles have created 'smart materials' which can be controlled externally and deliver the drugs, or act as active drug molecules, reducing the side effects of conventional medicine. The external stimuli can be magnetic field, ultrasound, radiofrequency wave, near infra-red light (NIR), x-rays, and electric field. The extensive *in vitro* and *in vivo* study has been performed for the demonstration of nanothernostic application of physically stimulated cancer nanomedicine.

This section briefly describes the use of externally stimulated nanothernostic for the treatment of breast cancer which is remotely controlled with magnetic field, electric field, NIR, x-rays, and sound waves and challenges for these nanothernostic strategies to be used for clinical trials.

3.2.1 Magnetic field

The delivery of the anticancer drugs at the targeted site is a major challenge not only in traditional treatment procedures but also in advanced nanothernostic treatment strategies, as the toxicity of drugs causes serious health hazards and wastes essential time during the treatment procedure. So, the construction of a drug delivery system (DDS) which can be modulated from outside the body is an essential requirement in targeted tumor cell treatment. For this purpose, the magnetic field has been extensively studied for the development of an effective DDS in the treatment of breast cancer, lung cancers and various other types of cancers.

For the stimulation of the body and blood circulation, a magnetic field is very effective. In tissues, small electric currents can be produced by alternating magnetic fields (AMF) or pulsed magnetic fields (PFM) which is directly proportional to the field frequency affecting the growth of the low oxygenated and low pH cells. AMF is also used for targeted drug delivery using large magnetic beads as DDS which can be activated through the oscillating magnetic field. Due to its small difference in the 'on' and 'off' state, this DDS system was not useful for the any clinical trials. With recent advancements, the PMF can be used for the activation of localized hyperthermia in cancer cells where magnetic nanoparticles (MNPs) are accumulated. Further, based on the AMF the magnetothermal responsive or magnetically triggered thermally-sensitive nanomaterials are designed for construction of magnetic DDS. In magnetic field-controlled DDS, the magnetic hyperthermia and thermal responsive or thermally rupturable layer are essential requirements.

Thorat *et al* developed the synergetic cancer therapy including magnetic fluid hyperthermia (MFH) guided by magnetic resonance imaging (MRI) [32]. The superparamagnetic $La_{0.7}Sr_{0.3}MnO_3$ nanoparticles (SPMNPs) with an oleic acid–polyethylene glycol (PEG) polymeric micelle (PM) structure was constructed with the anticancer cancer drug doxorubicin (DOX) in a high loading capacity ($\sim$60.45%) for *in vitro* delivery into breast cancer cells (MCF-7) (figure 3.2). Here the free DOX caused approximately 75% inhibition of cancer cells, whereas the SPMNPs with anticancer cargo affected approximately 90% of cancer cell inhibition on 24 h of incubation. The *in vivo* studies on mice showed effective distribution of the organs with no observable harm. From the study it was concluded that the polymeric micelle SPMNPs reported here can serve as a promising candidate for effective multimodal cancer theranostic such as in the combined chemotherapy–hyperthermia cancer therapy.

Chen *et al* developed a novel and facile strategy to fabricate highly stable iron nanoparticles with a silica nanocapsule and biocompatible polymeric layer as nanocarriers for pH triggered release for anticancer drug (doxorubicin, DOX) and MRI imaging in breast cancer cells [33]. The alginate polymeric capsule filled with DOX and imaging molecules was modulated by a magnetic core which was used to inhibit the breast cancer cells and imaging of the cancer cells using MRI (figure 3.3). Similarly, Brulé *et al* used magnetic hyperthermia and temperature responsive polymeric encapsulation for the effective targeted delivery of anticancer drugs in breast cancer cells. The magnetic alginate microbeads were encapsulated with DOX which was irradiated with AMF of frequency 700 kHz causing release of the DOX at the MCF-7 cells under magnetic field hyperthermia.

Hyperthermia combined with radiation reduces the total dose of radiation needed compared to radiation alone when a higher dose is needed to obtain the same effect [32]. Hyperthermia has been under active investigation as a cancer treatment for many years. Although it has been used as a single modality obtaining a high rate of complete response (38.5%) in chest wall recurrences, it is not recommended as a single therapy since only small and superficial tumors can be adequately heated and the response duration is short. So, extensive study is required to use the magnetic field for the recurrent breast cancer.

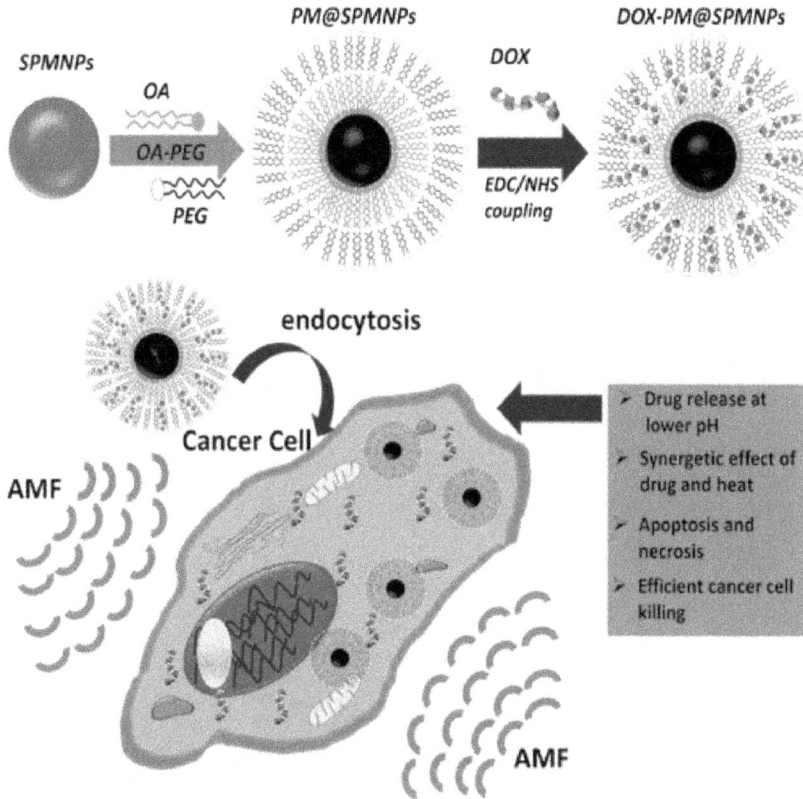

Figure 3.2. Schematics showing the overall mechanism of anticancer drug (DOX) conjugation into polymer encapsulated SPMNPs and the concept of killing cancer cells through synergetic therapy using an alternating magnetic field (AMF) [31], reproduced with permission © ESO.

3.2.2 Ultrasound field

Ultrasound (US) has been widely explored for its possible use in screening for diseases, assessing tissue condition and treatment in diseased state. The intrinsic property of ultrasound is being explored as an external controller for drug delivery from drug delivery cargo for pulsatile release. As the ultrasound method is non-invasive and the efficiency is based on the frequency of the ultrasound, it can be used to target the tissue in depth for soft tissue throughout the body [34]. The interaction between a drug carrier or drug delivery cargo and acoustic wave is essential in ultrasound assisted DDS. The design of the carrier is the main task in US assisted treatment which is responsive to ultrasound and non-toxic to human cells. These carriers should be able to carry large cargos and distribute evenly and remain in the cancer cells. By forming cavitation bubbles, localized tissue heating, and radiation force caused by the ultrasound in the biological environment is also used for controlled drug release from the nanocargo from the blood vessels to tumor site and to increase the penetration of the drug [35]. The ultrasound system can be used to

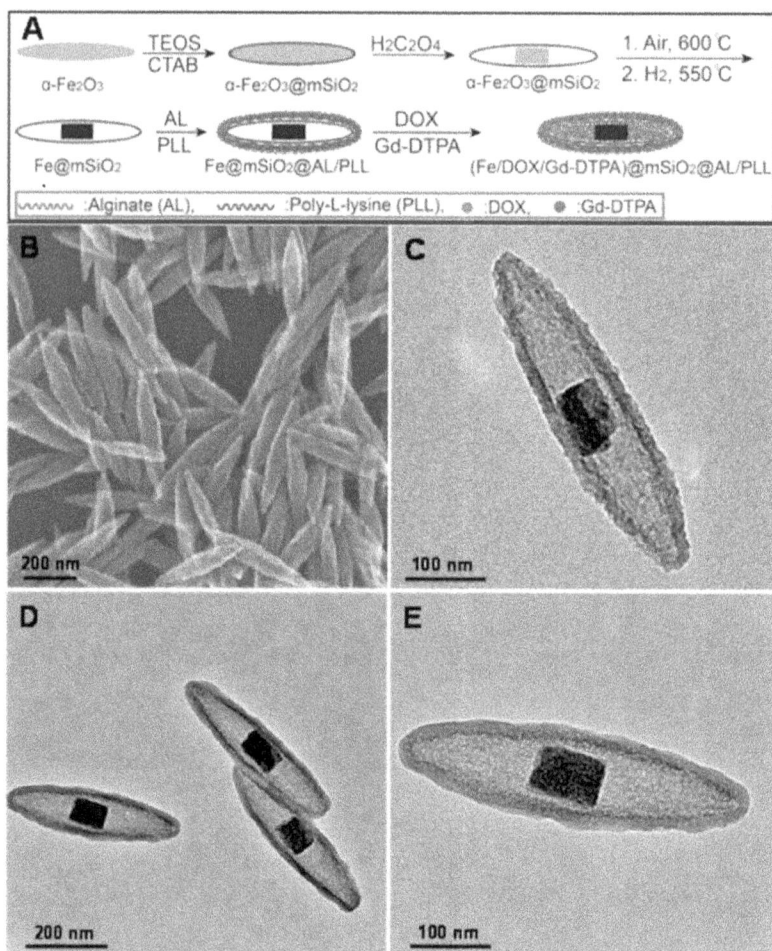

Figure 3.3. (A) Schematic showing the synthesis route of sodium alginate and poly-L-lysine coated nanoeye for pH triggered drug release, (B) SEM image of α-Fe$_2$O$_3$ nanorice, (C) TEM image of silica coated iron nanoeyes (Fe@SiO$_2$), (D) TEM image of sodium alginate and poly-L-lysine coated Fe@SiO$_2$ nanoparticles (Fe@SiO$_2$@AL/PLL) at low magnification, (E) TEM image of sodium alginate and poly-L-lysine coated Fe@SiO$_2$ nanoparticles (Fe@SiO$_2$@AL/PLL) at high magnification [33], reproduced with permission © ESO.

deliver temperature sensitive DDS which is responsive to small temperature changes at the tumor site.

Rapoport *et al* demonstrated the use of a technology that combined ultrasound imaging with ultrasound-mediated nanoparticle-based targeted chemotherapy where mixtures of drug-loaded polymeric micelles and perfluoropentane (PFP) nano/microbubbles were stabilized by the same biodegradable block copolymer [36]. The effect of the nano/microbubbles on the ultrasound-mediated cellular uptake of doxorubicin (Dox) in MDA MB231 breast tumors *in vitro* and *in vivo* (in mice bearing xenograft tumors) was studied with flow cytometry. Upon intravenous injection into mice and applying 20 kHz ultrasound externally, Dox-loaded micelles and nanobubbles extravasated selectively into the tumor interstitium, where the nanobubbles coalesced to produce microbubbles with a strong, durable ultrasound contrast.

Bai *et al* constructed the drug delivery cargo-based polymeric nanoparticles loaded with cancer resistance protein (ABCG2)-small interfering RNA (siRNA) for the treatment of Adriamycin resistant breast cancer cells which overexpresses the ABCG2 protein [37]. This cargo was delivered using ultrasound targeted micro-bubble destruction (UTMD) in breast cancer cells. It was found that the intracellular Adriamycin concentration was increased upon delivery of the ABCG2-small siRNA polymeric nanoparticles increasing the susceptibility of MCF-7/ADR (ADR resist-ant human breast cancer cells) towards Adriamycin. The siRNA-loaded NPs with UTMD + ADR showed better tumor inhibition effect and good safety *in vivo*. These results indicate that Adriamycin chemotherapy in combination with ABCG2-siRNA is an attractive strategy to treat breast cancer. Marino *et al* describe an innovative piezoelectric nanoplatform consisting of biocompatible piezoelectric barium titanate nanoparticles (BTNPs) functionalized with anti-HER2 antibody (Ab-BTNPs) (figure 3.4) [38]. Anti-proliferative effects of remote piezoelectric stimulation (US+Ab-BTNPs) on SK-R-3 breast cancer cells were verified in terms of cell cycle progression analysis, expression of proliferation markers, and metabolic activity of cell cultures. Since drug resistance of cancer cells can be counteracted by applying chronic electric stimulations, piezoelectric nanomaterials will be included in multifunctional nanosystems able to target cancer cells, locally release anticancer drugs, and remotely deliver electrical stimulations, in order to both inhibit cancer growth and enhance the cytotoxic effects of chemotherapy drugs. The use of virotherapy with focused ultrasound and polymeric nanoparticles for the inhibition

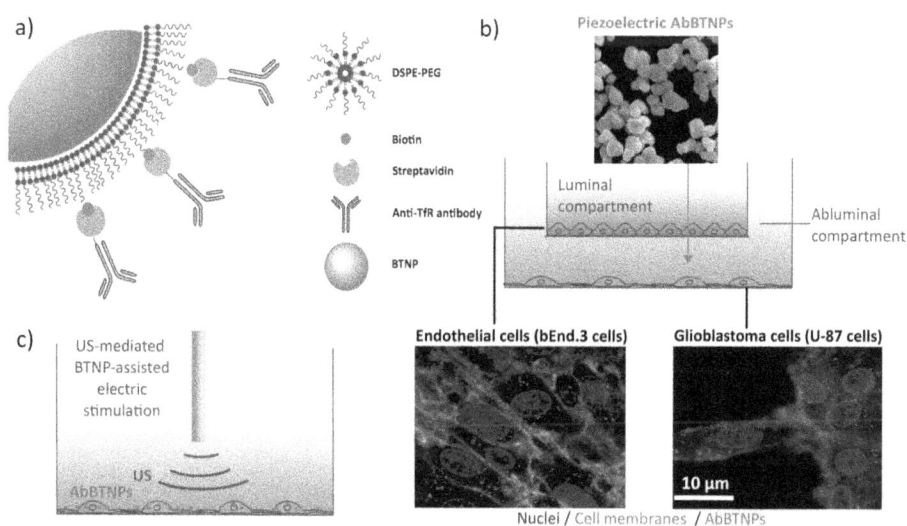

Figure 3.4. Schematic representation of functionalization of BTNP nanoparticles with antibodies against transferrin receptor (TfR) for ultrasound assisted drug delivery in glioblastoma cells [38], reproduced with permission © ESO.

of proliferation of breast cancer cells was demonstrated by Carlisle *et al* [39]. Combining drug stealthing and ultrasound-induced cavitation may ultimately enhance the efficacy of a range of powerful therapeutics, thereby improving the treatment of metastatic cancer.

3.2.3 Radiofrequency mediated hyperthermia

Radiofrequency mediated ablation uses heat made by radio waves to treat cancer. Radiofrequency ablation (RFA) provides an effective technique for minimally invasive tissue destruction. An alternating current delivered via a needle electrode causes localized ionic agitation and frictional heating of the tissue around the needle. The efficacy of the radiofrequency has been enhanced using gold and iron nanoparticles. With the use of stimuli responsive polymeric material and metal nanoparticles, external stimuli responsive DDS cargo carrying an anticancer drug can be constructed to deliver drugs at the targeted site and better thermal ablation. The nanomaterials which have the ability to heat up in the RF field are used for the RFA therapy. Metallic and non-metallic materials, such as gold, iron oxide, cobalt, carbon-based nanomaterials, and quantum dots (QDs) have been demonstrated for significant RF responsive heating in the *in vitro* experimental conditions (figure 3.5).

Cetuximab (C225) is a monoclonal antibody which is widely used to target epidermal growth factor receptor (EGFR) in many cancers. Glazer *et al* used C225 conjugated QDs such as cadmium–selenide and indium–gallium–phosphide to demonstrate concentration dependent RF heating *in vitro* in high EGFR^{+ve} pancreatic carcinoma cells (Panc-1), and low EGFR^{+ve} breast carcinoma cells,

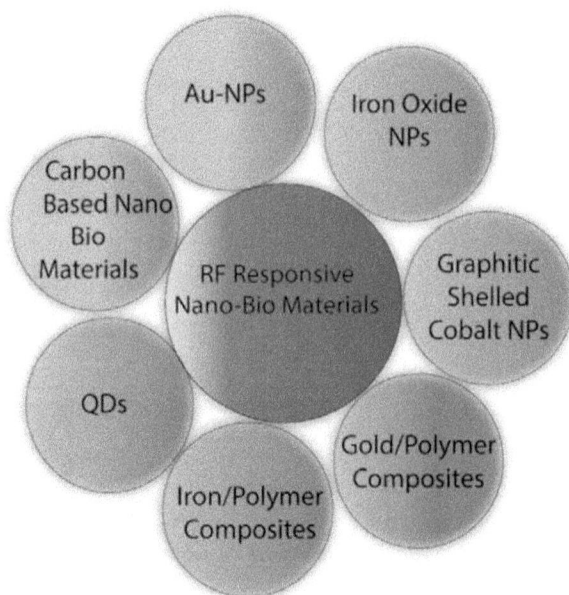

Figure 3.5. The major RF responsive metallic and non-metallic nanoparticles [42], reproduced with permission © ESO.

Cama-1 where selective RF response was observed based on EGFR concertation [40]. Karmakar and co-workers showed that carbon coated iron nanoparticles (C/Fe-MNPs) can be used as RF responsive nanocomposites for the destruction of cancer cells [41]. In the research work, to enhance the accumulation of the RF responsive nanocargo, the epidermal growth factor (EGF) was used for the functionalization of the C/Fe-MNPs which increased the uptake of the EGF functionalized C/Fe-MNPs to the EGF^{+ve} cancer cells such as MCF-7 (breast cancer cells) and Panc-1. A short time RF excitation of 10 min treatment, inhibited the growth of MCF-7 breast cancer cells showing approximately less than 3% cell viability after treatment with EGF modified C/Fe-MNPs (figure 3.6).

The most common and effective practice in the DDS is the use of polymeric materials carrying cargo responsive to external stimuli which can be remotely triggered to release the chemotherapeutic agents at the targeted sites. The RF responsive Au-NPs with MnO_2 nanorods for effective RF responsive cargo encapsulated in chitin polymer (ACM-TNGs) making a multicomponent nanogel for effective RF therapy [43]. On exposure to RF of 20–100 W, the heating effect of the ACM-TNGs increased compared to the cargo only containing MnO_2 or Au-NPs. Under 100 W of RF exposures for 2 min, the ACM-TNGs treated human ductal carcinoma cells (T47D) were found to be dead. For a better understanding of the toxicity of this RF responsive cargo a detailed toxicity profiling is essential with animal studies to properly understand its *in vivo* behavior. Rejinold and co-workers used curcumin, a natural phyto anticancer agent, for RF treatment encapsulated in thermo, pH, and RF responsible chitosan-graft-poly(*N*-vinyl caprolactam) NPs with

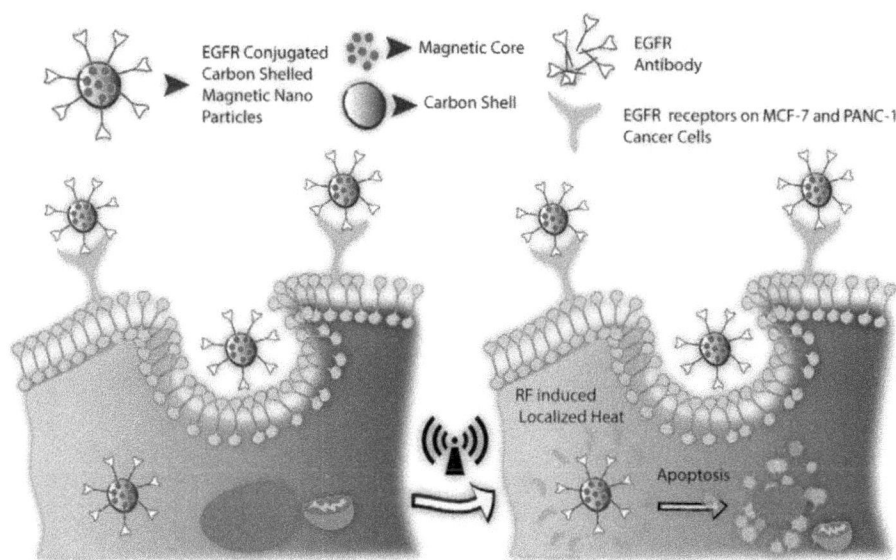

Figure 3.6. Schematic representation for the selective internalization of EGFR modified carbon shelled magnetic nanoparticles in EGFR^{+ve} cancer cells followed by apoptosis on RF excitation [42], reproduced with permission © ESO.

10 nm sized Au-NPs [44]. The physiological and RF responsive cargo designed for the targeted delivery of the curcumin and effective thermal ablation in breast cancer cells. On exposure for 5 min under RF condition of 40 W released curcumin in 4T1 mammary ductal carcinoma cells leading the induction of apoptosis. To analyze the response of the chitosan-graft-poly(N-vinyl caprolactam) NPs at low RF condition, the Au-NPs were replaced with Fe_3O_4 nanoparticles which was found to be useful in pH responsive cargo. Though researchers have demonstrated the *in vivo* accumulation of these NPs in 4T1 tumor models, the *in vivo* studies involving the release of the anticancer agent under RF condition from the drug delivery cargo would have to be clarified with application to the dimension of the NPs.

3.2.4 X-ray irradiation

X-ray irradiation is also an important remotely triggered treatment which can be non-invasive but can penetrate deep in the soft tissue for cancer treatment. The combination of the nanotechnology and irradiation technology has created tremendous opportunities for the development of rapid and effective diagnosis and therapeutic designs. In one study, EGFR-2 functionalized Au-NPs by conjugating trastuzumab (Herceptin) to 30 nm Au-NPs forming Au-T nanocomposite and administrated in athymic mice bearing subcutaneous MDA-MB-361 xenografts (breast cancer cells) which is further treated with a single dose of 11 Gy of 100 kVp x-rays 24 h after intra-tumoral injection of Au–T (~0.8 mg of Au) or no X-radiation (figure 3.7) [45]. The clonogenic survival of cells exposed to Au–T and x-rays was significantly lower than that of cells exposed to X-radiation alone, which translated to a dose enhancement factor of 1.6. In contrast, the survival of cells exposed to Au–P and x-rays versus X-radiation alone were not significantly different. *In vivo*, the combination of Au–T and X-radiation resulted in regression of MDA-MB-361 tumors by 46% as compared to treatment with X-radiation (16.0% increase in tumor volume). No significant normal tissue toxicity was observed. Radiosensitization of breast cancer to X-radiation with Au-NPs was successfully achieved with an optimized therapeutic strategy of molecular targeting of HER2 and intratumorally administration. With further detailed toxicity analysis and an increased mouse study sample size, the molecularly designed Au-NPs can be used for the clinical trials.

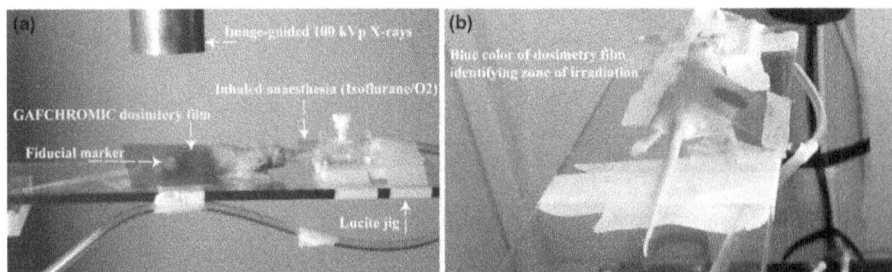

Figure 3.7. Experimental set-up for delivery of image-guided X-radiation. (a) anesthesia for the mice having HER2-overexpressing MDA-MB-361 human breast cancer xenografts which is immobilized on fiber table; (b) the placement of dosimetry film for the calculation of radiation dose for each mouse [45], reproduced with permission © ESO.

Kong and co-workers used the Au-NPs and modified Au-NPs having two kinds of functionalization, cysteamine (AET) and thioglucose (Glu) for the enhancement of cellular uptake and radiation cytotoxicity in a breast-cancer cell line (MCF-7) in comparison with a nonmalignant breast-cell line (MCF-10A) [46]. The transmission electron microscopy (TEM) study revealed the distribution of modified Au-NPs in MCF-7 cells where AET–Au-NPs were mostly bound to the MCF-7 cell membrane, while Glu–Au-NPs enter the cells. Upon irradiation with low-energy 200 kVp x-rays, the Au-NPs showed 30%–40% increased cytotoxicity. The detailed *in vivo* and *in vitro* studies are required for the clear understanding of toxicity mechanism which can be further used for the therapeutic application. Chattopadhyay *et al* demonstrated x-ray irradiation induced cytotoxicity of EGFR targeted trastuzumab factionalized PEG–Au-NPs in locally advanced breast cancer cells which increased 40% toxicity towards cancerous cells compared to naked Au-NPs.

HER-2 receptor mediated internalization Au-NPs could be trapped in endosomes or be deposited in the cytoplasm where they would interact with X-radiation to produce damaging secondary electrons. The fate of the EGFR receptor associated with Au-NPs is unknown but following internalization it could undergo degradation or be recycled back to the cell surface. Internalized Au-NPs irradiated with x-ray creating the Auger and photoelectrons enhancing the DNA damage of the cells. Further systematic and comparative studies are needed to achieve the best *in vivo* radiosensitization effect with advancement of the instruments which facilitates easy detection and treatment of breast cancer cells.

3.2.5 Phototriggered theranostics

One of the most explored, non-invasive, and externally controllable phototriggered theranostics treatments for cancer cells which has attracted enormous attention in recent years is easily focused and tunable with light irradiation enabling localized treatment. In recent decades, new designed nanomaterials have been incorporated in phototriggered theranostic modalities enhancing the performance of the synergistic theranostic properties which indeed will reduce the required multiple rounds of administration, and minimize overdoses and side effects. The various aspects of phototriggered theranostics like photodynamic therapy, photothermal therapy, and chemotherapy with phototriggered drug release involve various forms of nano-materials, i.e. photoresponsive or functionalized with photosensitive biomaterials for targeted delivery of the anticancer chemicals, reducing the side effects of the drugs to healthy cells and facilitating the early diagnosis and tracking the treatment progress (figure 3.8) [46]. The photodynamic therapy is now being used for the cancer treatments, whereas photothermal therapy, and chemotherapy with phototriggered drug release are still undergoing clinical trials for cancer theranostic application.

3.2.5.1 Photodynamic therapy
The photosensitizers (PSs) are used in photodynamic therapy (PDT) which are administrated topically or systemically where the PSs get excited from ground states to produce reactive oxygen species (ROS) which activate the destruction pathway

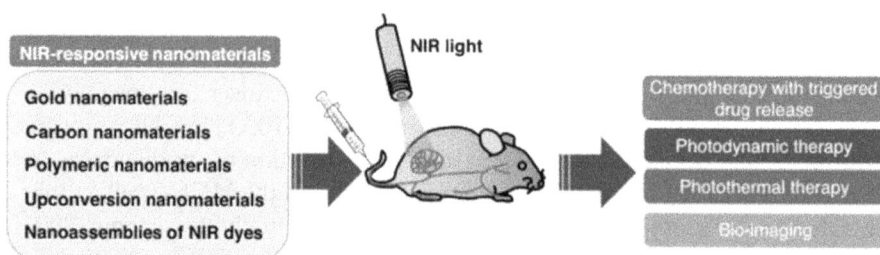

Figure 3.8. Schematic illustration of various kinds of near-infrared (NIR)-responsive nanomaterials investigated for cancer therapy, including for precisely controlled drug release, photodynamic therapy, photothermal therapy and bioimaging [46], reproduced with permission © ESO.

for cell death. The major role of the PSs is to absorb light energy in the form of photons and channel it towards the substrate. The targeted delivery of the PSs is affected due to the hydrophobicity of PSs, non-specific targeting, and toxicity which is overcome by using nanoparticles for the targeted delivery of PSs. Depending on the type of cancer, different forms of laser probe are used as light sources. Head, neck, ocular, and breast cancers can be treated with an external laser source, whereas intestinal or pancreatic cancers may be treated using a laser on the end of an endoscope.

Calavia *et al* used Au-NPs factionalized with a mixed monolayer of a zinc–phthalocyanine and a lactose derivative which utilizes the carbohydrate not only as stabilizer but also as the targeting agent for breast cancer cells [47]. The lactose functionalized Au-NPs were used for the targeting gelatin overexpressing breast cancer cells where the lactose functionalized Au-NPs on PDT studies *in vitro* showed approximately 97% cell inhibition. Lower growth inhibition was found for the low gelatin expressing cells in PDT studies using lactose functionalized Au-NPs. The use of lactose for the functionalization and target cancer cells has an advantage over the convectional antibody functionalized nanoparticles as cancer cells develop resistance towards antibodies. Further detailed toxicity analysis would require understanding the efficacy of the system to use it for clinical trials. Uppal and co-workers demonstrated that Rose Bengal, an anionic photosensitizer, modified silica nanoparticles having 3-amino propyl groups can be used for the treatment of breast cancers and oral cancer showing effective photoablation activity with low cytotoxicity [48].

The cancers cells are identified with the over-expression of several cell receptors as estradiol receptors, the human epidermal growth factor receptor 2, gonadotropin-releasing hormone receptors, and tisular factor VII receptors etc [49]. The researchers used these receptor molecules effectively to guide the PS-nanoparticles for targeted delivery, reducing the toxicity to normal cells and increasing the uptake of nanotherapeutic molecules, killing specifically cancer cells on red light exposure. Stuchinskaya *et al* used a 4-component nanoparticle conjugate containing a zinc–phthalocyanine derivative (photosensitiser), polyethylene glycol, and gold nanoparticle which is capable of specifically targeting and then photodynamically treating cancer cells. The PEG modified Au-NPs with monoclonal antibodies against ErbB-2 (HER2) to target HER2 overexpressing mammary epithelial cells

(MCF-10A) enhancing the PDT efficacy of the conjugate and lowering the toxicity of designed conjugates towards normal cells. Further investigation of the conjugates *in vivo* application would be useful for the development of the effective anticancer nanomedicine.

In the new era of nanomedicine, the use of upconversion nanoparticles (UCNPs) has increased interest for cancer treatment as these nanoparticles emit high energy photons with excitation by low-energy near infra-red light (NIR) [50, 51]. The UCNPs containing rare-earth elements exhibiting upconversion luminescence emission have applied for the biological labelling, sensing and imaging [52–55]. Also, now UCNPs have been demonstrated for nanotherapeutic application as the photons generated from the photosensitizer present on the nanoparticles killing cancer cells ROS generation, resonance energy transfer *etc* [56].

Wang *et al* used polymer coated UCNPs attached with chlorin e6 (Ce6) where Ce6 acted as a photosensitizer and the complex on exposure to NIR produced ROS to kill cervical cells and breast cancer cells, selectively affecting the growth of cancer cells after sensitization [56] (figure 3.9). *In vivo* studies of two groups of 4T1 tumor-bearing mice were irradiated to 660 nm and 980 nm and laser injected with the same concentration of Ce6 and UCNP–Ce6, respectively, at the optical power density (0.5 W cm^{-2}) for 30 min which inhibited the growth of cancer cells showing reduced tumor size. The biodistribution of the UCNP–Ce6 in the organs of the treated mice was studied extensively. Still further toxicity profiling towards mammalian cells of UCNP–Ce6 complex is required to be further used for clinical trials. Wang

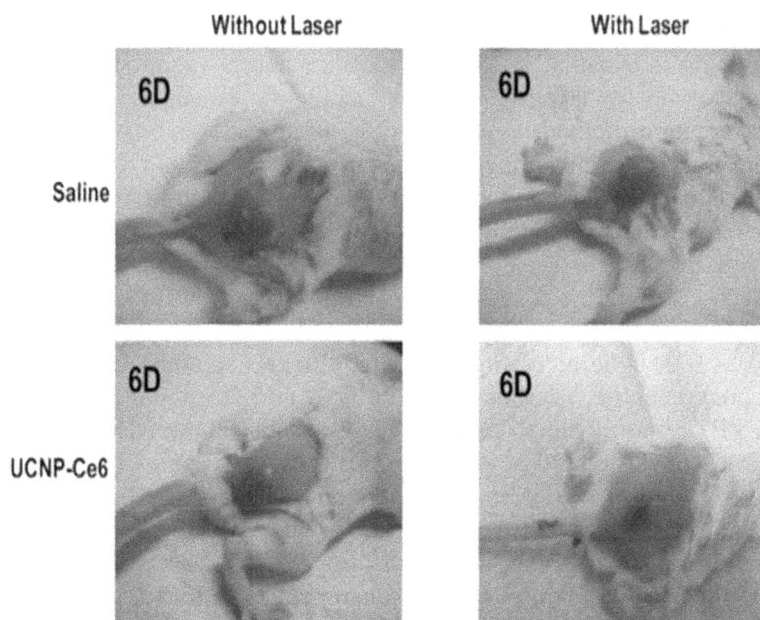

Figure 3.9. Photos of mice after tumor-injected with saline and UCNP–Ce6 treatments indicated on the sixth day with or without NIR irradiation [57], reproduced with permission © ESO.

and co-workers used a 'see and treat' strategy where multifunctional biodegradable polyacrylamide nanoplatforms were used for cancer theranostic combining photo-dynamic therapy and fluorescence imaging under NIR irradiation [58].

Zang *et al* used folic acid (FA)-targeted $NaGdF_4$:Yb/Tm@SiO_2@TiO_2 nano-composites for both *in vivo* MRI and NIR-responsive inorganic PDT, containing NIR-responsive UCL which converts UV responsive TiO_2 into a NIR-responsive one [59]. The presented composite showed good T1-weighted MRI contrast agents with high longitudinal relaxivity (r1) for *in vitro* studies and showed bright signals in the MCF-7 tumor of nude mice. The composite also demonstrated for NIR-responsive nanotherapeutic application where HeLa and MCF-7 cells exposed to the engineered composite and irradiated with 980 nm laser at the power density of 0.6 W cm^{-2} for 20 min which affected the growth of cancer cells. For *in vivo* PDT of MCF-7 tumor-bearing mice model the TiO_2 based composite was injected at the tumor site and the inhibition ratio was examined after treatment for 2 weeks, showing excellent inhibitory effect on breast cancer cells. Due to the biocompati-bility of the composite, it can be further studied for the clinical applications in both MRI and NIR-responsive PDT of cancers in deep tissues.

3.2.5.2 Photothermal therapy
The photothermal therapy (PPT) involves the use of electromagnetic radiation (Vis and NIR light) causing thermal ablation. The photoabsorber utilizes light to raise the temperature between 45–300 °C leading to cell death following various mechanisms [60]. The NIR light can penetrate deep inside the body, so the desired depth in the treatment procedure can be achieved using PTT. Like PDT, PTT also can be used to target diseased tissue without harming healthy tissue. The choice of laser irradiation is essential in PTT as specific cancers required specific types of irradiation method for effective thermal therapeutic outcomes. PTT utilizes two different irradiation methods consisting of continuous laser or pulsed laser irradi-ation where each has its advantages and applications depending on the nature of the target tissues. In PTT, the use of active photoresponsive molecules is essential, being non-toxic to tissues unless irradiated by specific light causing changes in the behavior of the targeted tissue.

The advancement in the field of nanotechnology has provided breakthroughs for PTT where the use of various engineered plasmonic nanoparticles has enhanced the absorption of NIR light increasing the desired therapeutic efficacy. Among different nanostructures, metallic nanoparticles like gold and silver nanocomposites show promising PPT response due to their strong absorbance in the NIR region with high photothermal conversion efficacy [61]. The NIR-responsive nanomaterials like gold nanorods [62], gold nanospheres [63], gold nanoshells [64], carbon nanotubes [65] and other nanostructures [66] were utilized for successful PTT application. These nanostructures are further modified with antigens, antibodies or specific ligands to increase the accumulation of the nanotherapeutic platform at the targeted site. Kang *et al* reported a novel nanocomplex composed of Au rods, porphyrin and trastuzumab called TGNs complex for PPT application against HER2-positive breast cancers [67]. The photothermal ablation by TGN complex on NIR laser irradiation inhibited the growth of HER2-positive cancer cells and prohibited the progress of tumor growth in the mouse model having HER2 over-expressed breast cancer xenograft. The reported complex studied for *in vivo* toxicity showed potentially less toxicity towards healthy tissues. The silica–gold nanoshells conjugated with anti-HER2 antibody applied for PTT of trastuzumab resistance breast tumors inhibiting growth of the breast cancer cells after exposing to 808 nm diode laser with 80 W cm^{-2} for 5 min [64]. The thermal treatment progress was tracked by fluorescence microscope signifying plausible application of nanoshells-mediated ablation to treat different types of breast cancer cells.

Graphene and its derivatives have demonstrated possible application for biosensors, tissue engineering, drug and gene cargo carrier and delivery with plasmonic application [68–70]. Yang and co-workers demonstrated reduced graphene oxide with magnetic iron oxide functionalized with polyethylene glycol can be applicable for multimodal imaging photothermal therapy against murine 4T1 breast cancer cells with *in vivo* and *in vitro* studies on irradiation by the 808 nm laser at 0.5 W cm^2 for 5 min [71]. Immuno molecule functionalized chitosan coated-copper sulfide nanoparticles were studied extensively for photothermal immunotherapy against breast cancer cells *in vivo* and *in vitro* where the composite breaks down to polymer coated-copper sulfide nanoparticles on exposure to NIR laser developing the thermal ablation at the tumor site (figure 3.10) [72]. Beqa *et al* reported Au nanopopcorn attached single walled carbon nanotube for targeted diagnosis and selective photothermal treatment targeting SK-BR-3 human breast cancer cell which sensed 10 cancer cells/mL level, using surface enhanced Raman scattering of single wall carbon nanotubes [73].

3.2.5.3 Chemotherapeutic release on light trigger

The major challenge in chemotherapy for any type of cancer is the action of the drug at the desired site where conventional drug delivery systems have failed to achieve target due to non-specific targeted delivery of the drugs. This leads to damage to healthy cells and lowers the effectiveness of the chemotherapeutic drugs. The use of remotely controlled nanomaterials for the targeted delivery of the drugs at the desired site is highly desirable for potentially lowering the systemic toxicity of the anticancer drugs and enhancing the efficacy of the treatment procedures. Light triggered drug release provides a major advantage as the treatment is possible for tissues at any depth. Goodman *et al* demonstrated DNA-based and protein-based hosts where the nanocomplex of Au nanoshells consisting of SiO$_2$ as core for the near-IR light triggered release of two drug molecules on NIR irradiation [74]. Here, two potential optically active drug delivery molecules, thiolated dsDNA and the protein human serum albumin (HSA), conjugated onto Au NS demonstrated for NIR triggered release of two breast cancer drugs, i.e. docetaxel (DTX) and lapatinib (LAP) molecules in SKBR3 breast cancer cells where the complex showed promising drug release at breast cancer cells increasing concentrations of the drug in the targeted site compared to non-phototriggered complex.

Wang and co-workers designed green nanospheres consisting of the anticancer drug, doxorubicine, with calcium carbonate @silica nanosphere, chemotherapeutic

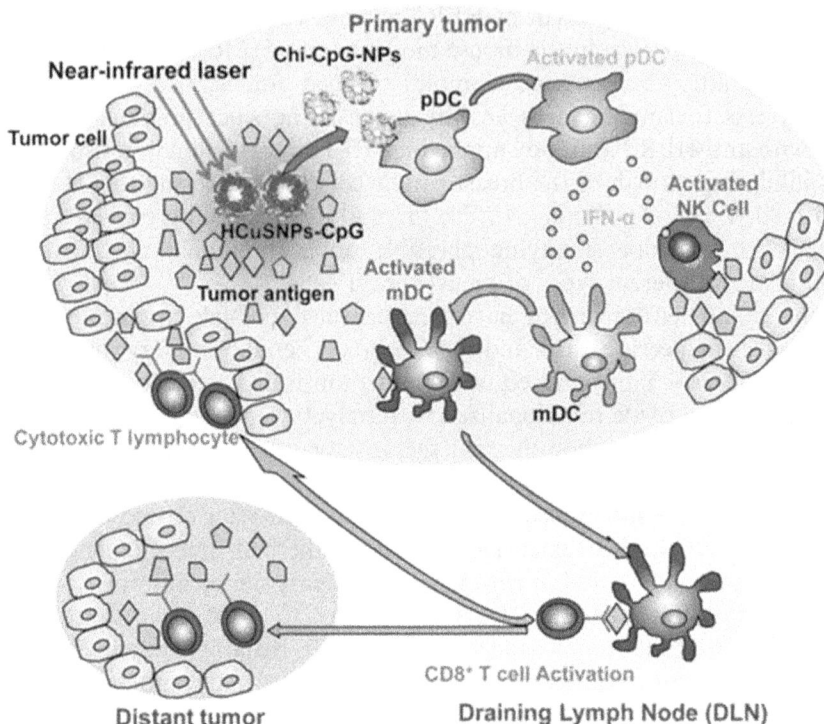

Figure 3.10. Schematics of HCuSNPs-CpG-mediated photothermal immunotherapy breast cancer tumors under near infrared laser irradiation [70], reproduced with permission © ESO.

release for drug resistant cancer therapy under infra-red irradiation against MCF-7/ADR breast cancer cells [75]. Manivasagan *et al* demonstrated folic acid-conjugated and doxorubicin-loaded chitosan oligosaccharide encapsulated gold nanorods for photochemical therapy which delivered the drugs on NIR irradiation for breast cancer cells [76]. The PTT was studied in MDA-MB-231 tumor-bearing nude mice exposed to an 808 nm laser (1.0 W cm^2) for 300 s at 4 h where after 18 days of treatment the tumor growth rate was reduced by 96.5% showing much less toxicity towards the cells. Though these results direct possible use in clinical trial, detailed study related to long-term toxicity needs to be studied thoroughly. There are several other designed phototriggerd chemotherapeutic molecules but these are still at early stages of experimentation and require long-term toxicity study.

3.3 Conclusion and future scope

Cancer nanomedicine has developed as an exciting branch of the therapeutic and diagnostic area which is rapidly gaining attention of researchers and physicians due to its advantages over the existing treatment procedures. The pharmacokinetics of conventional medicine is altered by nanotherapeutic moieties to be more effective at tumor sites. Massive development has occurred in nanotheranostics which is now utilizing biodegradable and biocompatible, multifunctional moieties encapsulating diagnostic agents or drugs [77–79]. These nanotheranostics have been designed to deliver biological, immunological, radiological, and genetic materials at specific sites. External stimuli responsive nanostructures have shown excellent promise for the treatment of cancer in preclinical models [80–83]. The ideal nanotheranostics deliver the cargo to the targeted tumor site on appropriate external triggers which allows the diagnosis and treatment of the cancer with tracking of the nanoparticles to record the response of cancer cells to the treatment molecule. To enhance the uptake or interaction of nanotheranostics with tumor cells, appropriate modifications are provided to nanotheranostics platforms. External stimuli mediated nanotheranostics are expected to deliver the diagnostic molecule or cancer nanomedicine without causing immunological response to tissue of interest, monitor the response, help to modulate the secondary treatment, and be compatible with healthy cells.

The development of new nanomaterials will be a crucial driver of progress in this field. The toxicity of the nanotherapeutic moieties is also a major concern in the clinical application of these engineered nanoparticles. Though these drug delivery molecules do not show any toxicity towards normal healthy cell lines for *in vitro* studies, their response in the human body must be analyzed in detail. In clinical trials, they may show adverse effects towards the human body. In breast cancer treatment locoregional procedures are involved where the specific site is easy to target compared to other cancer types. Great care must be taken by researchers to develop nanotherapeutic platforms which are not released in the systemic body and accumulate in other organs. The possible bioaccumulation or biomagnification of theses moieties and their long-term effects must be taken into consideration.

The locoregional recurrence of breast cancer is a major challenge for current cancer treatments where there is 4%–20% possibility of the breast cancer recurring after cancer therapy or radical mastectomy in women in phase I and II cancer [84]. The recurrence percentage increases if the patient is treated with a single therapy like chemotherapy or radiotherapy alone [85]. In the case of triple negative breast cancer and in primary inflammatory breast cancer the recurrence rate is very high with a high rate of mortality. These recurrences of breast cancer lead to severe health hazards, causing physical and mental illness. Very few studies have been performed using nanotherapeutic platforms against recurrent breast cancer. A comprehensive study must be done to treat recurrent breast cancer using external stimuli mediated nanotherapeutic moieties. The engineering advancement is not only required for DDS also for the external stimuli providing appliances.

Acknowledgments

Dr S V Otari acknowledges the 2015-KU Brain Pool Fellowship of Konkuk University. Dr Joanna Bauer and Dr Nanasaheb Thorat acknowledge that the project leading to this work has received funding from the European Union's Horizon 2020 research and innovation programme under the Marie Sklodowska-Curie grant agreement 751903.

References

[1] American Cancer Society 2017 *2011 American Cancer Society. Breast Cancer Facts & Figures 2017-2018* (Atlanta: American Cancer Society, Inc.)

[2] Anon 2006 Effect of screening and adjuvant therapy on mortality from breast cancer: Commentary Obstet *Gynecol. Surv.* **353** 1784–92

[3] Tran S, DeGiovanni P-J, Piel B and Rai P 2017 Cancer nanomedicine: a review of recent success in drug delivery *Clin. Transl. Med.* **6** 44

[4] Wang R, Billone P S and Mullett W M 2013 Nanomedicine in action: an overview of cancer nanomedicine on the market and in clinical trials *J. Nanomater.* **2013** 12

[5] Adair J H, Parette M P, Altinoğlu E I and Kester M 2010 Nanoparticulate alternatives for drug delivery *ACS Nano* **4** 4967–70

[6] Gao X, Cui Y, Levenson R M, Chung L W K and Nie S 2004 *In vivo* cancer targeting and imaging with semiconductor quantum dots *Nat. Biotechnol.* **22** 969–76

[7] Fan Z, Sun L, Huang Y, Wang Y and Zhang M 2016 Bioinspired fluorescent dipeptide nanoparticles for targeted cancer cell imaging and real-time monitoring of drug release *Nat. Nanotechnol.* **11** 388–94

[8] Frigerio B *et al* 2013 A single-chain fragment against prostate specific membrane antigen as a tool to build theranostic reagents for prostate cancer *Eur. J. Cancer* **49** 2223–32

[9] Lee G Y, Qian W P, Wang L, Wang Y A, Staley C A, Satpathy M, Nie S, Mao H and Yang L 2013 Theranostic nanoparticles with controlled release of gemcitabine for targeted therapy and MRI of pancreatic cancer *ACS Nano* **7** 732078–89

[10] Srinivasan S, Manchanda R, Fernandez-Fernandez A, Lei T and Mcgoron A J 2013 Near-infrared fluorescing IR820-chitosan conjugate for multifunctional cancer theranostic applications *J. Photochem. Photobiol. B Biol.* **119** 52–9

[11] Bwatanglang I B, Mohammad F, Yusof N A, Abdullah J, Hussein M Z, Alitheen N B and Abu N 2016 Folic acid targeted Mn:ZnS quantum dots for theranostic applications of cancer cell imaging and therapy *Int. J. Nanomed.* **11** 413–28

[12] Nagesetti A, Srinivasan S and McGoron A J 2017 Polyethylene glycol modified ORMOSIL theranostic nanoparticles for triggered doxorubicin release and deep drug delivery into ovarian cancer spheroids *J. Photochem. Photobiol. B Biol.* **174** 209–16

[13] Wicki A, Witzigmann D, Balasubramanian V and Huwyler J 2015 Nanomedicine in cancer therapy: Challenges, opportunities, and clinical applications *J. Control. Release* **200** 138–57

[14] Gabizon A, Shmeeda H and Barenholz Y 2003 Pharmacokinetics of pegylated liposomal doxorubicin: Review of animal and human studies *Clin. Pharmacokinet.* **42** 419–36

[15] Barenholz Y C 2016 Doxils—the first FDA-approved nano-drug: From basics via CMC, cell culture and animal studies to clinical use *Nanomedicines: Design, Delivery and Detection* ed M Braddock (RSC Drug Discovery Series) (London: Royal Society of Chemistry)

[16] O'Brien M E R *et al* 2004 Reduced cardiotoxicity and comparable efficacy in a phase III trial of pegylated liposomal doxorubicin HCl (CAELYXTM/Doxil®) versus conventional doxorubicin for first-line treatment of metastatic breast cancer *Ann. Oncol.* **15** 440–9

[17] Desai N *et al* 2006 Increased antitumor activity, intratumor paclitaxel concentrations, and endothelial cell transport of cremophor-free, albumin-bound paclitaxel, ABI-007, compared with cremophor-based paclitaxel *Clin. Cancer Res.* **12** 1317–24

[18] Lao J, Madani J, Puértolas T, Álvarez M, Hernández A, Pazo-Cid R, Artal Á and Antón Torres A 2013 Liposomal doxorubicin in the treatment of breast cancer patients: a review *J. Drug Deliv.* **2013** 456409

[19] Glantz M J *et al* 1999 A randomized controlled trial comparing intrathecal sustained-release cytarabine (DepoCyt) to intrathecal methotrexate in patients with neoplastic meningitis from solid tumors *Clin. Cancer Res.* **5** 3394–402

[20] Muggia F M *et al* 1997 Phase II study of liposomal doxorubicin in refractory ovarian cancer: Antitumor activity and toxicity modification by liposomal encapsulation *J. Clin. Oncol.* **15** 987–93

[21] Holme M N *et al* 2012 Shear-stress sensitive lenticular vesicles for targeted drug delivery *Nat. Nanotechnol.* **7** 536–43

[22] Jhaveri A, Deshpande P and Torchilin V 2014 Stimuli-sensitive nanopreparations for combination cancer therapy *J. Control. Release* **190** 352–70

[23] Tannock I F and Rotin D 1989 Acid pH in tumors and its potential for therapeutic exploitation *Cancer Res.* **49** 4373–84

[24] Obata Y, Tajima S and Takeoka S 2010 Evaluation of pH-responsive liposomes containing amino acid-based zwitterionic lipids for improving intracellular drug delivery *in vitro* and *in vivo J. Control. Release* **142** 267–76

[25] Wong C, Stylianopoulos T, Cui J, Martin J, Chauhan V P, Jiang W, Popovic Z, Jain R K, Bawendi M G and Fukumura D 2011 Multistage nanoparticle delivery system for deep penetration into tumor tissue *Proc. Natl Acad. Sci.* **108** 2426–31

[26] Tong R and Kohane D S 2012 Shedding light on nanomedicine *Wiley Interdiscip. Rev. Nanomed. Nanobiotechnol.* **4** 638–62

[27] Timko B P, Dvir T and Kohane D S 2010 Remotely triggerable drug delivery systems *Adv. Mater.* **22** 4925–43

[28] Needham D, Anyarambhatla G, Kong G and Dewhirst M W 2000 A new temperature-sensitive liposome for use with mild hyperthermia: Characterization and testing in a human tumor xenograft model *Cancer Res.* **60** 1197–201

[29] Ge J, Neofytou E, Cahill T J, Beygui R E and Zare R N 2012 Drug release from electric-field-responsive nanoparticles *ACS Nano* **6** 227–33

[30] Funkhouser J 2002 Reinventing pharma: The theranostic revolution *Curr. Drug Discov.* **2**

[31] Thorat N D, Bohara R A, Noor M R, Dhamecha D, Soulimane T and Tofail S A M 2017 Effective cancer theranostics with polymer encapsulated superparamagnetic nanoparticles: combined effects of magnetic hyperthermia and controlled drug release *ACS Biomater. Sci. Eng.* **3** 1332–40

[32] Hall E J and Giaccia A J 2006 *Radiobiology for the Radiologist* 6th edn (Philadelphia, PA: Lippincott Williams & Wilkins)

[33] Chen H, Sulejmanovic D, Moore T, Colvin D C, Qi B, Mefford O T, Gore J C, Alexis F, Hwu S J and Anker J N 2014 Iron-loaded magnetic nanocapsules for pH-triggered drug release and MRI imaging *Chem. Mater.* **26** 2105–12

[34] Ferrara K W 2008 Driving delivery vehicles with ultrasound *Adv. Drug Deliv. Rev.* **60** 1097–102

[35] Deckers R and moonen C T W 2010 Ultrasound triggered, image guided, local drug delivery *J. Control. Release* **148** 25–33

[36] Rapoport N, Gao Z and Kennedy A 2007 Multifunctional nanoparticles for combining ultrasonic tumor imaging and targeted chemotherapy *J. Natl. Cancer Inst.* **99** 1095–106

[37] Bai M, Shen M, Teng Y, Sun Y, Li F, Zhang X, Xu Y, Duan Y and Du L 2015 Enhanced therapeutic effect of Adriamycin on multidrug resistant breast cancer by the ABCG2-siRNA loaded polymeric nanoparticles assisted with ultrasound *Oncotarget* **6** 43779–90

[38] Marino A *et al* 2019 Piezoelectric barium titanate nanostimulators for the treatment of glioblastoma multiforme *J. Colloid Interface Sci.* **538** 449–61

[39] Carlisle R, Choi J, Bazan-Peregrino M, Laga R, Subr V, Kostka L, Ulbrich K, Coussios C C and Seymour L W 2013 Enhanced tumor uptake and penetration of virotherapy using polymer stealthing and focused ultrasound *J. Natl. Cancer Inst.* **105** 1701–10

[40] Glazer E S and Curley S A 2010 Radiofrequency field-induced thermal cytotoxicity in cancer cells treated with fluorescent nanoparticles *Cancer* **116** 3285–93

[41] Karmakar A, Xu Y, Mahmood M W, Zhang Y, Saeed L M, Mustafa T, Ali S, Biris A R and Biris A S 2011 Radio-frequency induced *in vitro* thermal ablation of cancer cells by EGF functionalized carbon-coated magnetic nanoparticles *J. Mater. Chem.* **21** 12761–9

[42] Rejinold N S, Jayakumar R and Kim Y C 2015 Radio frequency responsive nano-biomaterials for cancer therapy *J. Control. Release* **204** 85–97

[43] Rejinold N S, Ranjusha R, Balakrishnan A, Mohammed N and Jayakumar R 2014 Gold-chitin-manganese dioxide ternary composite nanogels for radio frequency assisted cancer therapy *RSC Adv.* **4** 5819–25

[44] Rejinold N S, Thomas R G, Muthiah M, Chennazhi K P, Park I K, Jeong Y Y, Manzoor K and Jayakumar R 2014 Radio frequency triggered curcumin delivery from thermo and pH responsive nanoparticles containing gold nanoparticles and its *in vivo* localization studies in an orthotopic breast tumor model *RSC Adv.* **4** 39408–27

[45] Chattopadhyay N, Cai Z, Kwon Y L, Lechtman E, Pignol J P and Reilly R M 2013 Molecularly targeted gold nanoparticles enhance the radiation response of breast cancer cells and tumor xenografts to X-radiation *Breast Cancer Res. Treat.* **137** 81–91

[46] Saneja A, Kumar R, Arora D, Kumar S, Panda A K and Jaglan 2018 Recent advances in near-infrared light-responsive nanocarriers for cancer therapy *Drug Discov. Today* **23** 1115–25

[47] Kong T, Zeng J, Wang X, Yang X, Yang J, McQuarrie S, McEwan A, Roa W, Chen J and Xing J Z 2008 Enhancement of radiation cytotoxicity in breast-cancer cells by localized attachment of gold nanoparticles *Small* **4** 1537–43

[48] Munnink T H O, Nagengast W B, Brouwers A H, Schröder C P, Hospers G A, de Hooge M N L, van der Wall E, van Diest P J and de Vries E G E 2009 Molecular imaging of breast cancer *Breast* **18** S66–73

[49] Stuchinskaya T, Moreno M, Cook M J, Edwards D R and Russell D A 2011 Targeted photodynamic therapy of breast cancer cells using antibody-phthalocyanine-gold nano-particle conjugates *Photochem. Photobiol. Sci.* **10** 822–31

[50] Auzel F 2004 Upconversion and anti-stokes processes with f and d ions in solids *Chem. Rev.* **104** 139–74

[51] Mai H X, Zhang Y W, Sun L D and Yan C H 2007 Highly efficient multicolor up-conversion emissions and their mechanisms of monodisperse NaYF4:Yb,Er core and core/shell-structured nanocrystals *J. Phys. Chem.* C **111** 13721–9

[52] Wang F, Han Y, Lim C S, Lu Y, Wang J, Xu J, Chen H, Zhang C, Hong M and Liu X 2010 Simultaneous phase and size control of upconversion nanocrystals through lanthanide doping *Nature* **463** 1061–5

[53] Sun P, Zhang H, Liu C, Fang J, Wang M, Chen J, Zhang J, Mao C and Xu S 2010 Preparation and characterization of Fe_3O_4/CdTe magnetic/fluorescent nanocomposites and their applications in immuno-labeling and fluorescent imaging of cancer cells *Langmuir* **26** 1278–84

[54] Jiang S and Zhang Y 2010 Upconversion nanoparticle-based FRET system for study of siRNA in live cells *Langmuir* **26** 6689–94

[55] Mader H S, Kele P, Saleh S M and Wolfbeis O S 2010 Upconverting luminescent nanoparticles for use in bioconjugation and bioimaging *Curr. Opin. Chem. Biol.* **14** 582–96

[56] Zhang P, Steelant W, Kumar M and Scholfield M 2007 Versatile photosensitizers for photodynamic therapy at infrared excitation *J. Am. Chem. Soc.* **129** 4526–7

[57] Wang C, Tao H, Cheng L and Liu Z 2011 Near-infrared light induced *in vivo* photodynamic therapy of cancer based on up-conversion nanoparticles *Biomaterials* **32** 6145–54

[58] Wang S, Kim G, Lee Y E K, Hah H J, Ethirajan M, Pandey R K and Kopelman R 2012 Multifunctional biodegradable polyacrylamide nanocarriers for cancer theranostics-A 'see and treat' strategy *ACS Nano* **6** 6843–51

[59] Zhang L, Zeng L, Pan Y, Luo S, Ren W, Gong A, Ma X, Liang H, Lu G and Wu A 2015 Inorganic photosensitizer coupled Gd-based upconversion luminescent nanocomposites for *in vivo* magnetic resonance imaging and near-infrared-responsive photodynamic therapy in cancers *Biomaterials* **44** 82–90

[60] Lindner U, Weersink R A, Haider M A, Gertner M R, Davidson S R H, Atri M, Wilson B C, Fenster A and Trachtenberg J 2009 Image guided photothermal focal therapy for localized prostate cancer: phase I trial *J. Urol.* **182** 1371–7

[61] Rai P, Mallidi S, Zheng X, Rahmanzadeh R, Mir Y, Elrington S, Khurshid A and Hasan T 2010 Development and applications of photo-triggered theranostic agents *Adv. Drug Deliv. Rev.* **62** 1094–124

[62] De Freitas L F, Zanelatto L C, Mantovani M S, Silva P B G, Ceccini R, Grecco C, Moriyama L T, Kurachi C, Martins V C A and Plepis A M G 2013 *In vivo* photothermal tumor ablation using gold nanorods *Laser Phys.* **23** 066003

[63] Lu W, Xiong C, Zhang G, Huang Q, Zhang R, Zhang J Z and Li C 2009 Targeted photothermal ablation of murine melanomas with melanocyte-stimulating hormone analog—conjugated hollow gold nanospheres *Clin. Cancer Res.* **15** 876–86

[64] Carpin L B, Bickford L R, Agollah G, Yu T K, Schiff R, Li Y and Drezek R A 2011 Immunoconjugated gold nanoshell-mediated photothermal ablation of trastuzumab-resistant breast cancer cells *Breast Cancer Res. Treat.* **125** 27–34

[65] moon H K, Lee S H and Choi H C 2009 *In vivo* near-infrared mediated tumor destruction by photothermal effect of carbon nanotubes *ACS Nano* **3** 3707–13

[66] Hussein E A, Zagho M M, Nasrallah G K and Elzatahry A A 2018 Recent advances in functional nanostructures as cancer photothermal therapy *Int. J. Nanomed.* **13** 2897–906

[67] Kang X *et al* 2017 Photothermal therapeutic application of gold nanorods-porphyrin-trastuzumab complexes in HER2-positive breast cancer *Sci. Rep.* **7** 42069

[68] Liu Z, Robinson J T, Sun X and Dai H 2008 PEGylated nanographene oxide for delivery of water-insoluble cancer drugs *J. Am. Chem. Soc.* **130** 10876–7

[69] Loh K P, Bao Q, Eda G and Chhowalla M 2010 Graphene oxide as a chemically tunable platform for optical applications *Nat. Chem.* **2** 1015–24

[70] He Q, Sudibya H G, Yin Z, Wu S, Li H, Boey F, Huang W, Chen P and Zhang H 2010 Centimeter-long and large-scale micropatterns of reduced graphene oxide films: Fabrication and sensing applications *ACS Nano* **4** 3201–8

[71] Yang K, Hu L, Ma X, Ye S, Cheng L, Shi X, Li C, Li Y and Liu Z 2012 Multimodal imaging guided photothermal therapy using functionalized graphene nanosheets anchored with magnetic nanoparticles *Adv. Mater.* **24** 1868–72

[72] Guo L, Yan D D, Yang D, Li Y, Wang X, Zalewski O, Yan B and Lu W 2014 Combinatorial photothermal and immuno cancer therapy using chitosan-coated hollow copper sulfide nanoparticles *ACS Nano* **8** 5670–81

[73] Beqa L, Fan Z, Singh A K, Senapati D and Ray P C 2011 Gold nano-popcorn attached SWCNT hybrid nanomaterial for targeted diagnosis and photothermal therapy of human breast cancer cells *ACS Appl. Mater. Interfaces* **3** 3316–24

[74] Goodman A M, Neumann O, Nørregaard K, Henderson L, Choi M-R, Clare S E and Halas N J 2017 Near-infrared remotely triggered drug-release strategies for cancer treatment *Proc. Natl Acad. Sci.* **114** 12419–24

[75] Peng H, Li K, Wang T, Wang J, Wang J, Zhu R, Sun D and Wang S 2013 Preparation of hierarchical mesoporous $CaCO_3$ by a facile binary solvent approach as anticancer drug carrier for etoposide *Nanoscale Res. Lett.* **8** 321

[76] Manivasagan P *et al* 2019 A multifunctional near-infrared laser-triggered drug delivery system using folic acid conjugated chitosan oligosaccharide encapsulated gold nanorods for targeted chemo-photothermal therapy *J. Mater. Chem.* B **7** 3811–25

[77] Masson J-F and Pelletier J N 2015 Will nanobiosensors change therapeutic drug monitoring? The case of methotrexate *Nanomedicine* **10** 521–4

[78] Bertrand N, Wu J, Xu X, Kamaly N and Farokhzad O C 2014 Cancer nanotechnology: The impact of passive and active targeting in the era of modern cancer biology *Adv. Drug Deliv. Rev.* **66** 2–25

[79] Ng K K, Lovell J F and Zheng G 2011 Lipoprotein-inspired nanoparticles for cancer theranostics *Acc. Chem. Res.* **44** 1105–13

[80] Benyettou F, Lalatonne Y, Chebbi I, Di Benedetto M, Serfaty J M, Lecouvey M and Motte L 2011 A multimodal magnetic resonance imaging nanoplatform for cancer theranostics *Phys. Chem. Chem. Phys.* **13** 10020–7

[81] Chen W, Xu N, Xu L, Wang L, Li Z, Ma W, Zhu Y, Xu C and Kotov N A 2010 Multifunctional magnetoplasmonic nanoparticle assemblies for cancer therapy and diagnostics (theranostics) *Macromol. Rapid Commun.* **31** 228–36

[82] Cole A J, Yang V C and David A E 2011 Cancer theranostics: The rise of targeted magnetic nanoparticles *Trends Biotechnol.* **29** 323–32

[83] Liao M Y, Lai P S, Yu H P, Lin H P and Huang C C 2012 Innovative ligand-assisted synthesis of NIR-activated iron oxide for cancer theranostics *Chem. Commun.* **48** 5319–21

[84] Clemons M, Danson S, Hamilton T and Goss P 2001 Locoregionally recurrent breast cancer: Incidence, risk factors and survival *Cancer Treat. Rev.* **27** 67–82

[85] Maluta S and Kolff M W 2015 Role of hyperthermia in breast cancer locoregional recurrence: a review *Breast Care* **10** 408–12

Chapter 4

Magnetically stimulated breast cancer nanomedicines

Shubhangi Shirsat, Vijaykumar V Jadhav and Rajaram S Mane

Breast cancer is a commonly diagnosed cancer in women. As there are already a number of therapeutic options, chemotherapy for metastatic breast cancer has become increasingly complex. The poor selectivity of the chemotherapy drugs can cause lethal damage to adjacent normal proliferating cells. On the other hand, many of the highly promising drugs have been dropped and ignored from the development pipeline due to their poor aqueous solubility as well as their side effects. Development of safe, efficient, and biocompatible drug nanomedicines with therapeutic properties is essential for clinical translation in breast cancer therapy. Targeted delivery of therapeutic agents inside the tumor tissue is an important aspect of nanomedicines. Magnetically stimulated drug delivery exhibits superiority for reversal of multiple drug resistance in breast cancer. Magnetically stimuli-responsive nanomedicine accumulates at specific sites and releases chemotherapeutic agents in response to external magnetic stimuli.

> *Magnetism, as you recall from physics class, is a powerful force that causes certain items to be attracted to refrigerators*
>
> —*Dave Barry*

4.1 Preface

Breast cancer is one of the most common malignant diseases found in women throughout the world. However, current drug therapy is far from optimal as indicated by the high death rate in breast cancer patients. Nanomedicine can be a promising option for breast cancer treatment. Researchers are trying to design and develop nanomedicine that is more tailored for breast cancer to achieve specificity, antitumorigenic, antimetastatic and drug resistance reversal effects. Currently magnetic field-targeted and induced hyperthermia has ushered in a new era for drug delivery by improving the therapeutic indices of the active pharmaceutical ingredients engineered within the nanoparticles. This chapter will help readers to understand the basic biology of breast cancer, available therapeutic options, multiple drug resistance in patients and use of magnetic field-induced targeting and hyperthermia in combination with chemotherapeutic agents in breast cancer to address the current issues in the treatment of breast cancer.

4.2 Introduction

Breast cancer is a commonly diagnosed and lethal cancer type in women as well as in animals (in some cases) worldwide [1]. Every year nearly 1.5 million new cases of breast cancer are being registered which makes it the second leading cause of mortality, just behind lung cancer, among females (figure 4.1). Breast cancer patients are diagnosed as palpable breast mass, nipple discharge, breast asymmetry, nipple inversion, breast skin erythema and thickening or by on screening mammogram [3]. High mortality rates are caused by unawareness of disease, lack of early detection followed treatment [4]. Factors responsible for breast cancer consist of BRCA 1 or 2 mutations, family history of the disease, reproductive factors that influence endogenous estrogen exposure, the use of exogenous hormones, alcohol drinking, excess body weight, and high-dose radiation to the chest, particularly at a young age [5, 6].

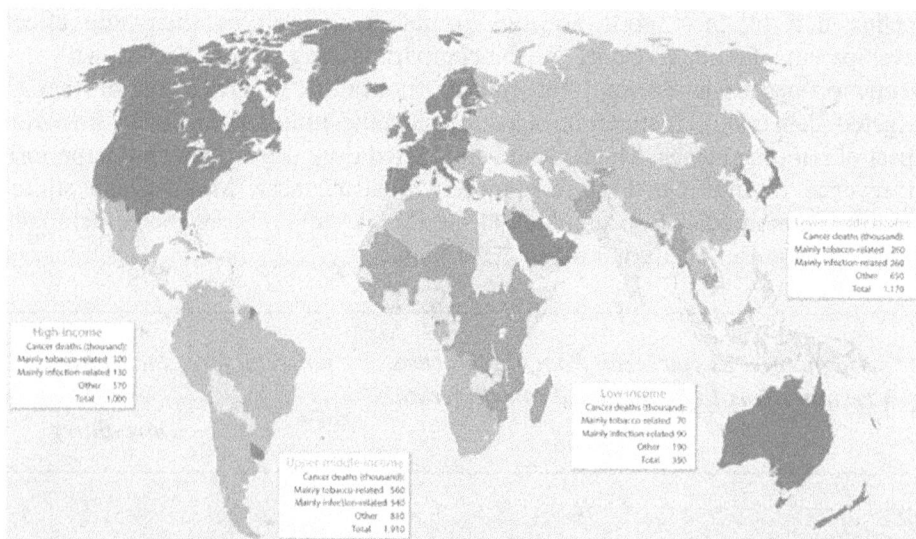

Figure 4.1. Global Breast Cancer Incidence in Women in 2012. *Note*: Values are estimated ASR per 100 000 women. ASR = age-standardized rate. Reproduced with permission from reference [2] under the terms of the Creative Commons Attribution Non-Commercial License. Copyright © 2015 International Bank for Reconstruction and Development/The World Bank.

Over the last few years considerable progress in the pathology and molecular biology of breast cancer has been evidenced where the classification of this heterogeneous disease is still complex [7]. The majority of breast cancer cases are invasive ductal carcinoma (50%–75% of patients), but other less-prevalent subtypes still draw attention because of their aggressiveness and occurrence in different patient subpopulations [7]. Stage I breast cancer has an anatomically defined tumor smaller than 2 cm and no lymph node involvement. In stages II and III it spreads to nearby tissues and lymph nodes (5%–15% of patients). Stage IV is known as distant metastatic stage in which cancer can spread to other sites in the body like the lung, liver, bone and brain [8]. Staging is crucial because once the tumor metastasizes the mortality rate dramatically increases [9].

Two main molecular targets in breast cancer pathogenesis have been identified. One is estrogen receptor alpha (ER), steroid hormone receptor and a transcription factor, upon activation initiates oncogenic pathways in the breast cancer cells. Expression of the closely related steroid hormone progesterone receptor (PR) is also a marker of ER signaling [8]. The second main molecular target is epidermal growth factor 2 (HER2), a transmembrane receptor which is associated with epidermal growth factor receptor family that can be amplified or overexpressed in approximately 20% of breast cancer. Depending upon molecular markers breast cancer is classified as luminal A, luminal B, HER2 type and triple negative subtypes [10]. Luminal A is the most common subtype in which the tumors are estrogen, progesterone receptor positive and are typically slow growing [10]. The luminal B subtype is estrogen receptor positive, progesterone negative and HER2 positive, they are generally fast growing. Approximately 20%–30% of patients with breast cancer have HER2 positive tumors [11]. Triple negative breast cancer, which makes up approximately 15% of all breast tumors, lacks the expression of molecular targets ER, PR, or HER2 [12, 13]. Triple negative breast tumors are more aggressive, highly metastasized and likely to occur in young, black, and/or Hispanic women [14]. The specific molecular pathophysiology of triple negative breast cancer remains unprecedented [15].

4.3 Tumor microenvironment and metastasis

Breast tumors consist of populations of epithelial-derived cancer cells and stromal components (non-cancerous cells) collectively known as the tumor microenvironment [16] (figure 4.2). Breast tumors have fibroblasts, immune cells, endothelial cells, cancer stem cells (CSC), cancer associated fibroblasts (CAS) infiltrating inflammatory cells, adipocytes as well as signaling molecules and extracellular matrix components (EMC) [18]. There is considerable heterogeneity in cellular and molecular components of the tumor microenvironment. Stromal cells influence the behavior of epithelial cells by secreting a range of extracellular matrix proteins, chemokines, cytokines and growth factors [19]. The various proteins secreted by stromal cells can aberrantly activate autocrine and paracrine loops, which affect the cell behavior in a paracrine or juxtacrine fashion. These interactions between stroma and tumor cells, along with underlying genetic defects of the tumor cells, dictate

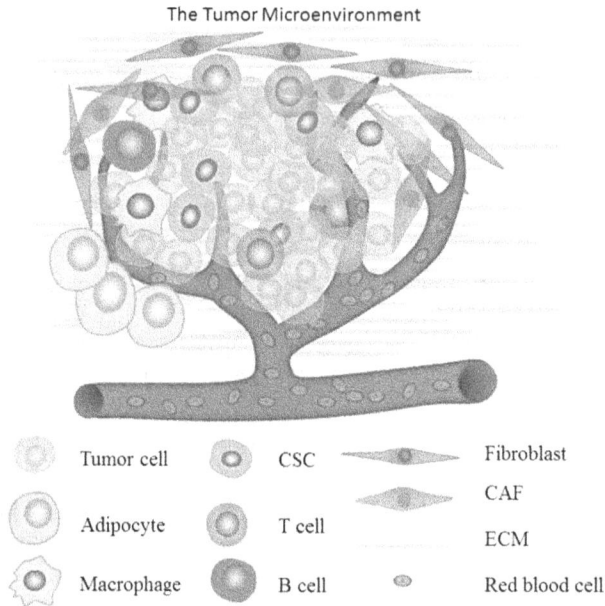

The Tumor Microenvironment

Figure 4.2. Tumor microenvironment. Reproduced with permission from reference [16]. Copyright © 2018, Springer Nature.

their growth characteristics, morphologies, and invasiveness [18] (figure 4.3). Tumor size attains a volume of 2 mm^3 and its core is exposed to hypoxic conditions due to the lack of oxygen. Therefore, new blood vessels formation is required to supply the nutrients and oxygen. The formation of new blood vessels is called angiogenesis [20].

4.4 Current trends and challenges in breast cancer treatment

Breast cancer patients, perhaps more than for any other solid tumor, are often given multimodality treatments such as surgery, radiation therapy and drug therapy, plus some optional complementary measures that range from acupuncture to diet management [21, 22]. Surgery and radiation therapy are two useful methods for eradicating tumors from the breast and regional lymph nodes. When the breast tumor progresses and becomes metastasized, the option of surgery and radiation therapy tends to decline [21]. Drug therapy is another very promising option for metastatic breast cancer treatment as it reduces the tumor burden by preventing/controlling the cancer cells [23]. Immunotherapy uses hormones or hormone-like drugs to suppress cancer cell proliferation and kill them with cytotoxic compounds. Recent breakthroughs in molecular biology and immunotherapy, specifically targeted drug therapies for pathophysiology of different breast cancer subtypes, are increasingly available. Trastuzumab (i.e. Herceptin) being the most famous targeted therapy uses humanized anti-HER2 monoclonal antibody for controlling the cancer cells [24]. Currently, metastatic breast cancer remains incurable in

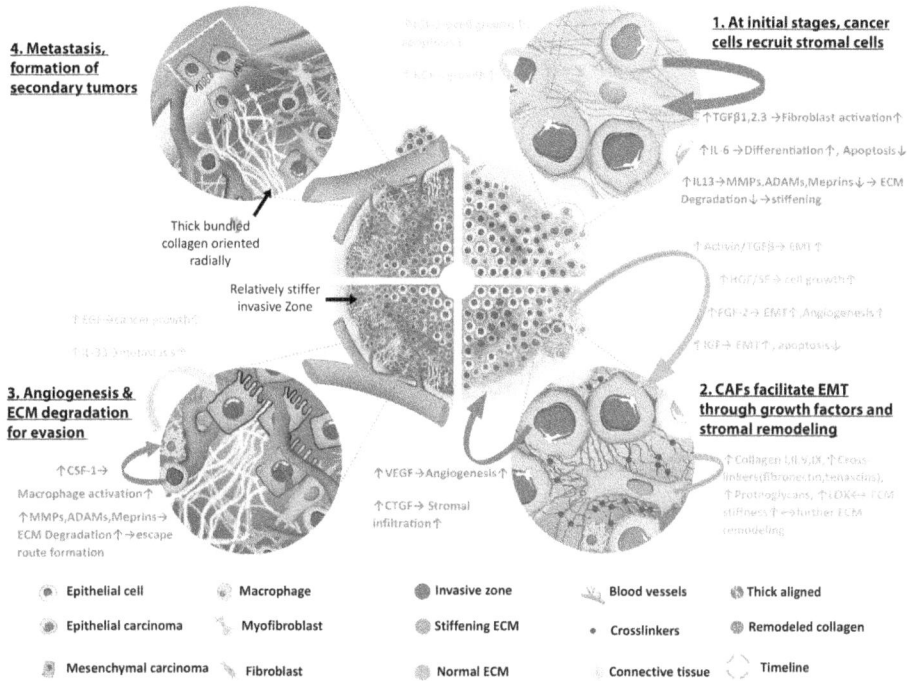

Figure 4.3. Metastatic progression of cancer. Reproduced with permission from reference [16]. Under the terms of the Creative Commons Attribution Non-Commercial License. Copyright © 2018 The Authors. Published by Elsevier B.V.

Table 4.1. Current standard drug therapy against breast cancer.

Type	Hormone therapy	Chemotherapy	Anti-HER2 (Trastuzumab)
Luminal A (low tumor burden)	Yes	No	No
Luminal A (high tumor burden)	Yes	Yes	No
Luminal B (HER2+)	Yes	Yes	Yes
Luminal B (HER2−)	Yes	Yes	No
Triple negative	No	Yes	No

virtually all affected patients. Different molecular subtypes have distinct risk profiles and treatment strategies. The optimal therapy for each patient depends on the tumor subtype, anatomic cancer stage, and patient preferences [25]. Table 4.1 summarizes the standard and known drug therapy options [23].

Application of basic scientific principles needs to be transformed for breast cancer treatments to reduce morbidity and mortality. There are several obstacles in drug therapy that reduce the current success rate [26–28]. Table 4.2 summarizes these challenges [23]. It is evident that the cancer stem cells are mostly responsible for the failure of chemotherapy and radiation therapy (box 4.1).

Table 4.2. Challenges in breast cancer drug therapy.

1. Insufficient specificity for breast cancer.
2. Inefficient access of drugs to metastatic sites such as brain and bone.
3. Undesirable pharmacokinetics such as quick clearance and short half-life.
4. Dose-limiting toxicity of the anticancer drugs or the recipients, for example, surfactants and organic co-solvents.
5. Drug resistance at cellular level, for example, increased drug efflux transport.
6. Drug resistance at tumor microenvironment level, for example, lower pH, hypoxia, cancer microenvironment crosstalk and so on.
7. Difficulty in eradicating cancer stem cells.
8. Undesirable pharmaceutical properties of the drugs, for example, low aqueous solubility, poor *in vivo* stability.
9. Suboptimal dosing schedule and sequence, especially when combinations of multiple drugs are involved.

Box 4.1 New target in breast cancer therapy—cancer stem cells

All cancerous cells in tumors are clones and represent the progeny of a single cell. A tumor is functionally heterogeneous, very few cells possess tumor-initiating function and are capable of maintaining tumor growth. These cells endow a high self-renewal capacity with proliferation rate, the ability to generate heterogenic lineages of cancer cells and metastatic potential, and are known as breast cancer stem cells. These cells are identified by specific markers such as CD44+/CD24−/low, and enhance a tumor's capacity for metastasis, invasion. It is the belief that therapy failure and therapy resistance are due to the presence of cancer stem cells because of (i) enhanced membrane transport, specific mechanisms of DNA repair, ROS scavenging systems and the ability to detoxify cytotoxic drugs; (ii) transcriptional factors, signaling pathways, and tumor suppressor genes act to maintain and amplify the cells; (iii) extensive interactions among cancer stem cells, their microenvironments, and other cells present initiate a cascade of growth factors and inducing elements, which in turn influence cancer stem cell function. Many researchers believe that cancer stem cells should be considered as possible therapeutic targets.

Multidrug resistance (MDR) is a severe issue associated with drug therapy which occurs shortly after the treatment. Molecular level changes in the tumor micro-environment, abnormal vascular, lymphatic systems induce high intratumoral pressure and hypoxia are the key factors in the development of MDR due to which 90% of treatments fail [28–30]. Table 4.3 represents the mechanism of MDR in cancer cells.

4.5 Cancer nanomedicines

The 21st century is the era of technology which is influencing every aspect of life including communication, space and underground technology, transportation,

Table 4.3. Mechanism of MDR in cancer cells.

(a) Overexpression of ATP-binding cassette pumps (e.g. P-glycoprotein, P-gp).
(b) Defective apoptotic mechanisms.
(c) Structural alterations of the drug targets.
(d) Repair of the damaged DNA.
(e) Detoxification of the drugs by certain enzymes.

medicine etc. It has also revolutionized the pharmaceutical and energy industries from relatively simple systems to highly sophisticated ones. An optimized cancer therapy would deliver drugs to target sites for localized control of the disease efficiently and with minimal toxicity. Extensive efforts in the past decade have been taken to produce a large arsenal of nanoplatforms with diversified capabilities for drug loading and releasing, and also for tumor targeting. Nanomedicine is an emerging method for treating cancer patients. There are several types of nano-materials that are being widely used for this process. Liposomes [31, 32], micelles [33, 34], gold nanoparticles [35, 36], magnetic nanoparticles [37, 38], carbon nanotubes [39, 40], dendrimers [41], mesoporous silica nanoparticles [42, 43] etc, are a few of them. Nanoparticles demonstrate 1–100 nm sizes due to which they approach quantum dimensions resulting in changes in their internal energy. They exhibit unique optical, chemical, electronic, magnetic and photochemical properties which can be utilized for cancer treatment [44–48].

The nanomaterials used in cancer research are monitored through their sizes, shapes, and surface characteristics for targeting specific tumors. Size is important to travel through the bloodstream and subsequent delivery of the nanocarriers to tumor tissues [49]. In passive targeting of tumors, abnormal vasculature is utilized for targeting the nanomedicine to the target site. Blood vessels in tumors are hetero-geneous with chaotic branching structures and are leaky due to the presence of an enlarged gap junction of around 100 nm to 2 μm [50, 51]. Increase of hypoxia and metastasis is caused by an accumulation of interstitial fluid pressure due to leakiness of the vessels and uneven distribution of blood, nutrients, oxygen etc. Tumors lack a well-defined lymphatic system, so they do not have efficient clearance of macro-molecules and drugs accumulated in the solid tumor tissues [52]. The enhanced vascular permeability and poor lymphatic drainage is collectively known as the enhanced permeability and retention effect, which is the gold standard for delivery of nanomedicine and other types of drugs and imaging agents [53]. Passive targeting can be achieved by utilizing the enhanced permeability and retention effect (figure 4.4(a)). Active targeting of nanocarriers is achieved by attaching a variety of targeting ligands on the surface of nanomaterials that can recognize a target within the tumor affected organs, tissues, cells or intracellular organelles [54]. Ligands have a high specificity to receptors and other cancer-specific targets which are overex-pressed on the surface of tumor cells, such as glycans [55]. These receptors must be expressed in high quantity and should be evenly distributed on the cancer cell's surface (figure 4.4(b)). Stimuli-responsive systems may reduce nonspecific exposure

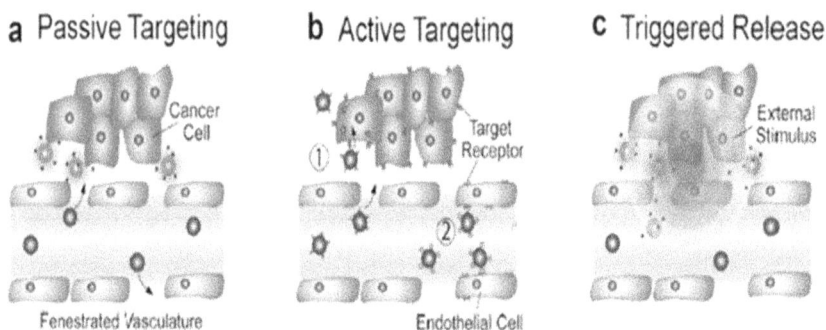

Figure 4.4. Types of targeting nanomedicine delivery to the tumor tissue. Reproduced from [54].

Table 4.4. Advantages of nanomedicines during cancer treatment.

1. Passive targeting and active targeting by nanomedicines to increase the tumor drug level and decrease the noncancer drug level.
2. Many nanomedicines improve brain and bone penetration of drugs.
3. Use of strategies such as PEGlyation to extend the circulation time.
4. Increased tumor specificity can control the drug release from nanocarriers by a solvent-surfactant-free nanoformulation.
5. Both passive and active targeting may enhance endocytosis as some nanoformulations may inhibit drug efflux mechanisms through co-delivery of agents that target the drug's resistance mechanism.
6. Targeting the tumor microenvironment.
7. Targeting cancer stem cells.
8. Many nanocarriers can achieve a drug solubilization thus protecting unstable drugs.
9. Careful optimization of dosing schedule and sequence.

to chemotherapeutic drugs. Internal and external stimuli can trigger the release of drugs by evoking a change in the nanocarriers. The pH, redox, ionic strength, and stress in target tissues are examples of internal stimuli, whereas temperature, light, ultrasound, magnetic force, and electric fields are examples of external (physical) stimuli [56] (figure 4.4(c)). Table 4.4 provides information on the advantages of nanomaterials in targeting the cancer therapeutic drug delivery [23].

4.6 Magnetic nanoparticles

Magnetic nanoparticles (MNs) offer great potential in the field of biology in medicine. They are usually composed of magnetic elements, such as iron, cobalt, nickel and their oxides. The size of MNs can be controlled in a wide range from a few nanometers up to tens of nanometers, as good as the dimensions of a cell (10–100 μm), a virus (20–450 nm), a protein (5–50 nm) or a gene (2 nm wide and 10–100 nm long) [57]. So they are very close to the biological entities of interest, MNs functionalized with drugs and bioactive agents like nucleic acid and peptides can form a discrete particulate system. These functionalized MNs can directly penetrate into the cells and tissues, offering organ-specific therapeutic and diagnostic modalities [58]. The concept of drug delivery using MNs is of great benefit. The developed stage of nanotechnology has made it possible not only to produce MNs in a very narrow size distribution range with superparamagnetic properties but also to engineer particle surfaces to provide site-specific drug delivery [59]. MNs obey Coulomb's law and can be manipulated by an external magnetic field gradient due to their strong magnetic properties. They are also used in biology and medicine for the magnetic separation of biological products and cells as well as magnetic guidance of particle systems for targeting specific sites. MNs resonantly respond to the varying magnetic field and acquire energy from the excited magnetic field. They can be heated up through an external magnetic field; thereby, they can be used as hyperthermia agents, delivering toxic amounts of thermal energy to tumors, or as chemotherapy and radiotherapy enhancement agents, where a moderate degree of tissue warming results in more effective malignant cell destruction [57]. In the last decade, activities in the clinical applications of magnetic carriers and magnetic particles have been increased greatly because of the need for better diagnosis procedures on one side and better treatment modalities at the other side. The most relevant biomedical applications for these particles may be targeted at healthcare and in theragnostics—the fusion of therapeutic and diagnostic technologies that target individualized medicine [60].

4.7 Magnetic field-induced breast tumor targeting

The major disadvantage of most chemotherapies is that they are relatively non-specific. Most chemotherapeutic drugs are administered intravenously leading to general systemic distribution and side effects on normal, healthy cells in addition to the target tumor cells. In 1970 researchers proposed the use of magnetic carrier to target specific sites (tumors) within the human body to reduce (i) the amount of systemic distribution of the cytotoxic drugs with minimal side effects and (ii) the dosage required by more efficient therapy [61–63]. Ineffective penetration of anticancer drugs in solid tumors is also a major issue in cancer chemotherapy. Even small sized drugs are able to penetrate only a few or several cell layers to the solid tumor due to vascular abnormalities and high hydrostatic pressure with tumors. Better drug penetration into solid tumors and the subsequent switchable release of the chemotherapeutic agent is of utmost importance for effective cancer treatment. Magnetically targeted and controlled release of chemotherapeutic drugs by an external magnetic field has received much attention for various potential applications [64–68]. In magnetic field-induced targeted therapy, the chemotherapeutic drug is attached to a biocompatible magnetic carrier to form a ferrofluid which then can be injected into the patient's circulatory system. When the magnetic drug carrier enters into the bloodstream, with the help of an external high gradient magnetic field, they accumulate at the target sites. All magnetic drug carriers respond to the external magnetic field and concentrate at target sites then the chemotherapeutic drug is released via enzymatic activity or changes in physiological conditions such as pH, temperature etc [69]. In this way the chemotherapeutic drug is taken up by the targeted cancerous cells and the effective cytotoxic activity of the drug is archived with reduced side effects (figure 4.5).

Varieties of MNs and microparticle carriers are being developed to deliver drugs to specific target sites *in vivo*. The size, surface chemistry, and charge on MNs and types of drug carrier are mostly important and strongly affect both the blood circulation time as well as the bioavailability of MNs within the body [70]. In addition, magnetic properties and internalization of drug carrier particles strongly depend on the sizes of MNs. Larger NPs (greater than 200 nm) show lower circulation times because of mechanical filtration by the spleen, eventually being removed by a cell of the phagocyte system. Smaller particles (less than 100 nm) can be rapidly removed through extravasations and renal clearance. Particles of 10–100 nm sizes demonstrate a prolonged blood circulation time and are optimal for intravenous injection. The particles in this size range are small enough both to evade the reticuloendothelial system of the body as well as to penetrate the very small capillaries within the body tissues and therefore, may offer the most effective distribution in tissues [71]. The magnetic component of the carrier particle is coated with a biocompatible polymer such as polyvinyl alcohol, dextran or inorganic porous silica to prevent agglomeration of MNs. Because of the large surface area-to-volume ratio, magnetic nanoparticles tend to agglomerate and adsorb plasma proteins. The apparent increased size due to agglomeration can cause a low circulation time and bioelimination by macrophages and reticuloendothelial system before it can reach the vicinity of the targeted sites. So carrier MNs are enclosed in biocompatible polymers [72, 73].

The coating is also functionalized by attaching carboxyl groups, biotin, avidin, carbo-diamide and other molecules, these molecules then act as attachment points for

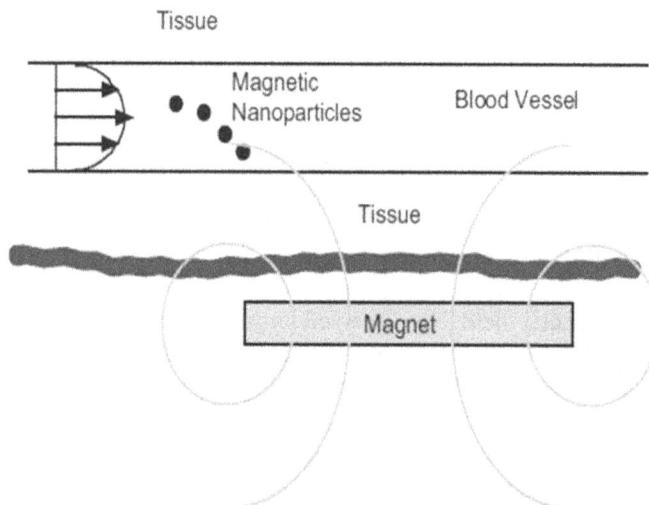

Figure 4.5. A hypothetical magnetic drug delivery system [57].

coupling of cytotoxic drugs or target antibodies to the nanocarrier complex [74–76]. Specific epitopes and receptors associated with cancer cells can be targeted using functional groups. Due to their mutated genotype, tumor cells often produce unique or overexpressed markers (antigens) which distinguish them from normal cells. These markers may then be exploited for targeting purposes. For example, the breast cancer marker HER2 is frequently overexpressed as a surface receptor in breast tumors and is therefore of targeting interest [77]. Another strategy involves targeting tumor-associated antigens such as vascular endothelial growth factor which is involved in angiogenesis [78]. For breast cancer (BT20) cells, PEG modification has facilitated the MNs internalization into the cells as PEG can dissolve in both polar and non-polar solvents and have a high solubility in cell membranes [79]. These carriers typically have two structural configurations; (i) a magnetic particle core (usually magnetite, Fe_3O_4, or maghemite, γ-Fe_2O_3) coated with a biocompatible polymer and (ii) a porous biocompatible polymer in which MNs are precipitated inside the pores [80] (figure 4.6). Recent work on carriers has largely focused the use of new polymeric or inorganic coatings on magnetite/maghemite nanoparticles [81–86], although noble metal-coatings such as gold are also considered [87]. Research also continues into alternative magnetic particles, such as iron, cobalt or nickel or yttrium aluminum iron garnet etc [88].

Curcumin (CUR) being an anticancer and chemo-preventive diphenol bioactive agent is obtained from rhizomes of *Curcuma longa*. It effectively upregulates p53, p27, p21, downregulates cycline E, decreases cell viability and anchorage independence in highly metastatic triple negative breast cancer cell line (MDA-MB-231). The cell cycle is arrested in the G1 phase and induced apoptosis through constitutive STAT3 signaling without affecting normal body calls [90–94]. However, its solubility, degradation in a physiological medium, and rapid metabolism hinder its clinical application [95]. To address this issue MN-coated nanocapsule containing β-cyclodextrin and pluronic F68 polymer (polyethylene oxide-*co*-polypropylene

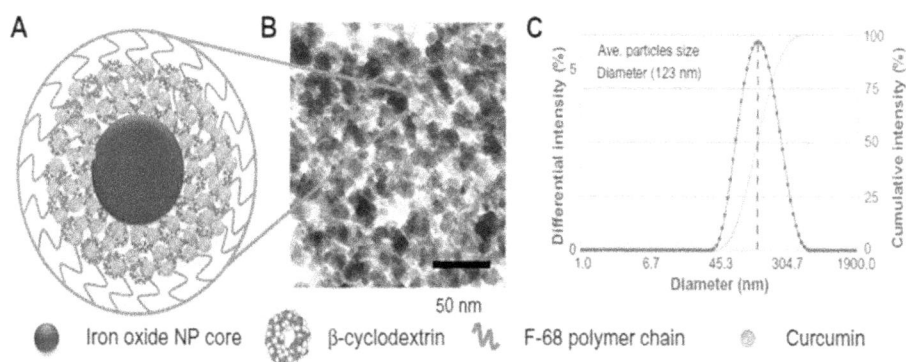

Iron oxide NP core   β-cyclodextrin   F-68 polymer chain   Curcumin

Figure 4.6. Physicochemical characterization of curcumin-loaded MNs formulation (MN-CUR). (A) Hypothetical schematic representation of MN-CUR. (B) Transmission electron microscopic image of MN-CUR. (C) Dynamic light scattering measurement of MN-CUR. Reproduced with permission from [89] under the terms of the Creative Commons Attribution Non-Commercial License. Copyright © 2012 Yallapu *et al*, publisher and licensee Dove Medical Press Ltd.

oxide-copolyethylene oxide) that allows loading of anticancer drugs were synthesized and examined for anticancer properties and magnetic targeting (figure 4.6).

The MDA-MB-231 cancer cells treated with MN-CUR were exposed to a neodymium external magnet for 3 h, after exposure cellular uptake of MN-CUR was visualized by a fluorescence microscope. Figure 4.7 shows that exposure to external magnets significantly increases the cellular uptake of MN-CUR and additional CUR internalization in malignant calls inducing more cellular apoptosis, thereby increasing the possible cytotoxicity of CUR. After 24 h treatment with MN-CUR indicated effective mitochondrial membrane damage, which is the key indicator of intrinsic apoptosis [96]. Higher reactive oxygen species (ROS) is also found in triple negative breast cancer cells (MDA-MB-231) after the treatment of MN-CUR because of internalization and delivery of CUR in their active form, while free CUR barely influences the production of ROS (figure 4.8). These results indicate that magnetically targeted CUR enhances the cytotoxic effect and effective delivery of CUR to the cancerous cells in its bioactive form [89].

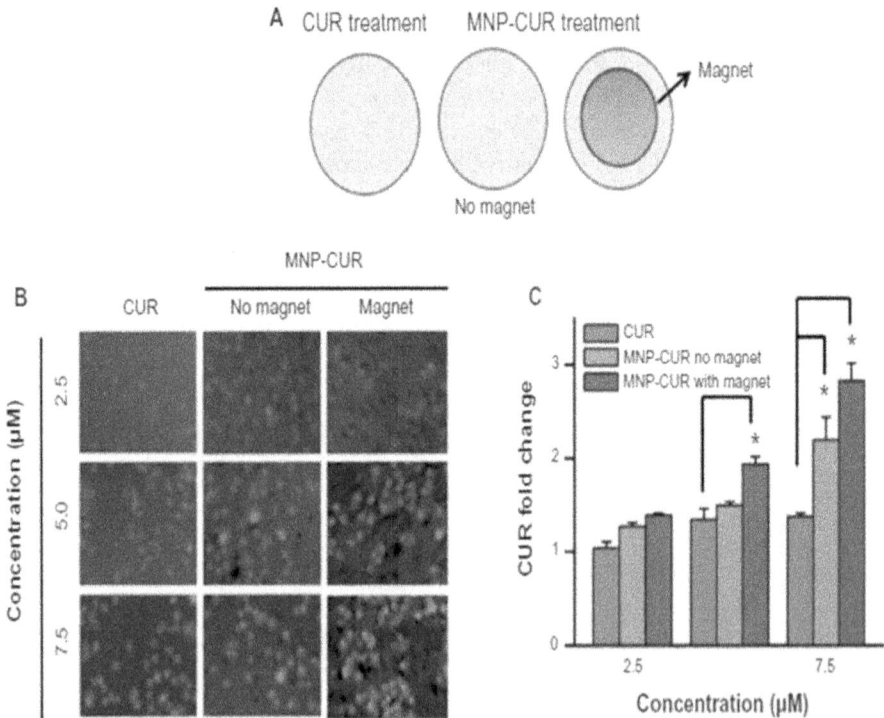

Figure 4.7. Magnetic targeting improves CUR uptake by MDA-MB-231 cells. (A) Schematic representation of experimental design for magnetic targeting. Cells (2×10^5) were seeded in six-well plates and treated for 3 h with (1) CUR, (2) MN-CUR, and (3) MN-CUR in the presence of a neodymium external magnet. (B) CUR uptake by cancer cells was visualized using a fluorescence microscope. (C) Internalized fluorescence levels. Reproduced with permission from [89] under the terms of the Creative Commons Attribution Non-Commercial License. Copyright © 2012 Yallapu *et al*, publisher and licensee Dove Medical Press Ltd.

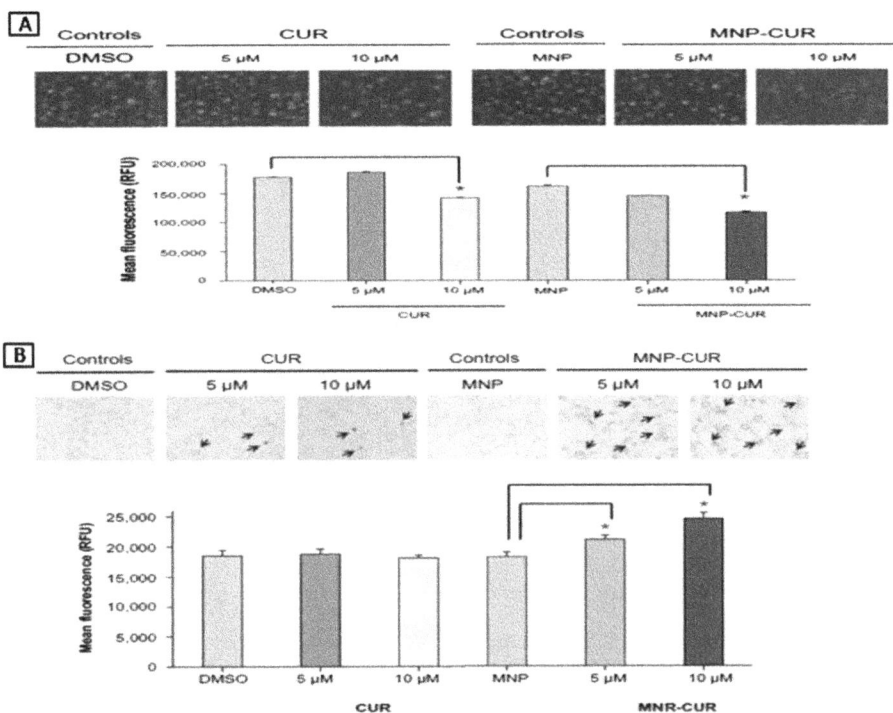

Figure 4.8. (A) Representative tetramethyl rhodamine ethyl ester staining image and quantitative estimation in MDA-MB-231 cells after treatment with CUR and MN-CUR. The less intense stain indicates loss of mitochondrial membrane potential. (B) Representative ROS generation images and quantitative estimation in MDA-MB-231 cells after treatment with CUR and (MN-CUR). Reproduced with permission from [89] under the terms of the Creative Commons Attribution Non-Commercial License. Copyright © 2012 Yallapu *et al*, publisher and licensee Dove Medical Press Ltd.

A novel hollow drug carrier is produced by immobilizing MNs of 10 nm (Fe_3O_4), anticancer drug, captothesin (hydrophobic)/doxorubicin (hydrophilic) into a silica shell (SiMNCs) (figure 4.9). The anticancer drug was loaded into the 100–150 nm diameter hollow capsules and its release was remotely activated by applied radio frequency field (at 100 kHz). A dramatic change in the amount of drug release is evidenced when the remote magnetic field is switched 'ON' and 'OFF'. The release of the hydrophobic drug to the environment after the initial release tends to slow down more than the hydrophilic drug. The magnetic field induces magnetic particle heating which causes the localized liquid temperature inside the nanosphere to rise. The resultant temperature gradient between the inside and outside of the sphere is likely to be the key element that stimulates accelerated diffusion and drug release [97].

Tumor penetration capability of the core of SiMNs was labeled with fluorescein isothiocyanate (FITC) green fluorescent tags [98, 99] and applied to colony aggregates of tumor cells under the influence of the magnetic field (~1200 Oe). Observation of confocal microscopy supported that upon application of the

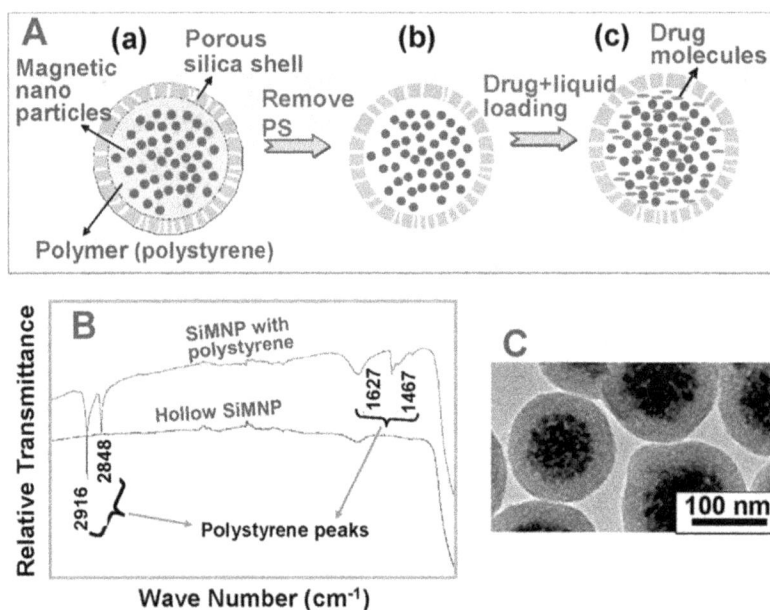

Figure 4.9. (A) Schematic illustration showing the preparation steps of drug containing hollow SiMNs with high saturation magnetization. (B) FT-IR data showing the removal of polystyrene from nanocapsule. (C) TEM micrograph showing trapped MNs in SiMNs. Reproduced with permission from [97]. Copyright © 2010 American Chemical Society.

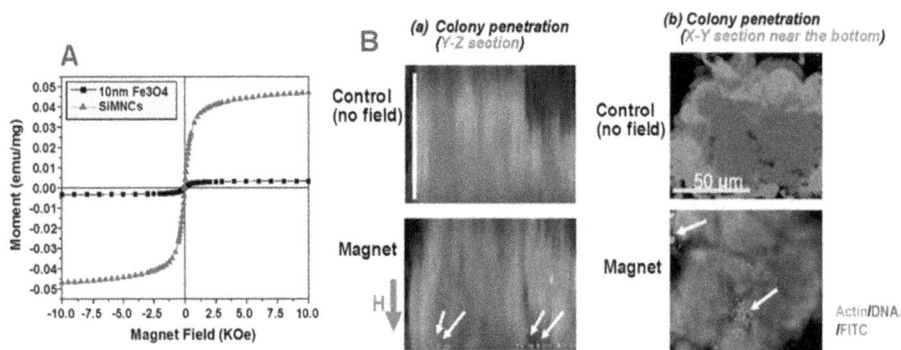

Figure 4.10. (A) M–H loops showing a significant increase in magnetic moment in SiMN configuration as compared with isolated MNPs. (B) Confocal X–Y sections (right-side images) at a location near the bottom of the colony show the presence of magnetic nanocapsules only when the gradient pulling force is applied by a magnet. Reproduced with permission from [97]. Copyright © 2010 American Chemical Society.

magnetic field, FITC green-marker-tagged magnetic nanocapsules are able to penetrate ~50 μm thickness which is equal to a thickness of ~10 cells. On the other hand, nanocapsule in the absence of applied magnetic field was washed away during the staining (Phalloidin-TRITC). Figure 4.10 shows magnetic hysteresis (M–H) loops in which a significant increase in magnetic moment in SiMNC configuration as compared with isolated 10 nm MNs of Fe_3O_4 is obtained. This is very good for effective penetration of SiMN in the tumor to reach the desired location. While confocal X–Y sections confirm the penetration and presence of MN under the influence of pulling force of stationary magnet (figure 4.10).

The *in vitro* anticancer potential of synthesized SiMNs was evaluated on the MT2 (breast cancer cell) in the presence and in the absence of radio frequency field activation with the help of MTT assay. As compared to the control group and cells treated with drug filled SiMNs without activation endows little change in tumor cell viability but the cells treated with drug filled SiMNCs and activated by radio frequency demonstrate significant variation in the tumor cell viability [97]. This is because the presence of radio frequency activation caused a substantially increased diffusional drug release rate and hence decreased the viability of tumor cells (figure 4.11(A)). The *in vivo* therapeutic potential of synthesized SiMNs was tested on tumor-bearing mice transplanted from a MMTV-ErbB2 transgenic mouse. The tumor-bearing mice were infused with 4×10^7 SiMNCs per 200 μL PBS of FITC-tagged magnetic nanocapsules. The mice were immobilized and the tumor sites were exposed to a constant magnet field (1200 Oe) for 2 h, after which the mice were euthanized and tumor tissue samples were collected for analysis. FITC imaging shows that tumor accumulates in nanocapsule only when a magnet is placed near the tumor tissues (green markers). The 4,6 diamidino-2-phenylindole (DAPI) image shows the damaged tumor structure through nuclei upon exposure to a magnet (figure 4.11(B)).

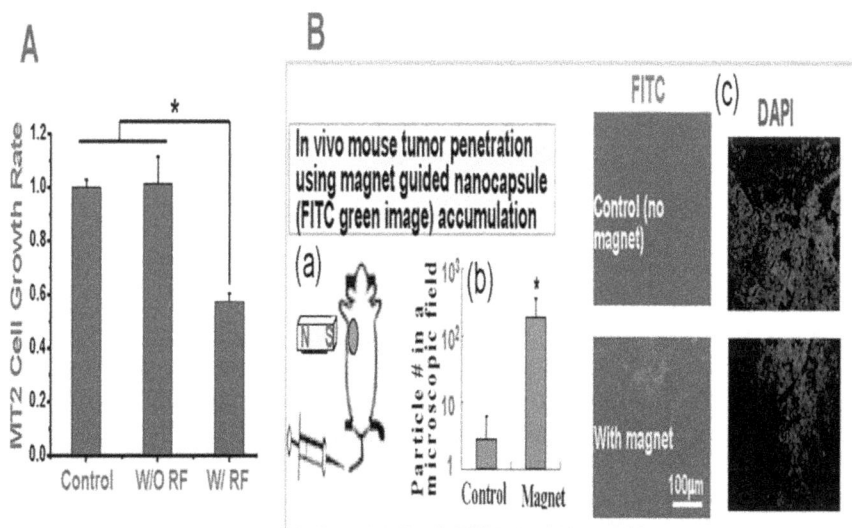

Figure 4.11. (A) Comparative growth rates (MTT assay) of MT2 (mouse breast cancer cell) in the presence versus the absence of activated SiMNs. (B) *In vivo* mouse tumor penetration of magnetic nanocapsules using a magnet near the induced mouse tumor (for 2 h). The red spot represents the tumor site. The FITC imaging shows the presence of accumulated capsules in the tumor tissue (green markers) for the case of a magnet placed nearby. The DAPI image shows the tumor structure via imaging of the nuclei in the tumor. Reproduced with permission from [97]. Copyright © 2010 American Chemical Society.

4.8 Mechanism of magnetic targeting

The physical principle of magnetic targeting therapy is magnetic separation. To accumulate the magnetic carrier at the tumor sites a strong magnetic gradient needs to be applied which can immobilize the magnetic carrier material via the magnetic force (equation (4.1)). The magnetic properties of both the injected particles and the ambient biomolecules in the bloodstream are illustrated by their different magnetic field response curves. The magnetic force should be high, so that it can overcome the hydrodynamic drag force acting on the magnetic particle in the flowing solution [57].

$$F_{d} = 6\pi\eta R_{m}\Delta v \tag{4.1}$$

where;

η—viscosity of the medium surrounding the cell (e.g. water, blood),
R_{m}—radius of the magnetic particle,
$\Delta v = v_{m} - v_{m}$ is the difference in velocities of the cell and surrounding liquid [7].
Equating the hydrodynamic drag and magnetic forces

$$V_{m} = \frac{4}{3}\pi R^{3}m \tag{4.2}$$

m—magnetic dipole

Velocity of the particle relative to the carrier fluid as

$$\Delta v = \frac{R^{2}m\chi\nabla(B^{2})}{9\mu 0\eta}$$

$$Or \tag{4.3}$$

$$\Delta v = \frac{\xi}{\mu_{0}}\nabla(B^{2}),$$

where

μ_{0}—permeability of free space,
B—magnetic induction,
χ—volumetric magnetic susceptibility.

Magnetophoretic mobility (ξ) describes how manipulable a magnetic particle is, for example magnetophoretic mobility of microsphere is greater than nanoparticles because of small size [57]. Several physical parameters like field strength, gradient, volumetric and magnetic properties of carrier molecule determine the effectiveness of the therapy as ferrofluids are administered intravenously with hydrodynamic parameters such as ferrofluid concentration, blood flow rate, infusion route and circulation time etc. Physiological parameters such as tissue depth i.e. distance from the magnetic field source, reversibility and strength of the drug/carrier binding, and tumor volume can also be taken into account while targeting the magnetic carrier at target sites [100].

4.9 Magnetic hyperthermia in breast cancer

Recently, hyperthermia directed specifically to cancer cells through the use of MNs heated with alternating magnetic field is increasingly being used in cancer treatment due to its advantages over other treatments [101]. The concept of magnetic hyperthermia use for the treatment of mammary carcinoma was proposed by Gordon *et al* in 1979 [102]. The MNs are able to convert electromagnetic energy into heat, which is most popularly utilized in magnetic hyperthermia for the destruction of tumor cells by heating them to their apoptosis threshold [57]. Generally, hyperthermia involves dispersing MNs throughout the tumor tissue, and then applying an AC magnetic field of sufficient strength and frequency to cause the particles to heat. The heat of MNs is immediately conducted into the tumor tissues if the temperature is above the therapeutic threshold of 42 °C for 30 min or more, tumor cells are destroyed. Magnetic hyperthermia is appealing as it ensures the localized heating of tumor tissue only. Effective treatment of cancer by hyperthermia is associated with delivery of an adequate quantity of the MNs to generate enough heat in the target using AC magnetic field conditions that are clinically acceptable. A variety of different types of MNs, different magnetic field strengths and frequencies and different methods encapsulation delivery of the particles are developed to achieve the effective hyperthermia [103, 104]. The amount of heat generated depends on the nature of magnetic material and of magnetic field parameters. Magnetic particles embedded around a tumor site and placed within an oscillating magnetic field will heat up to a temperature dependent on the magnetic properties of the material, the strength of the magnetic field, and the frequency of oscillation and the cooling capacity of the blood flow in the tumor sites [104].

Paclitaxel (PTX, taxol) is the anticancer drug which is mostly used for the treatment of ovarian, breast, lung, head, and neck cancer, and Kaposi sarcoma [105, 106]. It blocks cancer cells during cell division (mitosis) through reversible binding to tubulin, resulting in microtubule hyper-stabilization [106]. Such inhibition of microtubule dynamics activates the spindle assembly checkpoint, which prompts a persistent mitotic arrest in cancerous cells. Most breast cancer patients become resistant to the taxol in the first or second round of treatment, because cancer cells remain in mitosis until the drug clears, then grow and proliferate [107, 108]. Researchers have used the magnetic hyperthermia in overcoming the drug resistance and sensitivity of taxol in drug resistant tumor cells. Several experiments were performed to determine the effect of superparamagnetic iron oxide nanoparticles heat treatment alone or in combination with PTX in the MCF7 cells that are sensitive (wild type) and resistant to PTX. Combined treatment using PTX and MNPs hyperthermia kills cancerous cells more efficiently than PTX or hyperthermia treatments alone because hyperthermia forced mitotic exist and convert resistant cells to sensitive cells [109] (figure 4.12).

Breast cancer nodules BT474 (high HER2 expression) and SKOV3 (low HER2 expression) were introduced into *in vivo* mouse model. After sufficient growth of cancer nodules, the tumor temperature increased by applied alternating magnetic field to MNs to 45 °C. After the heat treatment tumor degeneration was observed

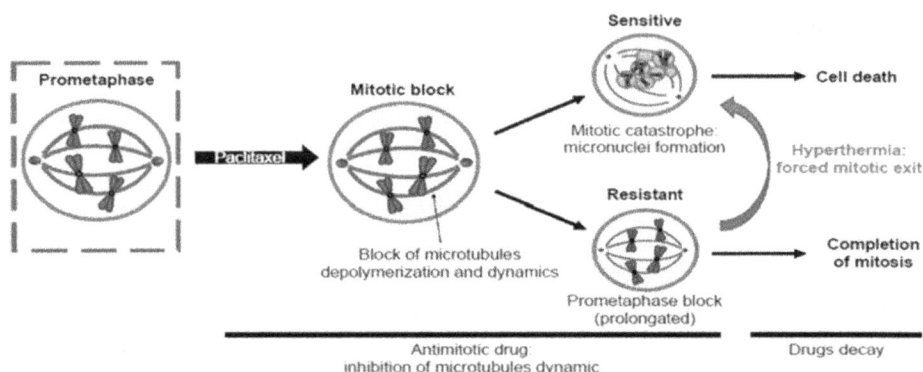

Figure 4.12. Paclitaxel induces a mitotic block in breast cancer cells. Sensitive cells die by mitotic catastrophe, while resistant cells remain in mitotic block longer and continue proliferation after drug decays. Mild hyperthermia triggers mitotic exit of PTX-pretreated cells, overcoming PTX resistance. Reproduced with permission from [109] under the terms of the Creative Commons Attribution Non-Commercial License. Copyright © 2018 Rivera-Rodriguez *et al* published and licensed by Dove Medical Press Limited.

and histological examination showed a coagulation necrosis in treated tumor cells [110]. In targeted magnetic hyperthermia, MNs were conjugated with anti-HER2 immunoliposomes and applied to breast cancer cells *in vitro*, subsequent application of an alternating magnetic field to MNs to achieve the hyperthermia. The combination of anti-HER2 antibody therapy with hyperthermia resulted in strong cytotoxic and antiproliferative effects [111].

Recently, iron oxide nanoparticles (IONPs) were synthesized by the co-precipitation method and coated with the fourth generation (G4) of polyamidoamine dendrimer (G4@IONPs) for magnetic hyperthermia and cytotoxic activity on human breast cancer cell line (MCF7). Synthesized G4@IONPs were stable without any aggregation and having a hydrodynamic size of 120 nm, maximum magnetization for IONPs and G4@IONPs at room-temperature (27 °C) were 63.4 and 40.6 emu g^{-1}, respectively. Hyperthermia was induced by a magnetic coil (12 kA m^{-1} and 300 kHz) for 120 min and cell viability was assessed by MTT assay immediately after the hyperthermia. The cell viability percentage in hyperthermia group decreased significantly while the viability in other groups was not significantly decreased compared with the control group. The terminal deoxynucleotidyl transferase dUTP nick end labeling (TUNEL) assay showed higher apoptotic activity in the hyperthermia group as compared to other groups. Prussian blue staining illustrated a high density of iron inside the MCF7 cells after 2 h incubation with G4@ IONPs [112] (figure 4.13).

4.10 Mechanism of hyperthermia

Heating of MNs in magnetic hyperthermia is due to hysteresis loss, so-called Brownian and Néel relaxation [113, 114]. In a low-viscosity liquid, MNs physically rotate in an alternating magnetic field while the surrounding medium opposes the rotation, resulting in energy loss and heat generation, known as Brownian losses.

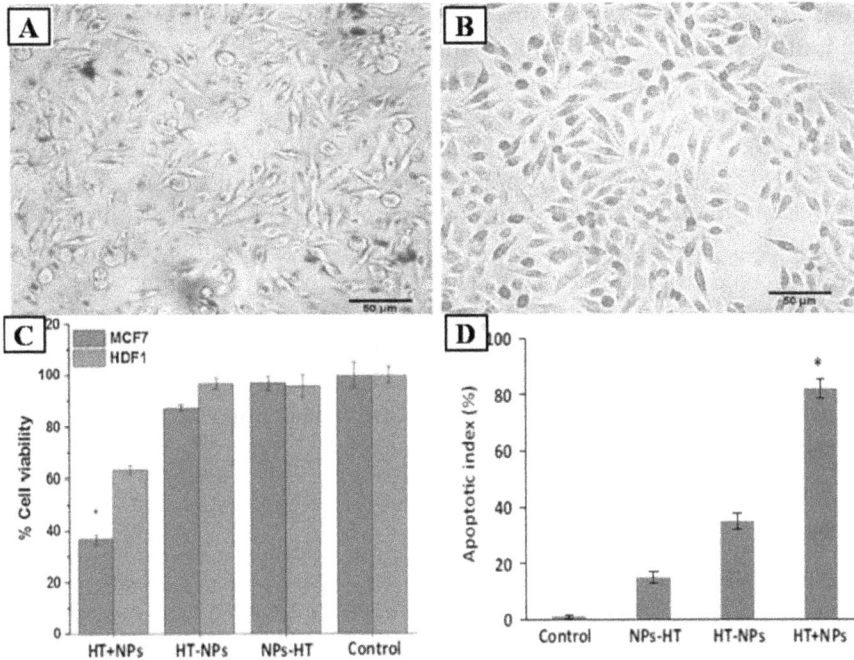

Figure 4.13. (A) Apoptotic index (TUNEL staining) of MCF7 cells. Prussian blue staining of MCF7 cells incubated C with and D without (control), (B) cell viability (MTT assay) of MCF7 and HDF1 cells after magnetic hyperthermia. Reproduced with permission from [112] under the terms of the Creative Commons Attribution Non-Commercial License. Copyright © The Author(s) 2018.

It is responsible for mechanically damaging the cells and depends on the hydro-dynamic properties of the fluid. The Néelian relaxation represents the rotation of the individual magnetic moment toward alternating magnetic field, determined by the magnetic anisotropy energy of MNs relative to the thermal energy [115]. When MNs are placed under an alternating magnetic field, heat generation in them depends upon hysteresis loops by plotting the magnetization versus applied magnetic field plots. When a ferromagnetic material is put into an oscillating/alternating magnetic field energy will be released as heat. The larger the magnetic hysteresis curve of the ferromagnet, the harder that ferromagnet is. Harder ferromagnets are used to create nanoparticles [57].

4.11 Conclusion and prospective

Substantial progress has been made in the treatment and diagnosis of breast cancer; there are several ways to treat breast cancer depending on the type and stage. With better understanding of the molecular biology of breast cancer nanomedicine has evolved considerably in the past two decades. Tumor targeting has been a central theme to nanomedicine that has also gradually evolved. The scientific community is concentrating on aspects of nanomedicine like EPR-based tumor uptake, nano-particle circulation half-lives, active and passive targeting. Stimuli-responsive nanomedicines provide a promising opportunity to trigger drug release in an on-demand manner and display various superiorities over conventional nanomedicines to combat multiple drug resistance in breast cancer. Active targeting of tumors via magnetic responsive nanomedicines could achieve more precise drug release, and avoid premature leakage during blood circulation. Combined magnetic targeting and control drug release inside the tumor tissue effectively kill the drug resistant tumor cells and increase the potential of chemotherapeutic drugs with decreases in their side effect. MNPs associated hyperthermia has the potential to directly kill the cancer cells or enhance their susceptibility to chemotherapy. Magnetically targeted drug delivery and hyperthermia have some limitations like embolization of the blood vessels by magnetic carrier, larger distances between the target site and the magnet, toxic responses to the magnetic carriers. These issues need to be addressed in the near future for the effective treatment of breast cancer using magnetically induced nanomedicine.

References

[1] Torre L A, Islami F, Siegel R L, Ward E M and Jemal A 2017 Global cancer in women: burden and trends *Cancer Epidemiol. Biomarkers Prev.* 26 444–57

[2] Anderson B O *et al* 2015 Breast cancer *Cancer: Disease Control Priorities* vol 3 3rd edn ed H Gelband, P Jha, R Sankaranarayanan and S Horton (Washington DC: World Bank Publications)

[3] Morrow M, Harris J R, Lippman M E, Morrow M and Osborne C K (ed) 2014 *Physical Examination of the Breast. Diseases of the Breast* (Philadelphia, PA: Wolters Kluwer Health)

[4] Waks A G and Winer E P 2019 Breast cancer treatment *J. Am. Med. Assoc.* **321** 288–300

[5] Bradbury A R and Olopade O I 2007 Genetic susceptibility to breast cancer *Rev. Endocr. Metab. Disord.* **8** 255–67

[6] Acolditz G and Bohlke K 2014 Priorities for the primary prevention of breast cancer *CA Cancer J. Clin.* **64** 186–94

[7] Dillon D, Guidi A J, Schnitt S J, Harris J R, Lippman M E, Morrow M and Osborne C K (ed) 2014 *Pathology of Invasive Breast Cancer. Diseases of the Breast* 5th edn (Philadelphia, PA: Wolters Kluwer Health)

[8] Joshi H, Press M F, Bland K I, Copeland E M, Klimberg V S and Gradishar W J (ed) 2018 Molecular oncology of breast cancer *The Breast* (Amsterdam: Elsevier)

[9] Hammond M F, Hayes D F and Dowsett M 2010 American Society of Clinical Oncology/College of American Pathologists guideline recommendations for immunohistochemical testing of estrogen and progesterone receptors in breast cancer *J. Clin. Oncol.* **28** 2784–95

[10] Sinn H P and Kreipe H 2013 A brief overview of the WHO classification of breast tumors *Breast Care* **8** 149–54

[11] Bardia A, Mayer I A and Diamond J R 2017 Efficacy and safety of anti-trop-2 antibody drug conjugate sacituzumab govitecan (IMMU-132) in heavily pretreated patients with metastatic triple-negative breast cancer *J. Clin. Oncol.* **35** 2141–8

[12] Carey L, Winer E, Viale G, Cameron D and Gianni L 2010 Triple-negative breast cancer: disease entity or title of convenience? *Nat. Rev. Clin. Oncol.* **7** 683–92

[13] Carey L A 2007 Triple-negative (basal-like) breast cancer: a new entity *Breast Cancer Res.* **9** S13

[14] Denkert C, Liedtke C, Tutt A and von Minckwitz G 2017 Molecular alterations in triple-negative breast cancer-the road to new treatment strategies *Lancet* **389** 2430–42

[15] Foulkes W D, Smith I E and Reis-Filho J S 2010 Triple-negative breast cancer *N. Engl. J. Med.* **363** 1938–48

[16] Huang L, Rob B, Yi Y H, Gordan H and Ruo-pan H 2018 New insights into the tumor microenvironment utilizing protein array technology *Int. J. Mol. Sci.* **19** E559

[17] Emon B, Bauer J, Jain Y, Jung B and Saif T 2018 Biophysics of tumor microenvironment and cancer metastasis—a mini review *Comput. Struct. Biotechnol. J.* **16** 279–87

[18] Hanahan D and Coussens L M 2012 Accessories to the crime: functions of cells recruited to the tumor microenvironment *Cancer Cell* **21** 309–22

[19] Lawler J 2009 Introduction to the tumour microenvironment review series *J. Cell Mol. Med.* **13** 1403–4

[20] Danhier F, Feron O and Préat V 2010 To exploit the tumor microenvironment: Passive and active tumor targeting of nanocarriers for anti-cancer drug delivery *J. Control. Release* **148** 135–46

[21] Greenlee H, DuPont Reyes M J and Balneaves L G *et al* 2017 Clinical practice guidelines on the evidence-based use of integrative therapies during and after breast cancer treatment *CA Cancer J. Clin.* **67** 194–232

[22] Masoud V and Pagès G 2017 Targeted therapies in breast cancer: New challenges to fight against resistance *World J. Clin. Oncol.* **8** 120–34

[23] Di W, Mengjie S, Hui Y and Wong H-L 2017 Nanomedicine applications in the treatment of breast cancer: current state of the art *Int. J. Nanomed.* **12** 5879–92

[24] Francisco E J, Vanessa H M, Jun T and Lajos P 2019 Immunotherapy and targeted therapy combinations in metastatic breast cancer *Lancet Oncol.* **20** e175–86

[25] Pan H, Gray R and Braybrooke J *et al* 2017 20-Year risks of breast-cancer recurrence after stopping endocrine therapy at 5 years *N. Engl. J. Med.* **377** 1836–46

[26] Zhang R X, Wong H L, Xue H Y, Eoh J Y and Wu X Y 2016 Nanomedicine of synergistic drug combinations for cancer therapy–Strategies and perspectives *J. Control. Release* **240** 489–503

[27] Majidinia M and Yousefi B 2017 Breast tumor stroma: a driving force in the development of resistance to therapies *Chem. Biol. Drug Des.* **89** 309–18

[28] Gottesman M M 2002 Mechanisms of cancer drug resistance *Annu. Rev. Med.* **53** 625–7

[29] Hennessy M and Spiers J 2007 A primer on the mechanics of P-glycoprotein the multidrug transporter *Pharmacol. Res.* **55** 1–15

[30] Hu C M J and Zhang L 2012 Nanoparticle-based combination therapy toward overcoming drug resistance in cancer *Biochem. Pharmacol.* **839** 1104–11

[31] Alamoudi K, Martins P, Croissant J G, Patil S, Omar H and Khashab N M 2017 Thermoresponsive pegylated bubble liposome nanovectors for efficient siRNA delivery via endosomal escape *Nanomedicine* **12** 1421–33

[32] Gao M H, Xu Y Z and Qiu L Y 2017 Sensitization of multidrug-resistant malignant cells by liposomes co-encapsulating doxorubicin and chloroquine through autophagic inhibition *J. Liposome Res.* **27** 151–60

[33] Muddineti O S, Kumari P, Ray E, Ghosh B and Biswas S 2017 Curcumin-loaded chitosan-cholesterol micelles: evaluation in monolayers and 3D cancer spheroid model *Nanomedicine* **12** 1435–53

[34] Huang S Y, Liu J Y, Zhu H, Hussain A, Liu Q and Li J 2017 PEGylated doxorubicin micelles loaded with curcumin exerting synergic effects on multidrug resistant tumor cells *J. Nanosci. Nanotechnol.* **17** 2873–80

[35] Wang Y, Zhang Z P, Xu S H, Wang F H, Shen Y Y and Huang S T 2017 pH, redox and photothermal tri-responsive DNA/polyethylenimine conjugated gold nanorods as nano-carriers for specific intracellular co-release of doxorubicin and chemosensitizer pyronaridine to combat multidrug resistant cancer *Nanomedicine* **13** 1785–95

[36] Song L, Jiang Q, Liu J, Li N, Liu Q and Dai L 2017 L,DNA origami/gold nanorod hybrid nanostructures for the circumvention of drug resistance *Nanoscale* **9** 7750–4

[37] Lee T, Son H Y, Choi Y, Shin Y, Oh S and Kim J 2017 Minimum hyaluronic acid (HA) modified magnetic nanocrystals with less facilitated cancer migration and drug resistance for targeting CD44 abundant cancer cells by MR imaging *J. Mater. Chem* B **5** 1400–407

[38] Cho M H, Kim S, Lee J M, Shin T H, Yoo D and Cheon J 2012 Magnetic tandem apoptosis for overcoming multidrug-resistant cancer *Nano Lett.* **16** 7455–60

[39] Farvadi F, Tamaddon A, Sobhani Z and Abolmaali S S 2017 Polyionic complex of single-walled carbon nanotubes and PEG-grafted-hyperbranched polyethyleneimine (PEG-PEI-SWNT) for an improved doxorubicin loading and delivery: development and *in vitro* characterization *Artif. Cells Nanomed. Biotechnol.* **45** 855–63

[40] Pai C L, Chen Y C, Hsu C Y, Su H L and Lai P S 2016 Carbon nanotube-mediated photothermal disruption of endosomes/lysosomes reverses doxorubicin resistance in MCF-7/ADR cells *J. Biomed. Nanotechnol.* **12** 619–29

[41] Wang M, Han M, Li Y, Jin Y and Gao J Q 2016 Chemosensitization of doxorubicin in multidrug-resistant cells by unimolecular micelles via increased cellular accumulation and apoptosis *J. Pharm. Pharmacol.* **68** 333–41

[42] Sun L J, Wang D G, Chen Y Y, Wang L Y, Huang P and Li Y P 2017 core–shell hierarchical mesostructured silica nanoparticles for gene/chemo-synergetic stepwise therapy of multidrug-resistant cancer *Biomaterials* **133** 219–28

[43] Liu J, Li Q L, Zhang J X, Huang L, Qi C and Xu L M 2017 Safe and effective reversal of cancer multidrug resistance using sericin-coated mesoporous silica nanoparticles for lysosome-targeting delivery in mice *Small* **13** 2567–81

[44] Krolikowska A, Kudelski A, Michota A and Bukowska J 2003 SERS studies on the structure of thioglycolic acid monolayers on silver and gold *Surf. Sci.* **532** 227–32

[45] Peto G, Molnar G L, Paszti Z, Geszti O, Beck A and Guczi L 2002 Electronic structure of gold nanoparticles depositated on SiO$_x$/Si(100) *Mater. Sci. Eng.* C **19** 95–9

[46] Chandrasekharan N and Kamat P V 2000 Improving photochemical performance of monostructured TiO$_2$ film by abortion of gold nanoparticles *J. Phys. Chem.* **B104** 10851–7

[47] Watson J H P, Ellwood D C, Soper A K and Charnock J 1999 Nanosized strongly-magnetic bacterially-produced iron sulfide materials *J. Magn. Magn. Mater.* **203** 69–72

[48] Kumar V and Yadav S K 2009 Plant-mediated synthesis of silver and gold nanoparticles and their applications *J. Chem. Technol. Biotechnol.* **84** 151–7

[49] Sinha R, Kim G J, Nie S and Shin D M 2006 Nanotechnology in cancer therapeutics: bioconjugated nanoparticles for drug delivery *Mol. Cancer Ther.* **5** 1909–17

[50] Jain R K 2005 Normalization of tumor vasculature: An emerging concept in antiangiogenic therapy *Science* **307** 58–62

[51] Hobbs S K, Monsky W L, Yuan F, Roberts W G, Griffith L, Torchilin V P and Jain R K 1998 Regulation of transport pathways in tumor vessels: Role of tumor type and micro-environment *Proc. Natl. Acad. Sci.* **95** 4607–12

[52] Byrne J D, Betancourt T and Brannon-Peppas L 2008 Active targeting schemes for nanoparticle systems in cancer therapeutics *Adv. Drug Deliv. Rev.* **60** 1615–26

[53] Albanese A, Tang P S and Chan W C 2012 The effect of nanoparticle size, shape, and surface chemistry on biological systems *Annu. Rev. Biomed. Eng.* **14** 1–16

[54] Peer D, Karp J M, Hong S, Farokhzad O C, Margalit R and Langer R 2007 Nanocarriers as an emerging platform for cancer therapy *Nat. Nanotechnol.* **2** 751–60

[55] Cho K, Wang X, Nie S, Chen Z G and Shin D M 2008 Therapeutic nanoparticles for drug delivery in cancer *Clin. Cancer Res.* **14** 1310–6

[56] Wicki A, Witzigmann D, Balasubramanian V and Huwyler J 2015 Nanomedicine in cancer therapy: challenges, opportunities, and clinical applications *J. Control. Release* **200** 138–57

[57] Pankhurst Q A, Connolly J, Jones S K and Dobson J 2003 Applications of magnetic nanoparticles in biomedicine *J. Phys. D: Appl. Phys.* **36** R167–81

[58] McCarthy J R 2007 Targeted delivery of multifunctional magnetic nanoparticles *Nanomedicine* **2** 153–67

[59] Yarar E, Karakas G, Rende D, Ozisik R and Malta S 2016 Influence of surface coating of magnetic nanoparticles on mechanical properties of polymer nanocomposites *APS March Meeting Abstract, Boston, MA, USA*

[60] Ozdemir V 2006 Shifting emphasis from pharmacogenomics to theragnostics *Nat. Biotechnol.* **24** 942–6

[61] Widder K J, Senyei A E and Scarpelli D G 1978 Magnetic microspheres: a model system for site specific drug delivery *in vivo Proc. Soc. Exp. Biol. Med.* **58** 141–6

[62] Senyei A, Widder K and Czerlinski C 1978 Magnetic guidance of drug carrying micro-spheres *J. Appl. Phys.* **49** 3578–83

[63] Mosbach K K and Schroder U 1979 Preparation and applicationnof magnetic polymers for targeting of drugs *FEBS Lett.* **102** 112–6

[64] Yallapu M M, Dobberpuhl M R, Maher D M, Jaggi M and Chauhan S C 2012 Design of curcumin loaded cellulose nanoparticles for prostate cancer *Curr. Drug Metab.* **13** 120–8

[65] Yallapu M M, Gupta B K, Jaggi M and Chauhan S C 2010 Fabrication of curcumin encapsulated PLGA nanoparticles for improved therapeutic effects in metastatic cancer cells *J. Colloid Interface Sci.* **351** 19–29

[66] Yallapu M M, Jaggi M and Chauhan S C 2010 Poly(β-cyclodextrin)/curcumin selfassembly: a novel approach to improve curcumin delivery and its therapeutic efficacy in prostate cancer cells *Macromol. Biosci.* **10** 1141–51

[67] Yallapu M M, Jaggi M and Chauhan S C 2010 Beta-cyclodextrin-curcumin selfassembly enhances curcumin delivery in prostate cancer cells *Colloids Surf. B Biointerfaces* **79** 113–25

[68] Yallapu M M, Othman S F, Curtis E T, Gupta B K, Jaggi M and Chauhan S C 2011 Multi-functional magnetic nanoparticles for magnetic resonance imaging and cancer therapy *Biomaterials* **32** 1890–905

[69] Alexiou C, Arnold W, Klein R J, Parak F G, Hulin P, Bergemann C, Erhardt W, Wagenpfeil S and Lubbe S A 2000 Locoregional cancer treatment with magnetic drug targeting *Cancer Res.* **60** 6641–8

[70] Chouly C, Pouliquen D, Lucet I, Jeune P and Pellet J J 1996 Development of super-paramagnetic nanoparticles for MRI: effect of particles size, charge and surface nature on biodistribution *J. Microencapsul.* **13** 245–55

[71] Pratsinis S E and Vemury S 1996 Particle formation in gases—a review *Powder Technol.* **88** 267–73

[72] Sun C, Lee J S and Zhang M 2008 Magnetic nanoparticles in MR imaging and drug delivery *Adv. Drug Deliv. Rev.* **60** 1252–65

[73] Li Z, Zhang C, Wang B, Wang H, Chen X, Mohwald H and Cui X 2014 Sonochemical fabrication of dual-targeted redox-responsive smart microcarriers *ACS Appl. Mater. Interfaces* **6** 22166–73

[74] Mehta R V, Upadhyay R V, Charles S W and Ramchand C N 1997 Direct binding of protein to magnetic particles *Biotechnol. Tech.* **11** 493–6

[75] Koneracka M, Kopcansky P, Antalk M, Timko M, Ramchand C N, Lobo D, Mehta V R and Upadhyay R V 1999 Immobilization of proteins and enzymes to fine magnetic particles *J. Magn. Magn. Mater.* **201** 427–30

[76] Koneracka M, Kopcansky P, Timko M, Ramchand C N, Sequeira A and Trevan M 2002 Direct binding procedure of proteins and enzymes to fine magnetic particles *J. Mol. Catal. B Enzym.* **18** 13–8

[77] Slamon D J 1990 studies of the HER-2/neu proto-oncogene in human breast cancer *Cancer Invest.* **8** 252–4

[78] Chen J, Wu H, Han D and Xie C 2006 Using anti-VEGF McAb and magnetic nanoparticles as double-targeting vector for the radioimmunotherapy of liver cancer *Cancer Lett.* **231** 169–75

[79] Yamazaki M and Ito M 1990 Deformation and instability of membrane structure of phospholipid vesicles caused by osmophobic association: mechanical stress model for the mechanism of poly (ethylene glycol)-induced membrane fusion *Biochemistry* **29** 1309–14

[80] Hans M L and Lowman A M 2002 Biodegradable nanoparticles for drug delivery and targeting *Curr. Opin. Solid State Mater. Sci.* **6** 319–27

[81] Rudge S R, Kurtz T L, Vessely C R, Catterall L G and Williamson D L 2000 Preparation, characterization, and performance of magnetic iron–carbon composite microparticles for chemotherapy *Biomaterials* **21** 1411–20

[82] Arias J L, Gallardo V, Gomez-Lopera S A, Plaza R C and Delgado A V 2001 Synthesis and characterization of poly(ethyl-2-cyanoacrylate) nanoparticles with a magnetic core *J. Control. Release* **77** 309–21

[83] Gomez-Lopera S A, Plaza R C and Delgado A V 2001 Synthesis and characterization of spherical magnetite/biodegradable polymer composite particles *J. Colloid Int. Sci.* **240** 40–7

[84] Mornet S, Grasset F and Portier J 2002 Maghemite/silica nanoparticles for biological applications *Eur. Cells Mater.* **3** 110–3

[85] Deng Y, Wang L, Yang W, Fu S and Elaissari A 2003 Preparation of magnetic polymeric particles via inverse microemulsion polymerization process *J. Magn. Magn. Mater.* **257** 69–78

[86] Santra S, Tapec R, Theodoropoulou N, Dobson J, Hebard A and Tan W 2001 Synthesis and characterization of silica-coated iron oxide nanoparticles in microemulsion: the effect of nonionic surfactants *Langmuir* **17** 2900–6

[87] Carpenter E 2001 Iron nanoparticles as potential magnetic carriers *J. Magn. Magn. Mater.* **225** 17–20

[88] Grasset F, Mornet S, Demourgues A, Portier J, Bonnet J, Vekris A and Duguet E 2001 Synthesis, magnetic properties, surface modification and cytotoxicity evaluation of $Y_3Fe_{5-x}Al_xO_{12}$ garnet submicron particles for biomedical applications *J. Magn. Magn. Mater.* **234** 409–18

[89] Yallapu M *et al* 2012 Curcumin-loaded magnetic nanoparticles for breast cancer therapeutics and imaging applications *Int. J. Nanomed.* **7** 1761–79

[90] Rowe D L, Ozbay T, O'Regan R M and Nahta R 2009 Modulation of the BRCA1 protein and induction of apoptosis in triple negative breast cancer cell lines by the polyphenolic compound curcumin *Breast Cancer* **3** 61–75

[91] Aggarwal B B, Shishodia S and Takada Y 2005 Curcumin suppresses the paclitaxel-induced nuclear factor-kappa B pathway in breast cancer cells and inhibits lung metastasis of human breast cancer in nude mice *Clin. Cancer Res.* **11** 7490–8

[92] Goel A and Aggarwal B B 2010 Curcumin, the golden spice from Indian saffron, is a chemosensitizer and radiosensitizer for tumors and chemoprotector and radioprotector for normal organs *Nutr. Cancer* **62** 919–30

[93] Yallapu M M, Maher D M, Sundram V, Bell M C, Jaggi M and Chauhan S C 2010 Curcumin induces chemo/radio-sensitization in ovarian cancer cells and curcumin nanoparticles inhibit ovarian cancer cell growth *J. Ovarian Res.* **3** 11–7

[94] Ravindran J, Prasad S and Aggarwal B B 2009 Curcumin and cancer cells: how many ways can curry kill tumor cells selectively? *AAPS J.* **11** 495–510

[95] Yallapu M M, Jaggi M and Chauhan S C 2012 Curcumin nanoformulations: a future nanomedicine for cancer *Drug Discov. Today* **17** 71–80

[96] Ly J D, Grubb D R and Lawen A 2003 The mitochondrial membrane potential (deltapsi (m)) in apoptosis; an update *Apoptosis* **8** 115–28

[97] Kong S D, Zhang W, Hee Lee J, Brammer K, Lal R, Karin M and Jin S 2010 Magnetically vectored nanocapsules for tumor penetration and remotely switchable on-demand drug release *Nano Lett.* **10** 5088–92

[98] Yu Z, Yiaoliang W, Xuman W, Hong X and Hongchen G 2007 Acute toxicity and irritation of water-based dextran-coated magnetic fluid injected in mice *J. Biomed. Mater. Res.* **85A** 582–7

[99] Shi D *et al* 2009 Fluorescent polystyrene-Fe_3O_4 composite nanospheres for *in vivo* imaging and hyperthermia *Adv. Mater.* **21** 1–4

[100] Lubbe A S, Bergemann C, Brock J and McClure D G 1999 Physiological aspects in magnetic drug-targeting *J. Magn. Magn. Mater.* **194** 149–55

[101] Jiang Q, Zheng S, Hong R, Deng S, Guo L and Hu R 2014 Folic acid-conjugated Fe_3O_4 magnetic nanoparticles for hyperthermia and MRI *in vitro* and *in vivo* *Appl. Surf. Sci.* **307** 224–33

[102] Gordon R T, Hines J R and Gordon D 1979 Intracellular hyperthermia: a biophysical approach to cancer treatment via intracellular temperature and biophysical alteration *Med. Hypotheses* **5** 83–102

[103] Hilger I, Fruhauf K, Andra W, Hiergeist R, Hergt R and Kaiser W A 2002 Heating potential of iron oxides for therapeutic purposes in interventional radiology *Acad. Radiol.* **9** 198–202

[104] Moroz P, Jones S K and Gray B N 2002 Magnetically mediated hyperthermia: current status and future directions *Int. J. Hyperthermia* **18** 267–84

[105] Dumontet C and Jordan M A 2010 Microtubule-binding agents: a dynamic field of cancer therapeutics *Nat. Rev. Drug Discov.* **9** 790–803

[106] Jordan M A and Wilson L 2004 Microtubules as a target for anticancer drugs *Nat. Rev. Cancer* **4** 253–65

[107] Bonneterre J, Spielman M and Guastalla J P 1999 Efficacy and safety of docetaxel (Taxotere) in heavily pretreated advanced breast cancer patients: the French compassionate use programme experience *Eur. J. Cancer* **35** 1431–9

[108] Crown J, O'Leary M and Ooi W-S 2004 Docetaxel and paclitaxel in the treatment of breast cancer: a review of clinical experience *Oncologist* **9** 24–32

[109] Rodriguez A R, Chiu-Lam A, Morozov V M, Ishov A M and Rinaldi C 2018 Magnetic nanoparticle hyperthermia potentiates paclitaxel activity in sensitive and resistant breast cancer cells *Int. J. Nanomed.* **13** 4771–9

[110] Hilger R, Hiergeist R, Hergt K, Winnefeld K, Schubert H and Kaiser W A 2002 Thermal ablation of tumors using magnetic nanoparticles: an *in vivo* feasibility study *Invest. Radiol.* **37** 580–6

[111] Ito A, Kuga Y and Honda H 2004 Magnetite nanoparticle-loaded anti-HER2 immuno-liposomes for combination of antibody therapy with hyperthermia *Cancer Lett.* **212** 167–75

[112] Salimi M, Sarkar S, Saber R and Delavari H 2018 Ali Mohammad Alizadeh and Mulder H T 2018 Magnetic hyperthermia of breast cancer cells and MRI relaxometry with dendrimer-coated iron-oxide nanoparticles *Cancer Nano* **9** 1–19

[113] Deatsch A E and Evans B A 2014 Heating efficiency in magnetic nanoparticle hyperthermia *J. Magn. Magn. Mater.* **354** 163–72

[114] Mamiya H and Jeyadevan B 2011 Hyperthermic effects of dissipative structures of magnetic nanoparticles in large alternating magnetic fields *Sci. Rep.* **1** 157

[115] Rosensweig R E 2002 Heating magnetic fluid with alternating magnetic field *J. Magn. Magn. Mater.* **252** 370–4

Chapter 5

Magneto-plasmonic stimulated breast cancer nanomedicine

Ahmaduddin Khan, S Arunima Rajan, R K Chandunika and N K Sahu

5.1 Introduction

Cancer is defined by abnormal growth of cells. Carcinogenesis is a multistage procedure in which the damage occurs to the genetic material of normal cells. Generally, tumors are categorized as malignant or benign. Malignant tumors invade rapidly unlike benign tumors which develop slowly and are localized. Some of the most common cancer types diagnosed in the United States (US) are colorectal cancer, breast cancer, bladder cancer, leukemia, lung cancer, prostate cancer etc. Breast cancer is one of the most prevalent types and approximately 0.3 million new invasive breast cancer cases are expected to be diagnosed in women and 2670 cases in men in the US by 2019; in addition, *in situ* breast lesions DCIS (ductal carcinoma *in situ*) or LCIS (lobular carcinoma *in situ*) cases are observed in 62 930 women as per the American Cancer Society report 2019. Nearly 42 260 deaths may occur in 2019 [1]. The estimated new leading cancer cases and related deaths in females are represented in figure 5.1 [1]. Risk factors that promote breast cancer are weight gain after menopause which leads to excessive estrogen due to higher fat accumulation [2], alcohol consumption [3], smoking [4], usage of oral contraceptives [5] etc. Treatment for breast cancer usually involves surgery which may be one of two types; (1) lumpectomy where the tumor is removed along with some cancer-free surrounding tissue and (2) mastectomy which involves removal of the entire breast. Radiation therapy is usually administered after lumpectomy to decrease the risk of breast cancer mortality [6]. Various other therapies such as chemotherapy and hormonal therapy are also practised but they have their own limitations. Some of the associated chemotherapeutic side effects are diarrhea, fatigue, hair loss, headache, nausea etc [7]. Similarly, two of the common side effects of hormonal therapy are hot flushes and sweats [8]. Since the conventional methods are associated with some

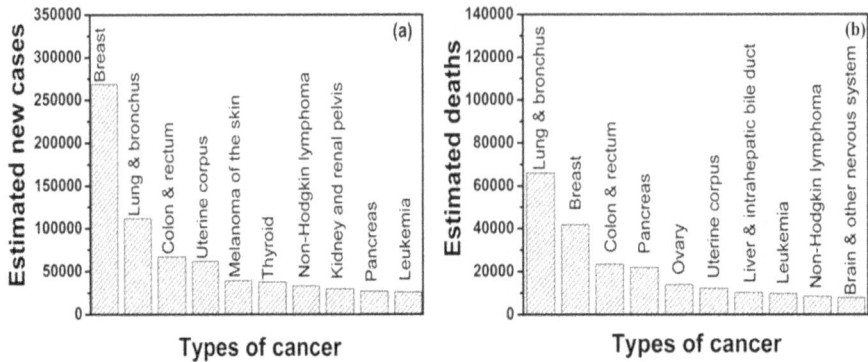

Figure 5.1. Estimated (a) new cancer cases in 2019 and (b) cancer related deaths for females in the United States. Concept adapted from [1].

serious side effects, e.g. damage to the immune system and other organs by the chemotherapeutics for non-specific targeting reasons [9], nanoparticles (NPs) mediated cancer therapy is the subject of extensive research. This chapter discusses breast cancer and related symptoms, existing therapeutic techniques, nanomaterial mediated therapy especially focussed on magneto-plasmonic stimulation, synthesis and functionalisation of nanomaterials, the mechanism of cellular uptake of nano-materials and the pros and cons of nanomedicine.

5.2 Breast cancer and its causes

Breast cancer is a malignant tumor or a collection of cancerous cells which occurs in the breast. Though it is most prevalent in women, it also occurs in men. According to the site of occurrences, breast cancer is of two types, i.e. non-invasive and invasive. Lobular carcinoma *in situ* (LCIS) and ductal carcinoma *in situ* (DCIS) are the two non-invasive types of breast cancer. LCIS grows in the breast lobules and doesn't grow outside the lobules. DCIS is one of the most common breast cancer types, its growth is also restricted to the breast duct [10]. There are many types of invasive breast cancer but most common are invasive lobular carcinoma (ILC) and invasive ductal carcinoma (IDC). ILC develops in the milk lobules and normally grows (metastasizes) and spreads to other parts of the body. IDC develops in milk ducts and penetrates the duct wall. It invades breast fatty tissue as well as other parts of the body. Some other less common breast cancer types are inflammatory breast cancer, Paget's disease of the nipple, Phylloides tumor, etc [11]. The exact cause of breast cancer is not yet fully known, however, some of the related risk factors are the use of oral contraceptives [5], family history, genetic risk factors such as mutation in BRCA1 and BRCA2 genes [12], obesity [2] etc.

Different types of signaling pathways for the development of breast cancer are reported in the literature such as ER signaling, HER signaling, Wnt signaling, CDKs signaling, Notch signaling, SHH signaling, BRK pathway etc [13, 14]. Among these, the major signaling pathways are ER and HER signaling pathways. In ER signaling, nuclear hormone receptors (ERα and ERβ) present in breast cancer cells form homo or heterodimer upon ligand binding and further translocate in the nucleus of the cell for regulation of transcription [13]. One of the mechanisms by which ERα initiates breast cancer cell growth is its interaction with cyclin D1 [15] which is an important activator of cyclin-dependent kinase (CDKs) 4 and 6. It helps in the transition of the cell cycle from G1 to S phase [13, 16]. Human epidermal growth factor receptors 1, 2, 3, 4 are members of the receptor family which show expression in many kinds of normal and cancer tissues [14]. In the human breast cancer cell line (HER signaling pathway), human epidermal growth factor receptor-2 (c-erbB2, HER2/neu) which is one among the above four tyrosine kinases shows amplification [17]. This leads to overexpression of HER2 protein that is associated with cancer progression and tumor cell proliferation [14]. Early symptoms of breast cancer include palpable breast mass or LN in the breast, discharge from nipple (which can be serious if malignant), organomegaly, nipple retraction, ulceration or erythema, skin edema and bone pain in late stages [18].

5.3 Existing breast cancer therapies

Generally, three types of cancer therapies such as surgical therapy, chemotherapy and endocrine therapy are usually practised by doctors. However, radiation therapy is frequently given along with surgical and chemotherapy as adjuvant therapy.

5.3.1 Surgical therapy

Surgery is one of the most effective and conventional methods to remove the tumor and its nearby tissues. Various surgery techniques such as diagnostic surgery, curative surgery, preventive surgery, staging surgery, debulking surgery, palliative surgery, supportive surgery, are practised by surgeons. Diagnostic surgery is used to examine whether the cells are cancerous or not. If a cancerous tumor is located in a particular area of the body curative surgery is adopted. However, if the tissue doesn't contain cancerous cells but has the ability to transform into malignant, preventive surgery is used to remove the tissue. Staging surgery is used to check the extent of the spread of tumor, whereas debulking surgery is used to remove some part of the tumor so as to avoid damage to the surrounding tissue. Supportive surgery is used along with other cancer treatments to increase efficiency. Restorative surgery helps to restore the function of an organ or a body part after a patient has undergone some kind of surgery. Patients with single lesions with sizes of less or equal to 4 cm should have breast conserving surgery also known as lumpectomy. Normally, removal of more than 15% of the breast volume results in poor cosmetic result and hence, the lesion should be removed with a margin of normal tissue. The National Institute of Health and Care Excellence (NICE) guidelines suggests 2 mm for DCIS, but for invasive cancer the optimum margin is not clear [19]. For DCIS, mastectomy is one of the most efficient treatments to prevent local recurrence. After mastectomy, the local recurrence rate is around 1% and the mortality rates is nearly zero as reported in the research [20]. Mastectomy is done in multicentricity case (multiple quadrants), whereas in the case of unicentric DCIS, the lesion is much too large to excise with clear margins and allowable cosmetic results [20]. Axillary staging procedures should be done for patients undergoing breast surgery for invasive cancer or extensive DCIS such as axillary lymph node dissection (ALND), sentinel lymph node biopsy (SLNB) and axillary node sampling [19].

5.3.2 Chemotherapy

Chemotherapeutic drugs obstruct the divisive ability of cancer cells and stop their proliferation. Either a single drug or a combination of drugs are used to kill the cancer cells. The drug can be administered via the bloodstream or through targeted mechanisms specifically onto the particular tumor sites. Adjuvant chemotherapy e.g. administration of drugs after surgery is practised to avoid recurrence. Drugs are prescribed in combination to increase their efficiency, reduce side effects and combat drug resistance [21]. Neoadjuvant chemotherapy (NAC) is a procedure for making a locally advanced and inoperable breast cancer resectable. More recently, it has been used to downstage the operable tumors such as the disease in axilla and the breast with the purpose of improving breast conservation and avoiding axillary lymph node dissection (ALND) in some cases [22, 23]. Conventional chemotherapy has some drawbacks which hinder its potential to kill cancer cells. These include: (a) limited aqueous solubility of chemotherapeutic drugs, as nearly all chemotherapeutics are synthetic or derived from plant source and hence are hydrophobic, thus requiring specific solvent and leading to toxicity, (b) lacking in selectivity of chemotherapeutics which harms the normal cells, and (c) multidrug resistance of the tumor due to the efflux pumps which eject the drugs out of the cell [24–27].

5.3.3 Endocrine therapy

Breast cancers with progesterone receptors and estrogen receptors are called *PR-positive* (or PR+) and ER-positive (ER+) cancer respectively. ER-negative, PR-positive cancers are unusually rare (0.3%) [28]. Nearly 80% of breast cancers are ER-positive (ER+) cancers, thus endocrine (also termed hormonal) therapy should be supportive to surgery in the majority of patients [29]. Adjuvant endocrine therapy can be executed before or after radiotherapy or surgery or along with chemotherapy. Current endocrine therapy for breast cancer consists of ovarian suppressor (OS), selective estrogen receptor modulator and downregulator (SERDs), aromatase inhibitor (AIs) [30]. Neoadjuvant endocrinal therapy can be done past the conventional month to permit further downstaging of the tumor [31].

5.3.4 Radiotherapy

Post-mastectomy radiotherapy is suggested in patients having a high risk of regional or local relapse. Chest wall radiotherapy is suggested for patients with large tumors (~5 cm) having four or more axillary lymph nodes, extensive lymphovascular invasion and/or surgical margins [32]. In conventional radiotherapy, a pair of tangential beams is provided across the chest to cover the remaining breast tissue after breast conservation or the full breast bed after mastectomy. The total radiation dose to obtaining locoregional control is 40–50 Gy while maintaining complications at an acceptable level and is given for 3–5 weeks [33, 34]. Oligometastatic breast cancer is a subtype of metastatic breast cancer having a limited number of metastatic sites. This forms around 20% of metastatic breast cancer [35]. Use of ablative radiotherapy for oligometstatic breast cancer management is gaining much attention because of its non-invasive nature, efficiency in gaining local control and outstanding safety profile. Additional research is required in genomic and molecular profiling for the characterization of metastatic breast cancer patients who will get advantage from such treatment [36].

These conventional therapies are associated with various side effects ranging from discomfort to the formation of secondary tumors and adverse clinical effects to normal surrounding cells. Hence, there is an urgent need to develop some new techniques that can overcome these drawbacks.

5.4 Nanomaterial aspect of breast cancer therapy

Nanomaterials are used extensively as the carriers of anticancer drugs in chemotherapy. They can be used intravenously or by mouth. The conventional chemotherapy is a major alternative for breast cancer patients but it has lot of drawbacks such as toxicity, function loss due to drug resistance, irregular bio-distribution etc. Nanomedicine has the potential to overcome these challenges. In a recent report, the anticancerous property of biosynthesized gold nanoparticles (AuNPs) was demonstrated in combination with AuNPs–curcumin, AuNPs–turmeric, AuNPs–quercetin and AuNPs–paclitaxel and then compared with the pristine drugs [37]. The conjugates in combination are most effective therapeutically for the inhibition of breast cancer cell proliferation, angiogenesis, apoptosis, spheroid formation and colony formation showing synergistic effect as compared to their individual analogue. NPs are also conjugated with targeting agents like folic acid for targeting folate receptors which are overexpressed in various cancer cell types. In a report, β-lactoglobulin NPs (BNPs) were attached to folic acid and then loaded with anticancerous agent doxorubicin (DOX) [38]. Results show that the folic acid modified BNPs represents enhanced cell proliferation inhibition in both MCF-7 and MDA-MB-231 cells at lower dosage in comparison to free DOX. Another active targeting of NPs based on vascular endothelial growth factor (VEGF) is demonstrated by Semkina *et al* [39]. PEGylated magnetic nanoparticles (MNP) were coated with bovine serum albumin (BSA) and conjugated with anti-VEGF antibodies and IgG (non-specific immunoglobulin). MNP–VEGF exhibited the highest percentage of accumulation in comparison to MNP–IgG and bare MNP as analyzed by flow cytometry. The median survival time of mice having 4T1 tumors cells was nearly 50% more when treated with MNP–VEGF@DOX in comparison to pure DOX. VEGF acts as a target for the active targeting of DOX and also as a target for antiangiogenic therapy [39]. Further nanomaterials are also used to enhance radiosensitization. In a report, three kinds of AuNPs were synthesized in the presence of three different capping/stabilizing agents such as citrate (Au–CIT), glutathione (Au–GSH) and DEAE-dextran (Au–DEX) to increase cellular uptake in the cancer cells [40]. These NPs are localized in cytoplasmic vesicles when treated with MDA-MB-231 breast cancer cells. Western blot and flow cytometry analysis proves that only

AuNPs alone did not affect cell cycle. However, with the application of high energy radiation of 6 MV of dose 4 Gy, the radiation itself caused an S phase arrest after 6 and 8 h of irradiation, and G2/M arrest after 24 and 48 h of irradiation [40].

Immunotherapy is a kind of cancer therapy which uses the immune system of a body to fight against cancer. In a report, polymeric NPs consisting PLGA–PEG was loaded with breast cancer antigen ovalbumin and delivered to the dendritic cells (DCs) [41]. Liu *et al* reported a dual pH-responsive multifunctional NPs system based on poly(L-histidine) (PHIS) and hyaluronic acid (HA) for the combination of chemo- and immunotherapy to treat breast cancer [42]. A pH responsive HA–DOX/PHIS/R848 conjugate NPs was formed by encapsulating PHIS with antitumor immune regulator (R848) to form PHIS/R848 nanocore and by attaching DOX with HA through acid labile hydrazone linkage which forms HA–DOX polymer prodrug. At pH 6.5, PHIS is ionized and changed from hydrophobic to hydrophilic which released R848 to provide immunoregulatory action [42]. Zhang *et al* demonstrated the delivery of CpG (cytosine phosphate guanine) via (3-aminopropyl) triethoxysilane modified Fe_3O_4 (FeNPs)[43]. CpG is an agonist of toll-like receptor (TLR9) which can stimulate Th1 response in tumor. Free CpG has some drawbacks, e.g. it is prone to nuclease mediated degradation under biological conditions, it lacks specificity and has poor cellular uptake after administration, and the route of administration affects its toxicity and antitumor ability. FeNP/CpG have enhanced cellular uptake rate comparable to free CpG and have significant antitumor effects in the tumor model of 4T1 breast cancer with inhibitory rate of 64.3% [43]. Besides the use of nanomaterials in chemo-, radio- and immunotherapy, they are exclusively used in magnetic hyperthermia and photothermal therapy of cancer.

Elimination and restraint of specific cancer cells by utilizing heat is known as thermotherapy and is a probable future technique for efficient cancer therapy [44]. As the characteristics of the tumor microenvironment are acidic, hypoxic and nutrient-deficient, tumor cells are more prone to heat compared to normal tissues [45]. The elevated temperature promotes the cell death via denaturation of intracellular protein and/or the disruption of cellular membrane. This intracellular heat can be generated by sending suitable NPs (magnetic or plasmonic) to the tumor sites and subsequently applying either alternating magnetic field or electromagnetic wave (photon). Hence, the performance of thermotherapy can be tuned by the power density of the external source, the energy absorption and subsequent thermal behavior of the biological environment. Thermotherapy may possess minimal side effects compared to conventional chemotherapy and radiotherapy [46].

5.4.1 Principle of magnetic hyperthermia

Hyperthermia (HPT) is a thermotherapy which elevates temperature in the range of 41–43 °C for the selective destruction of tumor tissues. Conventionally, various energy sources such as lasers, radio frequency waves, infrared waves, microwaves and ultrasonic waves have been used for inducing heat in tumor tissues [47–49]. These types of heating can often lead to non-uniform heat distribution causing adverse effects on healthy tissues as heat distribution and thermal dosage are the crucial factors to be monitored during this therapy. Magnetic hyperthermia (MHT), also termed magnetic fluid hyperthermia (MFT), utilizes suitable magnetic nanoparticles (MNPs) to dissipate heat upon exposure to an external alternating current magnetic field (ACMF). Localized heat delivery to deep seated tumor tissues makes MHT advantageous over traditional hyperthermia methods. A maximum field-frequency of $H*f < 5 \times 10^8$ A m s^{-1} (Atkinson–Brezovich limit) is often monitored for safe clinical MHT [50]. The heat dissipation is quantified by a specific absorption rate (SAR) or specific loss power (SLP).

Heat dissipation of MNPs under an external magnetic field may mainly be attributed to two different mechanisms: (a) hysteresis losses and (b) relaxation losses. Heat loss mechanisms differ with respect to size of the NPs. In the single domain region ferro/ferromagnetic materials, there exist two types of magnetism such as ferro/ferrimagnetism and superparamagnetism. Hysteresis heat losses are exhibited by ferro/ferrimagnetic particles as a result of irreversible magnetization processes in hysteresis loops. The applied magnetic energy absorbed by MNPs aligns all atomic moments with respect to the magnetic field and demagnetizes partially upon removal of the field. This causes a change in magnetisation in a different path providing a hysteresis loop. This magnetisation–demagnetization dissipates energy in the form of heat [51]. However, superparamagnetic (SPM) NPs are often preferred for bio-applications [52]. SPM NPs are characterized by a single giant magnetic moment which provides for the total magnetisation of the material. Such MNPs exhibit high magnetisation upon an external magnetic field and remain zero after the removal of field. This fluctuation in magnetisation reduces magnetic dipolar interactions thereby reducing any adverse effects in the human body. Heat dissipation in SPM NPs is mainly due to two relaxation processes such as Néel relaxation and Brownian relaxation [53]. Heat loss due to Néel relaxation occurs when the magnetic moment inside each domain rotates irrespective of its particle rotation by overcoming the anisotropy energy barrier. Brownian relaxation occurs by physical rotation of particles in the suspended fluid with the applied field. The suspended medium opposes this rotational movement resulting in heat loss as a result of frictional forces. Both relaxation mechanisms contribute to the effective heat release which can be calculated in terms of effective relaxation time [54] as:

$$\tau_{\text{effective}} = \frac{\tau_B \tau_N}{\tau_B + \tau_N}$$

where, τ_N and τ_B are the Néel relaxation time and Brownian relaxation time respectively given by,

$$\tau_N = \frac{\tau_0}{2}\sqrt{\pi \frac{k_B T}{KV}}\, e^{\frac{KV}{k_B T}} \qquad \text{and} \qquad \tau_B = \frac{3\eta V_H}{k_B T}$$

where τ_0 is the attempt time scale ranging values from 10^{-13} to 10^{-9} s, k_B is the Boltzmann's constant, T is temperature, K is anisotropy constant, V is volume of MNP, V_H defines the hydrodynamic volume of the MNP and η is the viscosity of the medium. Figure 5.2(a) represents a schematic of heat loss mechanism due to above

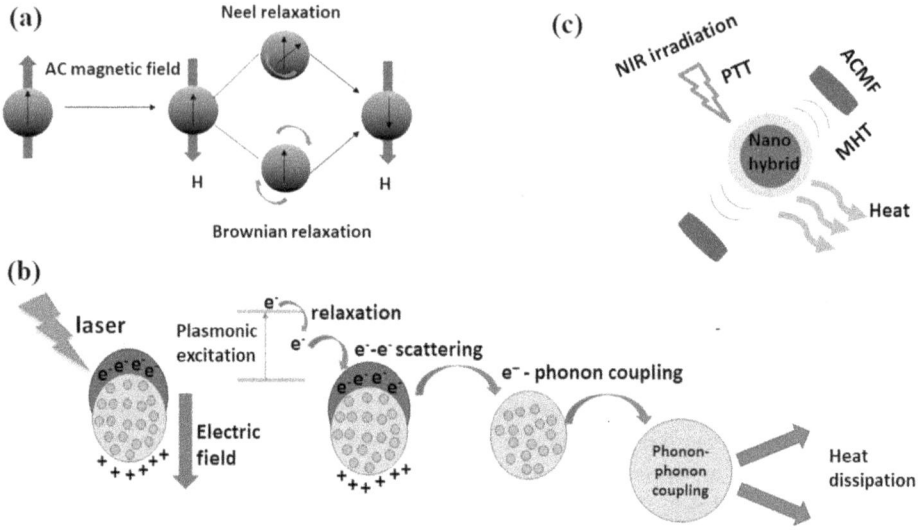

Figure 5.2. Schematic representation of (a) heat generation mechanism in SPM NPs, (b) NIR irradiation on plasmonic NPs and (c) magneto-plasmonic combined therapy.

relaxations. Accordingly, the total power dissipated (thermal energy) by the MNPs upon an external magnetic field considering the relaxation losses is calculated by [54],

$$P_{\text{dissipated}} = \frac{1}{2}\mu_0\chi_0\omega H^2\frac{\omega\tau_{\text{effective}}}{1 + (\omega\tau_{\text{effective}})^2}$$

where χ_0 is the initial static susceptibility; ω and H are the angular frequency ($\omega = 2\pi f$, where f is the applied frequency) and magnitude of the applied magnetic field. The heating ability of MNPs is defined as the ratio of thermal power dissipation ($P_{\text{dissipated}}$) to the mass of MNPs (M_{mnp}) is in terms of SAR [54] as

$$\text{SAR} = \frac{P_{\text{dissipated}}}{M_{\text{mnp}}}$$

Calorimetric method is the most commonly used method for measuring SAR in terms of temperature rise in colloidal MNPs exposed to AC magnetic field [55] as

$$\text{SAR} = C_s \cdot \frac{\Delta T}{\Delta t} \cdot \frac{1}{M_{\text{mnp}}}$$

where, C_s is the specific heat capacity of the solvent where MNPs are dispersed, M_{mnp} is the mass of the MNP, and $\frac{\Delta T}{\Delta t}$ is the temperature-time dependent slope.

5.4.2 Principle of photothermal therapy

Phototherapy, a combination of both photothermal therapy (PTT) and photo-dyanamic therapy (PDT), is a new technique for malignant tissue treatment and has gained much interest. In comparison to the traditional therapies, PTT has various advantages such as low toxicity, minimal invasion, temporal and spatial control. PTT utilizes plasmonic NPs which can convert absorbed near-infrared (NIR) irradiation into heat energy by photothermal agents for effective killing of specific tissues. Although tissues and cells possess inherent natural photothermal agents such as hemoglobin, cytochromes etc, their absorption efficiency is too low. Hence plasmonic NPs with higher absorption efficacy as photothermal agents are often considered as potentiating for PTT. Even though plasmonic resonances in the near Vis–NIR range are exhibited by noble metals such as silver (Ag), gold (Au), platinum (Pt) and copper (Cu), Au remains the prime choice for PTT applications [56]. This can be attributed to the fact that Au is the only noble metal which can exhibit localized surface plasmon resonance (LSPR) in the NIR region advantageous for cancer therapy. The phenomenon surface plasmon resonance (SPR) is defined as the collective excitation of conduction electrons at the interface of NPs stimulated by incident light. This collective excitation can induce strong surface fields on exposure to electromagnetic irradiation. The excited electrons relax rapidly thereby inducing heat beneficial for cancer cell destruction [56]. For sufficient heating, light radiation interacting with NPs should have a frequency which strongly overlaps with the NP SPR absorption band [57]. A schematic representing NIR irradiation on plasmonic NPs is shown in figure 5.2(b). Energy absorbed by the surface NPs can be released by either the emission of photons of different frequency (luminescence) or by the emission of phonons which are responsible for heat dissipation. In addition, part of the incident radiation scatters with the same frequency in all directions. Absorption rate is higher, whereas scattering rate is lower for smaller NPs. Upon increasing the size of NPs, scattering rate increases, whereas absorption rate decreases [58]. SPR peak of AuNPs includes both absorption and scattering components. NPs with large absorption efficiency (η_{abs}) result in significant light-to-heat energy conversion favorable for efficient PTT and it can be calculated as:

$$\eta_{abs} = \frac{\text{Number of absorbed photons}}{\text{Total number of incident photons interacted with the NPs}}$$

The absorbed light rapidly converts to heat thereby forming a thermal metallic lattice via two major processes such as (a) electron–electron relaxation and (b) electron–phonon relaxation [59]. The first relaxation is the electron thermalization resulting in the formation of hot electrons of several thousand kelvins. This occurs within femtoseconds due to electron–electron scattering. The second relaxation is due to the exchange of energy between the electrons and the lattice through electron–phonon coupling. The heat release from hot electrons to the lattice causes the lattice temperature to rise in the order of a few tens of degrees within picoseconds [60]. The lattice then cools off by phonon–phonon relaxation within ~100 ps during which heat is dissipated from the NPs into the surroundings [61]. Heat generated by NPs can be calculated as [62]:

$$Q_{NP} = C_{abs}I$$

where, C_{abs} is the absorption cross-section area of the NPs and I is the laser irradiance. In a sample of NPs with certain concentration (C), SAR can be expressed as [62]:

$$SAR = Q_{NP} \times C = C_{abs}IC$$

It is well known that biological tissues absorb light radiation in the NIR wavelength range of 650–900 nm [63]. NIR PTT is favorable for *in vivo* applications as NIR irradiation penetrates tissue optimally. The major factors to be focused on during selection of NPs in PTT are (a) wavelength of the maximal absorption, (b) spectral bandwidth and (c) size of the NPs. As per the laser dosage, type of laser and irradiation time, PTT induced cell death occurs either by apoptosis or by direct necrosis [12]. When PTT is used with continuous wave (CW) laser irradiation, plasmonic NPs accumulated in the cytoplasm are more efficient for induced cell death than NPs localized at the nucleus, whereas high energy pulse laser enables the NPs at the nucleus for cell necrosis. Zharov *et al* experimented and inferred that a nanosecond pulse of energy of 2–3 J cm^{-2} induced cell death when cells are accumulated with 10–15 AuNPs [64]. Such laser pulses induce cell damage for several reasons such as thermal expansion of particles, photoinduced breakdown or plasmon generation of AuNPs. CW laser and nanosecond pulsed laser trigger cell death through apoptosis and necrosis, respectively. Generally, a CW laser is used for sufficient heating in PTT [65]. These findings reveal that the cell death mechanism depends on the type of laser used.

5.4.3 Nanomaterials used for MHT and PTT

Metals like iron (Fe), cobalt (Co), nickel (Ni), zinc (Zn), manganese (Mn), gadolinium (Gd), magnesium (Mg) as well as their alloys and oxides are certain MNPs which are used commonly for MHT. As iron oxide is approved by the US Food and Drug Administration (FDA) for clinical studies, iron oxide-based NPs (IONP) are most widely utilized for MHT [66]. *In vivo* study by Hilger *et al* proved that magnetite NPs on human breast adeno-carcinomas under an ACMF of field strength 6.5 kA m^{-1} and frequency of 400 kHz with exposure time of 4 min can generate sufficient heat for cell death [67]. In a similar work, IONP coated with fourth generation (G^4) polyamidoamine dendrimer tested on MCF-7 cell, viability get reduced under an ACMF of 300 kHz and field intensity of 12 kA m^{-1} [68]. Chen *et al* performed *in vitro* studies in MCF-7 breast cancer cell lines under an ACMF (230 kHz, 28 A) which exhibited maximum cell inhibition rate of 78% and the inhibition effect was found to be concentration dependent of Fe$_3$O$_4$ NPs [69]. In a study by Almaki *et al*, a tumor-targeting magnetic core–shell based nanocarrier (Fe$_3$O$_4$–PEG–TRA) was evaluated for both *in vitro* and *in vivo* MHT in various breast cancer cells. The results indicated that MHT at 230 kHz increased the inhibition growth rate in cells which depend on time and temperature. The sensitivity was higher in SKBR-3 cell line (HER2+) compared to the other cells such as HER2- cell lines (HSF 1184, MDA-MB-231 and MDA-MB-468) [70].

Hong *et al* reported that PEGylated Au nanoclusters (AuNCs) conjugated with AS1411 achieves cancer targeted PTT and anti-proliferation activity [71]. Such NPs showed specific cellular uptake in MDA-MB-231 breast cancer cells via selective binding of AS1411 to nucleolin in the plasma membrane. A decreased cell viability to 64% in NIH 3T3 cells through NIR light (power density 0.7 W cm^{-2} and wavelength of 808 nm) irradiation for 3 min exposure/treatment time was found due to its elevated cellular uptake which further mediated NIR light responsive photo-thermal treatment in specific cancer cells. In another report by Liu *et al*, a novel phototherapeutic agent IABDP possessing the property of NIR absorption exhibited excellent PTT on increasing the power density from 0.4 to 0.8 W cm^{-2} and at a wavelength of 635 nm for an exposure time of 5 min, tumor suppress efficiency was reported about 97% for intratumor injection upon irradiation [72]. Park *et al* reported the targeting efficacy of cooperative nanomaterials such as DOX loaded magnetic nanoworms (NWs) and liposomes (LPs) to the local microenvironment of MDA-MB-435 xenograft tumor that can be enhanced by photothermally activated gold nanorods [73]. Kirui *et al* prepared a gold/iron oxide nanohybrid by the thermal decomposition method, functionalized with a single chain antibody bonded to colorectal cancer antigen A33 [74]. Cellular uptake and laser irradiation studies have experimented on colorectal cancer cell lines (SW1222 and HT 29 cells) using 808 nm CW laser. When the intensity of the laser was increased from 5.1 W cm^{-2} to 31.5 W cm^{-2}, 50% of cell destruction occurred within 6 min. A report by Wang *et al* proved that PS@CS@Au–Fe$_3$O$_4$–FA/ICG nanocomposite under NIR laser (808 nm 1.0 W cm^{-2} for 300 s) exhibits significant cell destruction in both *in vitro* and *in vivo* studies using HeLa cells [75].

Iron–platinum (FePt) NPs due to their high energy absorption and excellent superparamagnetic behavior have been recently recognized as multifunctional materials for bio-application. Chen *et al* synthesized folate conjugated FePt NPs to target breast cancer cells EMT-6 for PTT [76]. It was observed that a sudden rupture of plasma membrane occurs under NIR irradiation for a femtosecond. For a comparative analysis, AuNPs on the similar EMT-6 cells under the same exper-imental conditions were performed and found that FePt NPs showed superior activity to AuNPs for cell death. Similarly, Phan *et al* developed FePt NPs functionalized with polypyrrole (PPy) which displayed high NIR absorbance and excellent biocompatibility in biological studies [77]. When MDA-MB-231 breast cancer cells are incubated with FePt@PPy NPs *in vitro* and exposed to the NIR laser at a power density of 1 W cm^{-2} for 5 min, there is temperature rise of 39.3 °C and 71 °C for 20 μg ml^{-1} and 120 μg ml^{-1} concentration of NPs, respectively. Carbon-based nanomaterials have also been investigated in biomedical areas due to their high porosity, biocompatibility and large specific surface area. Kam *et al* [78] reported single walled carbon nanotube (SWCNT) with strong optical absorbance and performed *in vitro* PTT studies on HeLa cells. Cell studies were carried out in a concentration of 25 mg l^{-1} NPs which on exposure to NIR laser of wavelength 808 nm with a power density of 1.4 W cm^{-2} for 2 min resulted in an increased temperature of 70 °C. Yang *et al* [79] fabricated PEG coated nanographene sheets (NGS) by fluorescent labeling method. *In vivo* PTT carried out on 4T1 tumor model on mice and found that when exposed to 808 nm NIR laser (power density of 2 W cm^{-2}), temperature increased to ~50 °C within 5 min NGS showed a strong optical absorbance in NIR region thereby causing efficient tumor cell destruction.

5.4.4 Combination therapy

One of the most interesting factors to focus on nanomaterials is the symbiosis, i.e. its ability to combine with other nanomaterials to obtain a synergistic effect. In the biomedical field, there exists a need to obtain single nanoentities with multifunctional properties. MNPs with the aid of external magnetic field and plasmonic NPs via near-infrared (NIR) radiation dissipate heat lead to the combined therapy. A nanohybrid of IONPs coated with gold (Au) has been explored for a combined therapy of MHT and PTT [80]. Such nanohybrids possess both diagnostic and therapeutic functions for cancer therapy. A schematic of the combinational therapy is depicted in figure 5.2(c).

Das *et al* [81] reported an increment in SAR of 210 W g^{-1} for silver/magnetite (Ag/Fe$_3$O$_4$) nanoflowers when subjected to laser irradiation (442 nm, 0.93 W cm^{-2}, 5 min) and magnetic field (200 Oe). Another study by Espinosa *et al* [82] reported the duality of Fe$_3$O$_4$ nanocubes as magneto-plasmonic agents. Upon combined exposure of NIR laser irradiation (808 nm, 0.3 W cm^{-2}, 10 min) and alternating magnetic field (900 kHz, 250 Oe), *in vitro* studies tested on PC3 (prostate cancer), SKOV3 (ovarian cancer) and A431 (epidermoid cancer) cells exhibited an increased heating effect with a specific loss power (SLP) of 5000 W g^{-1}. In addition, *in vivo* studies were performed on mice bearing solid tumors by injecting 14 mg of Fe$_3$O$_4$ nanocubes three times at 24 h intervals which showed an enhanced dual therapy towards the cancer cell. A comparison of temperature rise in SKOV3 cells for different size NPs due to MHT alone and combined MHT–PTT is shown in figure 5.3(a). Average tumor growth in non-injected mice and iron oxide nanocube-injected mice with A431 human epidermoid carcinoma cells when exposed to MHT, PTT and combined MHT and PTT is shown in figure 5.3(b). It has been seen that with application of combined MHT and PTT, the tumor growth is completely suppressed. Schematic of MNPs coated with plasmonic NPs and injected intratumorally to mice which subsequently subjected to combined MHT–PTT is shown in figure 5.3(c).

Targeted chemo-PTT is also employed as a synergistic therapy for effective cancer treatment. Multifunctional magnetic gold NPs conjugated with PEG and DOX were treated on human breast cancer cells (MCF-7) for achieving a combined chemo-PTT [84]. CW laser irradiation of power density 2 W cm^{-2} for a duration of 10 min showed significant inhibition of cancer cells. The SPR effect of Au raises the local temperature and accelerates the release of the drug resulting in significant inhibition. Yao *et al* encapsulated DOX with mesoporous silica nanoparticles (MSNs) modified with graphene quantum dots (GQDs) and investigated the synergistic therapeutic efficacy on 4T1 breast cancer cells [85]. The results showed an improved heat generation upon NIR irradiation of power density 2.5 W cm^{-2} for a duration of 3 min with a simultaneous delivery of chemotherapeutic drug. No cytotoxicity was

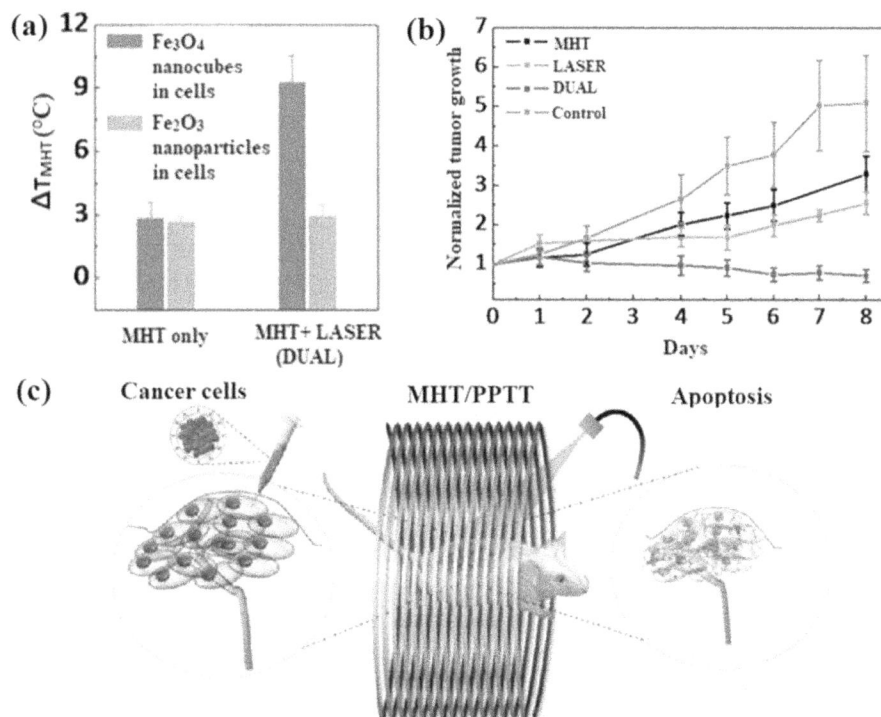

Figure 5.3. (a) Temperature rise in MHT alone and combined MHT–PTT contribution in SKOV3 cells for 20 nm iron oxide nanocubes and 9 nm NPs [82], (b) Average tumor growth in non-injected mice upon control, and in iron oxide nanocube-injected mice exposed to MHT, PTT and combined therapy for 8 days [82] and (c) schematic of MNPs coated with plasmonic NPs injected intratumorally to mice and subsequently subjected to combined MHT–PTT [83]. Figures reprinted with permission from American Chemical Society.

exhibited when GQD–MSNs were internalized by 4T1 cells. Gao *et al* reported a similar work where Rhodamine B (RhB) was loaded instead of DOX in the previous work and studied the photothermal release of RhB. RhB was found to release when the NPs were exposed to a NIR laser of wavelength 808 nm at 2.5 W cm^{-2} [86]. In a report by Li *et al*, a zwitterionic nanogel loaded with indocyanine green (ICG) and DOX was developed and synergistic photothermal and chemotherapeutic cytotoxicity was assessed *in vitro* [87]. Upon exposure to NIR radiation (808 nm, 1.5 W cm^{-2}), there was a decrease in nanogel diameter and improved photothermal effects were found. Furthermore, *in vivo* studies on H22 cell-bearing mice exhibited half tumor death. A biphasic system of Ce^{3+} sensitized GdPO$_4$:Tb^{3+} with IONP have been reported for *in vitro* chemo-thermal therapy of HeLa and MCF-7 cells [88]. There was 50%–55% of cell deaths in both the cell lines after 24 h incubation with DOX loaded biphasic system which increased up to 90% upon application of ACMF (250 kHz, 460 Oe) for a period of 25 min.

Xu *et al* reported pH and NIR responsive drug release platform composed of HA functionalized gold nanorods (GNRs) for the synergistic photo-chemotherapy of breast cancer [89]. HA was functionalized with dopamine, adipic acid dihydrazide and folate and then attached to GNRs by Au–catechol bonds. DOX was attached to HA by acid labile hydrazone bond to form GNRs–HA–FA–DOX. It manifests a remarkably higher ability to deliver DOX and GNRs into MCF-7 cells by folate receptor mediated endocytosis and can effectively kill the cells by apoptosis using NIR radiation. The combination of chemotherapy and PTT shows a synergistic effect in both *in vivo* and *in vitro*. Mahmoodzadeh *et al* reported novel theranostic NPs made by attaching polymer HS-poly (ε-caprolactone)-block-poly(*N*-isopropy-lacrylamide)-block-poly(acrylic acid) (HS-PCL-*b*-PNIPAAm-*b*-PAA) and gold NPs to form GNPs@polymer and further loaded with DOX [90]. The GNPs@polymer–DOX was tested against MCF-7 cells and the results prove that the chemo-photothermal therapy has an enhanced therapeutic effect compared to chemo and PTT alone. In a report by Machado *et al* [91], a new drug delivery system based on curcumin nanoemulsion (CNE) is analyzed for its use as a photosensitizing agent in photodyanamic therapy in MCF-7 and HFF-1 cells. Cell viability decreases to 57% after irradiation upon CNE treatment. 28% MCF-7 and 52% HFF-1 viable cells were observed after two doses of irradiation. After an extra CNE dose, 90% phototoxicity was observed for MCF-7 and 65% for HFF-1 cells. The CNE shows good efficiency as a photosentizing agent as shown by more ROS generation in both the cell lines, phototoxic effect and increment in caspase mediated apoptosis in tumor cell. The above findings revealed the potential therapeutic efficiency of NPs for synergistic chemo-thermal cancer therapy.

5.4.5 Synthesis and functionalization strategy of nanomaterial for therapeutic application

Many efficient synthesis strategies to produce stable and biocompatible magnetic and plasmonic NPs have been reported. The most common methods include co-precipitation, hydrothermal, sol–gel, thermal decomposition, bacterial (micro-organism) synthesis, and electrochemical synthesis. Nanomaterials have to be surface functionalized to obtain stability in a physiological environment to promote their bioactivity. Silica is a widely used inorganic coating material on NPs as it is able to avoid aggregation of NPs and helps in conjugating therapeutic molecules [92, 93]. In addition to this, polymers like dextran [94], chitosan [95], PEG [96], PEI [97], PVA [98] etc, are often used to improve the monodispersity of the particles, prolonging the circulation time, enhancing biodegradability and biocompatibility.

Co-precipitation is a common method for synthesizing narrow sized particles at room or elevated temperature [99]. Typically, ferric and ferrous salts are added in alkaline aqueous solution under an inert atmosphere. For the synthesis of IONP, the Fe^{2+}/Fe^{3+} ratio, reaction temperature, pH of the solution and ionic strength medium are the main factors influencing the monodispersity and quality of MNPs [100]. Ghosh *et al* [101] synthesized PEG coated Fe_3O_4 NPs by the co-precipitation method using ferrous sulfate ($FeSO_4 \cdot 7H_2O$) and ferric chloride ($FeCl_3 \cdot 6H_2O$) as precursors, PEG as surfactant and NH_4OH as reducing agent. *In vitro* studies were done in MCF-7 human breast cancer cells exhibited SAR of 33.5 W g^{-1} for a minimal concentration efficient for hyperthermia. Liu *et al* [102] synthesized 4 nm citrate coated Ag NPs and have performed *in vitro* (U251 glioblastoma cells) and *in vivo* tests which exhibited significant improvement in antiglioma effects upon a combination of Ag NPs and radiotherapy. Similarly, Parsian *et al* [103] synthesized chitosan modified IONPs and found enhanced cytotoxicity on SKBR and MCF-7 breast cancer cells by XTT assay.

Hydrothermal synthesis of NPs is often employed for controlled particle size and morphology [104]. The reaction is carried out at a controlled pressure of 0.3–4 MPa within a temperature of 100 °C–240 °C for a duration of a few hours to a few days [105]. An experimental investigation by Cai *et al* [106] on synthesizing Fe_3O_4–PEI NPs with tunable sizes via this method have been carried out for *in vitro* studies on both KB cells (a human epithelial carcinoma cell line) and raw 264.7 cells (a type of mouse macrophage cell line). The synthesis was carried out at 134 °C with a pressure of 2 bar for 3 h using iron chloride ($FeCl_2 \cdot 4H_2O$), ammonium hydroxide (NH_4OH), and polyethylenimine (PEI) as precursors. By varying the mass ratio of these reactants, the size was tuned from 11–22 nm. The obtained NPs were found to be highly stable and exhibited R_2 relaxivity in the range of 130–160 $mM^{-1} s^{-1}$. Cytocompatibility studies were done at different concentrations and hence found that these NPs can be multi-functionalized for cancer imaging and therapeutics. Thermal decomposition is yet another method for the synthesis of monodispersed NPs. This method is carried out at a higher temperature of 150–350 °C in high boiling organic solvents using organometallic precursors [107]. Generally, the ratios of organometallic precursors, surfactant and solvent, reaction temperature and time are the critical parameters to be focussed on for the controlled size and shape of NPs [108]. Xu *et al* [109] followed this method for synthesizing fcc-FePt NPs using Fe (acac)$_3$ and Pt(acac)$_2$ as metal precursor, oleylamine and oleic acid as surfactant and benzyl ether as solvent. The synthesis was carried out at 300 °C for a duration of 2 h. These NPs show significant tumor cell inhibition. Sadhukha *et al* [110] synthesized superparamagnetic NPs having a magnetite core of 12 nm via this high temperature method and have performed MHT in A549 and MDA-MB-231 tumor cells. These studies exhibited acute necrosis in some cells and also induced some ROS mediated cell death. Sol–gel method, a suitable wet chemical route, provides for the synthesis of nanostructured metal oxides [111]. Saldiver *et al* [112] synthesized $Mg_{0.4}Ca_{0.6}Fe_2O_4$ NPs via sol–gel method using ferric nitrate and ethylene glycol as precursors. The gel was formed at 80 °C within 2 h and later the xerogel was kept at a temperature range of 300 °C–600 °C for 30 min. Magnetic hyperthermia was performed on MG63 cell lines and found null toxicity and a sudden decrease of cellular viability up to 14%. For biomedical applications, the NPs should be dispersible in hydrophilic solvents. NPs synthesized using organic solvents need to be made water soluble using phase transfer agents such as trimethylammonium hydroxide (TMAOH), dimercaptosuccinic acid–dimethyl sulfoxide (DMSA–DMSO), 1-ethyl-3-(3 dimethylaminopropyl) carbodiimide hydrochloride–*N*-hydroxysuccinimide (EDC–NHS) etc via the ligand exchange process. In a study by Venkatesha *et al*, $MnFe_2O_4$–Fe_3O_4 core–shell structure synthesized using organic solvents (oleic acid, oleylamine) were made water dispersible using DMSA and DMSO through the ligand exchange mechanism. For this process, 1:1 ratio of $MnFe_2O_4$–Fe_3O_4 NPs stored in hexane and combination of

DMSA and DMSO solution made the NPs hydrophilic [113]. Similarly, TMAOH is also used for ligand exchange oleic acid and oleylamine coated NPs. Also, Maceira *et al* reported an effective method by transferring oleic acid/oleylamine-capped FePt NPs into hydrophilic using TMAOH. The phase transfer was achieved without altering its crystalline and mono disperse nature of the NPs [114]. Xie *et al* reported that hydrophobic PEG diacid coated magnetite NPs via EDC–NHS surface modification were modified to be hydrophilic [115].

5.5 Mechanism of cellular uptake and accumulation of NPs in tumors

NPs entry into cells depends mainly on the interfacial and physical characteristics of NPs, cell membrane properties, and their interaction with the biological environment [116]. All the body cells use endocytosis to communicate with the biological environment. The cells internalize signaling molecules and nutrients to interact with other cells and to get energy, respectively. Subtypes of endocytosis are phagocytosis, clathrin and caveolae mediate endocytosis, macropinocytosis and pinocytosis [117]. NPs phagocytosis normally starts by the process of opsonisation in which opsonins such as blood protein are adsorbed onto the surface of NPs [118]. These opsonized NPs are recognized and further attached to the phagocytes by ligand receptor interaction which triggers the actin assembly by initializing a signaling cascade, cell surface extension formation and further engulfing and internalization of particles [119, 120]. Clathrin and caveolae mediated endocytosis is a receptor mediated endocytosis. In clathrin mediated endocytosis, clathrin coated endocytotic vesicles are formed with diameters of 100 nm [121]. The uptake process starts by binding of ligand to its receptor. Caveolae are the vesicles made by cell membrane invagination having diameters of 50–100 nm [122] which contains caveolin, cholesterol and sphingolipids and forms microdomains by binding to associated proteins. This microdomain which has cationic lipids, mediates the endocytosis of anionic NPs [116, 123–125]. Macropinocytosis is an endocytic process driven by actin which internalizes a significant volume of liquid by large vesicles called macropinosome [126]. Pinocytosis absorbs biological fluid from the cell's external environment [127].

A mechanism was reported by Maeda and Matsumura where they described that in passive targeting, there is a preferential accumulation of macromolecules (including nanoparticles) in the neoplastic tissue because of the enhanced permeability and retention effect (EPR) [128–130]. Combination of poor lymphatic drainage and leaky vasculature tends to the EPR effect. NPs having sizes smaller than the fenestra can enter the interstitium and be trapped in the tumor [130]. When the NPs are loaded with a particular drug, it brings the drug towards a leaky zone which ultimately releases the drug because of degradation of carrier. Active targeting uses the surface conjugated targeting ligands for the increased delivery of NPs. The most obvious way to confer targeting abilities to a non-specific drug is by conjugating the drug with another molecule called the vector molecule or targeting moiety. Different kinds of targeting moieties can be used such as lectins, antibodies and fragments, hormones, charged molecules, mono, oligo and polysaccharides, lipoproteins, ligands having low molecular weight like folate etc [131]. In a report, biodegradable PLGA–lipid NPs encapsulated with Indocyanin green and folic acid that acts as a targeting agent (FA–ICG–PLGA–lipid NPs) were synthesized [132]. Specific cell targeting was evaluated comparatively by testing on MCF-7 and A549 cells which are folate receptor positive and negative, respectively. Targeting efficiency FA–ICG–PLGA–lipid NPs were more effective in MCF-7 than in A549 as proven by cellular uptake experiments [132].

5.6 Current status of clinical trials of nanomedicine based on MHT and PTT

AuroLase®, developed by Nanospectra biosciences (a medical device company) is a gold NP which can absorb light to ablate tumors thermally. This is under clinical trial for MRI, ultrasound fusion imaging and biopsy in combination with NP directed focal therapy for ablation of prostate tissue (NCT02680535) on human subjects. Previously it was under trial for primary and metastatic lung tumor treatment (NCT01679470) (www.clinicaltrials.gov/). Since no active drug is utilized, AuroLase is activated externally on the target tissue thereby minimizing the side effects on normal cells [133]. First generation Aurimune platform nanotherapy (CYT-6091) carries TNF (tumor necrosis factor) via gold NPs into the tumors for the disruption of blood vessels followed by chemotherapy for tumor penetration and killing of cancer cells (www.cytimmune.com/). In a preclinical study, combination of colloidal gold, rhTNF and thiolayted-PEG construct (CYT-6091) shows that when AuNPs are PEGylated, it reduces the uptake by the reticuloendothelial system and lowers the toxicity of rhTNF thus nanomedicine gets sequestered in solid tumors [134, 135]. In a phase I trial on humans, 50 µg m^{-2} to 600 µg m^{-2} dose of 27 nm colloidal gold NPs which was conjugated simultaneously with rhTNF and thiolayted-PEG was well tolerated and no MTD (maximum tolerated dose) was observed. The dose of rhTNF with AuNPs could be 3-fold more in comparison to its normal dose with no toxic effect [135].

So far as the clinical trial of magnetic thermoablation is concerned, magnablate I (iron NPs) was under phase 0 study which utilizes the magnetic field for ablation (NCT02033447) (www.clinicaltrials.gov/). Currently an active trial is taking place for the drug NU-0129 for the evaluation of its safety in its first in-human trial. It is based on spherical nucleic acid (SNA). SNA consists of nucleic acid which is arranged on the surface of AuNPs. Bcl2L12 is the gene present in glioblastoma associated with growth of tumor since it prevents apoptosis. Targeting Bcl2L12 by NU-0129 can reduce the growth of cancer cells (NCT03020017) (www.clinicaltrials. gov/). Magnetic fluid MFL-AS which contains water dispersed superparamagnetic IONPs having conc. 112 mg ml^{-1} was developed by Nanotherm, MagForce Nanotechnologies AG, Berlin, Germany and is applied for magnetic fluid hyperthermia studies. The core of NPs is covered by a shell made up of aminosilane and having a size nearly 15 nm in diameter. After three clinical studies in glioblastoma multiforme, recurrent and residual tumor (non-resectable pretreated), and recurrent prostate carcinoma, its feasibility is proven [136].

5.7 Toxicity of nanomaterials

Nanomaterial toxicity is a major concern for biological applications. Due to their smaller size and higher surface area-to-volume ratio and surface functionalisation, NPs show greater biological and chemical reactivity. It can have more risk of toxicity than its bulk counterpart. In a report by Goodman *et al*, cationic gold nanospheres having a diameter of 2 nm are toxic at certain concentrations [137]. However, the same NPs with negative charge on the surface were non-toxic on the same cell line at the same concentrations. The interaction ability of cationic gold NPs with the cell membrane which is negatively charged cause membrane disruption [137]. The smaller the particle size, the more possibility there is of interaction with subcellular and cellular components such as lysosomes, nucleus and genetic material. Pan *et al* reported that gold NPs having diameters of 1.4 nm coated with triphenylphosphine monosulfate (Au1.4MS) is more cytotoxic than Au15MS [138]. Cell death occurs mainly by necrosis indicating mitochondrial permeability transition and oxidative damage [138]. Mahmoudi *et al* reported that bare SPION possesses higher toxicity compared to a PVA (polyvinyl alcohol) coated one [139]. Wozniak *et al* reported the shape dependent toxicity of AuNPs [140]. They performed the experiment with different shapes and sizes of AuNPs such as spherical (~10 nm), nanorod (~41 nm), nanoprism (~160 nm), nanostar (~240 nm), nano-flower (~370 nm). Results show that Au nanorods and nanospheres showed toxicity depending upon the concentration and time interval of incubation. Nanoprism, nanoflower and nanostar were relatively unharmful when tested on HeLa cells. The highest concentration of nanoflower was lethal after 72 h incubation time [140]. Feng *et al* [141] reported that PEI coated IONPs are internalized more effectively in both macrophage and cancer cells compared to the PEGylated IONPs in the same size range probably because of affinity between the cationic NPs and the negative phospholipid head groups or protein domains present on the cellular membrane. It can cause acute cytotoxicity by ROS generation and apoptosis. 10 nm PEGylated IONP has more cellular uptake compared to the 30 nm IONP and can induce autophagy thus can act as a protection against cytotoxicity of IONP. Toxicity studies show that PEI coated IONP at a dose of 2.5 mg kg^{-1} or more can cause death of an animal due to capillary blockage because of large aggregate formation or rapid hemolysis. In a report, Wistar rats were given Fe_3O_4 suspension intravenously. After examining all parameters, exposure of Fe_3O_4 causes respiratory toxicity, cytotoxic damage and it induced some inflammatory response. After 28 days by some defense mechanism, the Fe_3O_4 NPs is eliminated through the respiratory tract [142]. Neha *et al* reported SPION functionalized with various organic surfactants such as L-arginine, succinic acid, oxalic acid, glutamic and citric acid. The NPs are biocompatible, however, oxalic acid coated NP at 100 μg ml^{-1} concentration show appreciable toxicity to HeLa cells [143]. Sargent *et al* demonstrated a 15-day inhalation study of MWCNT (5 mg m^{-3}, 5 h day^{-1}) following intraperitoneal injection of methylcholanthrene (MCA), which acts as initiator and found that MWCNT increases the tumor rate in the MCA exposed lung [144].

5.8 Conclusion

Since conventional methods for breast cancer therapy limits its effectiveness, NP mediated therapy appears to be a promising route to overcome the drawbacks. However, many challenges need to be overcome regarding the synthesis and functionalisation of NPs, cellular uptake, biological toxicity, loading and releasing of anticancer cargo and on-site therapy. Hyperthermia due to MHT/PTT is a potential strategy for tumor therapy with minimal side effects. But, research is still in progress to carry out the synergism in the breast tumor tissue based on magneto-plasmonic nanomedicine due to various pros and cons. Based on the early preclinical studies, magneto-plasmonic nanomedicine may provide a new platform for breast cancer treatment.

Acknowledgments

The authors gratefully acknowledge the financial support from DST-SERB (Project No. ECR/2016/000301).

References

[1] American Cancer Society 2019 *Cancer facts & figures 2019* (Atlanta, GA: American Cancer Society) pp 1–71
[2] Bhardwaj P, Au C C, Benito-Martin A, Ladumor H, Oshchepkova S, Moges R and Brown K A 2019 Estrogens and breast cancer: Mechanisms involved in obesity-related development, growth and progression *J. Steroid Biochem. Mol. Biol.* **189** 161–70
[3] Scoccianti C, Lauby-Secretan B, Bello P-Y, Chajes V and Romieu I 2014 Female breast cancer and alcohol consumption: a review of the literature *Am. J. Prev. Med.* **46** S16–25
[4] Dianatinasab M, Fararouei M, Mohammadianpanah M, Zare-bandamiri M and Rezaianzadeh A 2017 Hair coloring, stress, and smoking increase the risk of breast cancer: a case-control study *Clin. Breast Cancer* **17** 650–9
[5] Zolfaroli I, Tarín J J and Cano A 2018 Hormonal contraceptives and breast cancer: clinical data *Eur. J. Obstet. Gynecol. Reprod. Biol.* **230** 212–6
[6] Giannakeas V, Sopik V and Narod S A 2018 Association of radiotherapy with survival in women treated for ductal carcinoma *in situ* with lumpectomy or mastectomy *JAMA Netw. Open* **1** e181100
[7] Aslam M S, Naveed S, Ahmed A, Abbas Z, Gull I and Athar M A 2014 Side effects of chemotherapy in cancer patients and evaluation of patients opinion about starvation based differential chemotherapy *J. Cancer Ther.* **05** 817–22
[8] Gupta A 2018 Hormone therapy–related hot flashes and their management *JAMA Oncol.* **4** 595
[9] Sutradhar K B and Amin M L 2014 Nanotechnology in cancer drug delivery and selective targeting *ISRN Nanotechnol.* **2014** 1–12
[10] Akram M, Iqbal M, Daniyal M and Khan A U 2017 Awareness and current knowledge of breast cancer *Biol. Res.* **50** 1–23
[11] Sharma G N, Dave R, Sanadya J, Sharma P and Sharma K K 2010 Various types and management of breast cancer: an overview *J. Adv. Pharm. Technol. Res.* **1** 109–26

[12] Rezaee A, Buck A, Raderer M, Langsteger W and Beheshti M 2018 Breast Cancer *PET/CT in Cancer: An Interdisciplinary Approach to Individualized Imaging* ed M Beheshti *et al* (Amsterdam: Elsevier) pp 43–63

[13] Feng Y *et al* 2018 Breast cancer development and progression: Risk factors, cancer stem cells, signaling pathways, genomics, and molecular pathogenesis *Genes Dis.* **5** 77–106

[14] Nwabo Kamdje A H, Seke Etet P F, Vecchio L, Muller J M, Krampera M and Lukong K E 2014 Signaling pathways in breast cancer: therapeutic targeting of the microenvironment *Cell Signal.* **26** 2843–56

[15] Said T K, ConNéely O M, Medina D, O'Malley B W and Lydon J P 1997 Progesterone, in addition to estrogen, induces cyclin D1 expression in the murine mammary epithelial cell, *in vivo Endocrinology* **138** 3933–9

[16] Lundberg A S and Weinberg R A 1999 Control of the cell cycle and apoptosis *Eur. J. Cancer* **35** 1886–94

[17] King C R, Kraus M H and Aaronson S A 1985 Amplification of a novel v-erbB-related gene in a human mammary carcinoma *Science* **229** 974–6

[18] Murshed H 2019 Breast cancers *Fundamentals of Radiation Oncology* 3rd edn ed H Murshed (New York: Academic) pp 317–49

[19] Osborn G 2013 Management of breast cancer: basic principles *Surgery – Oxf. Int. Ed.* **31** 27–31

[20] Sibbering M and Courtney C-A 2016 Management of breast cancer: basic principles *Surgery – Oxf. Int. Ed.* **34** 25–31

[21] Barrett S V 2010 Breast cancer *J. R. Coll. Physicians Edinb.* **40** 335–8

[22] Moo T A, Sanford R, Dang C and Morrow M 2018 Overview of breast cancer therapy *PET Clin.* **13** 339–54

[23] van der Hage J A, van de Velde C J, Julien J P, Tubiana-Hulin M, Vandervelden C and Duchateau L 2001 Preoperative chemotherapy in primary operable breast cancer: results from the European Organization for Research and Treatment of Cancer trial 10902 *J. Clin. Oncol.* **19** 4224–37

[24] Chidambaram M, Manavalan R and Kathiresan K 2011 Nanotherapeutics to overcome conventional cancer chemotherapy limitations *J. Pharm. Pharm. Sci.* **14** 67–77

[25] Kwon G S 2003 Polymeric micelles for delivery of poorly water-soluble compounds *Crit. Rev. Ther. Drug Carrier Syst.* **20** 357–403

[26] Luo Y and Prestwich G D 2002 Cancer-targeted polymeric drugs *Curr. Cancer Drug Targets* **2** 209–26

[27] Stavrovskaya A A 2000 Cellular mechanisms of multidrug resistance of tumor cells *Biochemistry (Mosc.)* **65** 95–106

[28] Purdie C A, Quinlan P, Jordan L B, Ashfield A, Ogston S, Dewar J A and Thompson A M 2014 Progesterone receptor expression is an independent prognostic variable in early breast cancer: a population-based study *Br. J. Cancer* **110** 565–72

[29] Lumachi F, Santeufemia D A and Basso S M 2015 Current medical treatment of estrogen receptor-positive breast cancer *World J. Biol. Chem.* **6** 231–9

[30] Lumachi F, Brunello A, Maruzzo M, Basso U and Basso S M M 2013 Treatment of estrogen receptor-positive breast cancer *Curr. Med. Chem.* **20** 596–604

[31] Lobo-Cardoso R, Magalhães A T and Fougo J L 2017 Neoadjuvant endocrine therapy in breast cancer patients *Porto Biomed. J.* **2** 170–3

[32] Flatley M J and Dodwell D J 2016 Adjuvant treatment for breast cancer *Surgery – Oxf. Int. Ed.* **34** 43–6

[33] Violet J A and Harmer C 2004 Breast cancer: improving outcome following adjuvant radiotherapy *Br. J. Radiol.* **77** 811–20

[34] Romestaing P, Lehingue Y, Carrie C, Coquard R, Montbarbon X, Ardiet J M, Mamelle N and Gerard J P 1997 Role of a 10-Gy boost in the conservative treatment of early breast cancer: results of a randomized clinical trial in Lyon, France *J. Clin. Oncol.* **15** 963–8

[35] Jain S K, Dorn P L, Chmura S J, Weichselbaum R R and Hasan Y 2012 Incidence and implications of oligometastatic breast cancer *J. Clin. Oncol.* **30** e11512

[36] Possanzini M and Greco C 2018 Stereotactic radiotherapy in metastatic breast cancer *Breast* **41** 57–66

[37] Vemuri S K, Banala R R, Mukherjee S, Uppula P, Gpv S, Reddy A V G and Malarvilli T 2019 Novel biosynthesized gold nanoparticles as anti-cancer agents against breast cancer: Synthesis, biological evaluation, molecular modelling studies *Mater. Sci. Eng.: C* **99** 417–29

[38] Kayani Z, Bordbar A-K and Firuzi O 2018 Novel folic acid-conjugated doxorubicin loaded β-lactoglobulin nanoparticles induce apoptosis in breast cancer cells *Biomed. Pharmacother.* **107** 945–56

[39] Semkina A S *et al* 2018 Multimodal doxorubicin loaded magnetic nanoparticles for VEGF targeted theranostics of breast cancer *Nanomedicine* **14** 1733–42

[40] Hanžić N, Horvat A, Bibić J, Unfried K, Jurkin T, Dražić G, Marijanović I, Slade N and Gotić M 2018 Syntheses of gold nanoparticles and their impact on the cell cycle in breast cancer cells subjected to megavoltage X-ray irradiation *Mater. Sci. Eng.: C* **91** 486–95

[41] Zupančič E, Silva J, Videira M A, Moreira J N and Florindo H F 2014 Development of a novel nanoparticle-based therapeutic vaccine for breast cancer immunotherapy *Procedia Vaccinol.* **8** 62–7

[42] Liu Y, Qiao L, Zhang S, Wan G, Chen B, Zhou P, Zhang N and Wang Y 2018 Dual pH-responsive multifunctional nanoparticles for targeted treatment of breast cancer by combining immunotherapy and chemotherapy *Acta Biomater.* **66** 310–24

[43] Zhang X, Wu F, Men K, Huang R, Zhou B, Zhang R, Zou R and Yang L 2018 Modified Fe(3)O(4) magnetic nanoparticle delivery of CpG inhibits tumor growth and spontaneous pulmonary metastases to enhance immunotherapy *Nanoscale Res. Lett.* **13** 1–12

[44] Wust P, Hildebrandt B, Sreenivasa G, Rau B, Gellermann J, Riess H, Felix R and Schlag P M 2002 Hyperthermia in combined treatment of cancer *Lancet Oncol.* **3** 487–97

[45] Mendes R, Pedrosa P, Lima J C, Fernandes A R and Baptista P V 2017 Photothermal enhancement of chemotherapy in breast cancer by visible irradiation of gold nanoparticles *Sci. Rep.* **7** 1–8

[46] van der Zee J 2002 Heating the patient: a promising approach? *Ann. Oncol.* **13** 1173–84

[47] Sivakumar B, Aswathy R G, Romero-Aburto R, Mitcham T, Mitchel K A, Nagaoka Y, Bouchard R R, Ajayan P M, Maekawa T and Sakthikumar D N 2017 Highly versatile SPION encapsulated PLGA nanoparticles as photothermal ablators of cancer cells and as multimodal imaging agents *Biomater. Sci.* **5** 432–43

[48] Bleehen N M 1982 Hyperthermia in the treatment of cancer *Br. J. Cancer Suppl.* **5** 96–100

[49] Storm F K, Morton D L, Kaiser L R, Harrison W H, Elliott R S, Weisenburger T H, Parker R G and Haskell C M 1982 Clinical radiofrequency hyperthermia: a review *Natl. Cancer Inst. Monogr.* **61** 343–50

[50] Jordan A, Wust P, Fahling H, John W, Hinz A and Felix R 1993 Inductive heating of ferrimagnetic particles and magnetic fluids: physical evaluation of their potential for hyperthermia *Int. J. Hyperthermia* **9** 51–68

[51] Dennis C L and Ivkov R 2013 Physics of heat generation using magnetic nanoparticles for hyperthermia *Int. J. Hyperthermia* **29** 715–29

[52] Bañobre-López M, Teijeiro A and Rivas J 2013 Magnetic nanoparticle-based hyperthermia for cancer treatment *Rep. Pract. Oncol. Radiother.* **18** 397–400

[53] Deatsch A E and Evans B A 2014 Heating efficiency in magnetic nanoparticle hyperthermia *J. Magn. Magn. Mater.* **354** 163–72

[54] Kumar C S and Mohammad F 2011 Magnetic nanomaterials for hyperthermia-based therapy and controlled drug delivery *Adv. Drug Deliv. Rev.* **63** 789–808

[55] Suto M, Hirota Y, Mamiya H, Fujita A, Kasuya R, Tohji K and Jeyadevan B 2009 Heat dissipation mechanism of magnetite nanoparticles in magnetic fluid hyperthermia *J. Magn. Magn. Mater.* **321** 1493–6

[56] Huang X and El-Sayed M A 2011 Plasmonic photo-thermal therapy (PPTT) *Alexandria J. Med.* **47** 1–9

[57] Huang X and El-Sayed M A 2010 Gold nanoparticles: Optical properties and implementations in cancer diagnosis and photothermal therapy *J. Adv. Res.* **1** 13–28

[58] Jain P K, Lee K S, El-Sayed I H and El-Sayed M A 2006 Calculated absorption and scattering properties of gold nanoparticles of different size, shape, and composition: applications in biological imaging and biomedicine *J. Phys. Chem.* B **110** 7238–48

[59] Sivan Y, Un I W and Dubi Y 2019 Assistance of metal nanoparticles in photocatalysis – nothing more than a classical heat source *Faraday Discuss.* **214** 1–24

[60] Ekici O, Harrison R K, Durr N J, Eversole D S, Lee M and Ben-Yakar A 2008 Thermal analysis of gold nanorods heated with femtosecond laser pulses *J. Phys. D Appl. Phys.* **41** 1–11

[61] Furube A and Hashimoto S 2017 Insight into plasmonic hot-electron transfer and plasmon molecular drive: new dimensions in energy conversion and nanofabrication *NPG Asia Mater.* **9** 1–24

[62] Qin Z and Bischof J C 2012 Thermophysical and biological responses of gold nanoparticle laser heating *Chem. Soc. Rev.* **41** 1191–217

[63] Dreaden E C, Alkilany A M, Huang X, Murphy C J and El-Sayed M A 2012 The golden age: gold nanoparticles for biomedicine *Chem. Soc. Rev.* **41** 2740–79

[64] Zharov V P and Lapotko D O 2005 Photothermal imaging of nanoparticles and cells *IEEE J. Sel. Top. Quantum Electron.* **11** 733–51

[65] Huang X, Kang B, Qian W, Mackey M A, Chen P C, Oyelere A K, El-Sayed I H and El-Sayed M A 2010 Comparative study of photothermolysis of cancer cells with nuclear-targeted or cytoplasm-targeted gold nanospheres: continuous wave or pulsed lasers *J. Biomed. Opt.* **15** 058002-1–7

[66] Revia R A and Zhang M 2016 Magnetite nanoparticles for cancer diagnosis, treatment, and treatment monitoring: recent advances *Mater. Today* **19** 157–68

[67] Hilger I, Hiergeist R, Hergt R, Winnefeld K, Schubert H and Kaiser W A 2002 Thermal ablation of tumors using magnetic nanoparticles: an *in vivo* feasibility study *Invest. Radiol.* **37** 580–6

[68] Salimi M, Sarkar S, Saber R, Delavari H, Alizadeh A M and Mulder H T 2018 Magnetic hyperthermia of breast cancer cells and MRI relaxometry with dendrimer-coated iron-oxide nanoparticles *Cancer Nanotechnol.* **9** 1–19

[69] Chen D, Tang Q, Li X, Zhou X, Zang J, Xue W-q, Xiang J-y and Guo C-q 2012 Biocompatibility of magnetic Fe_3O_4 nanoparticles and their cytotoxic effect on MCF-7 cells *Int. J. Nanomed.* **7** 4973–82

[70] Hamzehalipour Almaki J, Nasiri R, Idris A, Nasiri M, Abdul Majid F A and Losic D 2017 Trastuzumab-decorated nanoparticles for *in vitro* and *in vivo* tumor-targeting hyperthermia of HER2+ breast cancer *J. Mater. Chem.* B **5** 7369–83

[71] Hong E J, Kim Y-S, Choi D G and Shim M S 2018 Cancer-targeted photothermal therapy using aptamer-conjugated gold nanoparticles *J. Ind. Eng. Chem.* **67** 429–36

[72] Liu Y, Song N, Li Z, Chen L and Xie Z 2019 Near-infrared nanoparticles based on aza-BDP for photodynamic and photothermal therapy *Dyes Pigm.* **160** 71–8

[73] Park J H, von Maltzahn G, Xu M J, Fogal V, Kotamraju V R, Ruoslahti E, Bhatia S N and Sailor M J 2010 Cooperative nanomaterial system to sensitize, target, and treat tumors *Proc. Natl. Acad. Sci. U. S. A.* **107** 981–6

[74] Kirui D K, Rey D A and Batt C A 2010 Gold hybrid nanoparticles for targeted phototherapy and cancer imaging *Nanotechnology* **21** 1–10

[75] Wang Y, Liu X, Deng G, Wang Q, Zhang L, Wang Q and Lu J 2017 Multifunctional PS@CS@Au–Fe_3O_4–FA nanocomposites for CT, MR and fluorescence imaging guided targeted-photothermal therapy of cancer cells *J. Mater. Chem.* B **5** 4221–32

[76] Chen C L, Kuo L R, Lee S Y, Hwu Y K, Chou S W, Chen C C, Chang F H, Lin K H, Tsai D H and Chen Y Y 2013 Photothermal cancer therapy via femtosecond-laser-excited FePt nanoparticles *Biomaterials* **34** 1128–34

[77] Phan T T V, Bui N Q, Moorthy M S, Lee K D and Oh J 2017 Synthesis and *In Vitro* performance of polypyrrole-coated iron-platinum nanoparticles for photothermal therapy and photoacoustic imaging *Nanoscale Res. Lett.* **12** 1–12

[78] Kam N W, O'Connell M, Wisdom J A and Dai H 2005 Carbon nanotubes as multifunctional biological transporters and near-infrared agents for selective cancer cell destruction *Proc. Natl. Acad. Sci. U. S. A.* **102** 11600–5

[79] Yang K, Zhang S, Zhang G, Sun X, Lee S-T and Liu Z 2010 Graphene in mice: ultrahigh *in vivo* tumor uptake and efficient photothermal therapy *Nano Lett.* **10** 3318–23

[80] Espinosa A, Bugnet M, Radtke G, Neveu S, Botton G A, Wilhelm C and Abou-Hassan A 2015 Can magneto-plasmonic nanohybrids efficiently combine photothermia with magnetic hyperthermia? *Nanoscale* **7** 18872–7

[81] Das R, Rinaldi-Montes N, Alonso J, Amghouz Z, Garaio E, García J A, Gorria P, Blanco J A, Phan M H and Srikanth H 2016 Boosted hyperthermia therapy by combined AC magnetic and photothermal exposures in Ag/Fe_3O_4 nanoflowers *ACS Appl. Mater. Interfaces* **8** 25162–69

[82] Espinosa A, Di Corato R, Kolosnjaj-Tabi J, Flaud P, Pellegrino T and Wilhelm C 2016 Duality of iron oxide nanoparticles in cancer therapy: amplification of heating efficiency by magnetic hyperthermia and photothermal bimodal treatment *ACS Nano* **10** 2436–46

[83] Di Corato R, Bealle G, Kolosnjaj-Tabi J, Espinosa A, Clement O, Silva A K, Menager C and Wilhelm C 2015 Combining magnetic hyperthermia and photodynamic therapy for tumor ablation with photoresponsive magnetic liposomes *ACS Nano* **9** 2904–16

[84] Elbialy N S, Fathy M M, Al-Wafi R, Darwesh R, Abdel-dayem U A, Aldhahri M, Noorwali A and Al-ghamdi A A 2019 Multifunctional magnetic-gold nanoparticles for efficient combined targeted drug delivery and interstitial photothermal therapy *Int. J. Pharm.* **554** 256–63

[85] Yao X, Tian Z, Liu J, Zhu Y and Hanagata N 2017 Mesoporous silica nanoparticles capped with graphene quantum dots for potential chemo–photothermal synergistic cancer therapy *Langmuir* **33** 591–9

[86] Gao Y, Zhong S, Xu L, He S, Dou Y, Zhao S, Chen P and Cui X 2019 Mesoporous silica nanoparticles capped with graphene quantum dots as multifunctional drug carriers for photo-thermal and redox-responsive release *Micropor. Mesopor. Mater.* **278** 130–7

[87] Li F, Yang H, Bie N, Xu Q, Yong T, Wang Q, Gan L and Yang X 2017 Zwitterionic temperature/redox-sensitive nanogels for near-infrared light-triggered synergistic thermo-chemotherapy *ACS Appl. Mater. Interfaces* **9** 23564–73

[88] Sahu N K, Singh N S, Pradhan L and Bahadur D 2014 Ce3+ sensitized $GdPO_4$:Tb_{3+} with iron oxide nanoparticles: a potential biphasic system for cancer theranostics *Dalton Trans.* **43** 11728–38

[89] Xu W, Qian J, Hou G, Suo A, Wang Y, Wang J, Sun T, Yang M, Wan X and Yao Y 2017 Hyaluronic acid-functionalized gold nanorods with pH/NIR dual-responsive drug release for synergetic targeted photothermal chemotherapy of breast cancer *ACS Appl. Mater. Interfaces* **9** 36533–47

[90] Mahmoodzadeh F, Abbasian M, Jaymand M, Salehi R and Bagherzadeh-Khajehmarjan E 2018 A novel gold-based stimuli-responsive theranostic nanomedicine for chemo-photo-thermal therapy of solid tumors *Mater. Sci. Eng.: C* **93** 880–9

[91] Machado F C, Adum de Matos R P, Primo F L, Tedesco A C, Rahal P and Calmon M F 2019 Effect of curcumin-nanoemulsion associated with photodynamic therapy in breast adenocarcinoma cell line *Bioorg. Med. Chem.* **27** 1882–90

[92] Alhmoud H, Delalat B, Elnathan R, Cifuentes-Rius A, Chaix A, Rogers M-L, Durand J-O and Voelcker N H 2015 Porous silicon nanodiscs for targeted drug delivery *Adv. Funct. Mater.* **25** 1137–45

[93] Vogt C, Toprak M S, Muhammed M, Laurent S, Bridot J-L and Müller R N 2010 High quality and tuneable silica shell–magnetic core nanoparticles *J. Nanopart. Res.* **12** 1137–47

[94] Berry C C, Wells S, Charles S and Curtis A S 2003 Dextran and albumin derivatised iron oxide nanoparticles: influence on fibroblasts *in vitro Biomaterials* **24** 4551–7

[95] Khor E and Lim L Y 2003 Implantable applications of chitin and chitosan *Biomaterials* **24** 2339–49

[96] Gupta A K and Curtis A S G 2004 Surface modified superparamagnetic nanoparticles for drug delivery: Interaction studies with human fibroblasts in culture *J. Mater. Sci.: Mater. Med.* **15** 493–6

[97] Karimzadeh I, Aghazadeh M, Doroudi T, Ganjali M R and Kolivand P H 2017 Superparamagnetic iron oxide (Fe_3O_4) nanoparticles coated with PEG/PEI for biomedical applications: a facile and scalable preparation route based on the cathodic electrochemical deposition method *Adv. Phys. Chem.* **2017** 1–7

[98] Shan G-b, Xing J-m, Luo M-f, Liu H-z and Chen J-y 2003 Immobilization of Pseudomonas delafieldii with magnetic polyvinyl alcohol beads and its application in biodesulfurization *Biotechnol. Lett.* **25** 1977–81

[99] Gupta A K and Gupta M 2005 Synthesis and surface engineering of iron oxide nanoparticles for biomedical applications *Biomaterials* **26** 3995–4021

[100] Xu Y and Zhu Y 2016 Synthesis of magnetic nanoparticles for biomedical applications *Nano Adv.* **1** 25–38

[101] Ghosh R *et al* 2011 Induction heating studies of Fe_3O_4 magnetic nanoparticles capped with oleic acid and polyethylene glycol for hyperthermia *J. Mater. Chem.* **21** 13388–98

[102] Liu P, Jin H, Guo Z, Ma J, Zhao J, Li D, Wu H and Gu N 2016 Silver nanoparticles outperform gold nanoparticles in radiosensitizing U251 cells *in vitro* and in an intracranial mouse model of glioma *Int. J. Nanomed.* **11** 5003–14

[103] Parsian M, Unsoy G, Mutlu P, Yalcin S, Tezcaner A and Gunduz U 2016 Loading of Gemcitabine on chitosan magnetic nanoparticles increases the anti-cancer efficacy of the drug *Eur. J. Pharmacol.* **784** 121–8

[104] Hayashi H and Hakuta Y 2010 Hydrothermal synthesis of metal oxide nanoparticles in supercritical water *Materials* **3** 3794–817

[105] Laurent S, Forge D, Port M, Roch A, Robic C, Vander Elst L and Muller R N 2008 Magnetic iron oxide nanoparticles: synthesis, stabilization, vectorization, physicochemical characterizations, and biological applications *Chem. Rev.* **108** 2064–110

[106] Cai H, An X, Cui J, Li J, Wen S, Li K, Shen M, Zheng L, Zhang G and Shi X 2013 Facile hydrothermal synthesis and surface functionalization of polyethyleneimine-coated iron oxide nanoparticles for biomedical applications *ACS Appl. Mater. Interfaces* **5** 1722–31

[107] Sun S, Zeng H, Robinson D B, Raoux S, Rice P M, Wang S X and Li G 2004 Monodisperse MFe_2O_4 (M = Fe, Co, Mn) nanoparticles *J. Am. Chem. Soc.* **126** 273–9

[108] Sun S and Zeng H 2002 Size-controlled synthesis of magnetite nanoparticles *J. Am. Chem. Soc.* **124** 8204–5

[109] Xu C, Yuan Z, Kohler N, Kim J, Chung M A and Sun S 2009 FePt nanoparticles as an Fe reservoir for controlled Fe release and tumor inhibition *J. Am. Chem. Soc.* **131** 15346–51

[110] Sadhukha T, Niu L, Wiedmann T S and Panyam J 2013 Effective elimination of cancer stem cells by magnetic hyperthermia *Mol. Pharm.* **10** 1432–41

[111] Laurent S, Dutz S, Häfeli U O and Mahmoudi M 2011 Magnetic fluid hyperthermia: Focus on superparamagnetic iron oxide nanoparticles *Adv. Colloid Interface Sci.* **166** 8–23

[112] Saldivar-Ramirez M M, Sanchez-Torres C G, Cortes-Hernandez D A, Escobedo-Bocardo J C, Almanza-Robles J M, Larson A, Resendiz-Hernandez P J and Acuna-Gutierrez I O 2014 Study on the efficiency of nanosized magnetite and mixed ferrites in magnetic hyperthermia *J. Mater. Sci., Mater. Med.* **25** 2229–36

[113] Venkatesha N, Pudakalakatti S M, Qurishi Y, Atreya H S and Srivastava C 2015 $MnFe_2O_4$–Fe_3O_4 core–shell nanoparticles as a potential contrast agent for magnetic resonance imaging *RSC Adv.* **5** 97807–15

[114] Salgueiriño-Maceira V, Liz-Marzán L M and Farle M 2004 Water-based ferrofluids from Fe_xPt_{1-x} nanoparticles synthesized in organic media *Langmuir* **20** 6946–50

[115] Xie J, Xu C, Kohler N, Hou Y and Sun S 2007 Controlled PEGylation of monodisperse Fe_3O_4 nanoparticles for reduced non-specific uptake by macrophage cells *Adv. Mater.* **19** 3163–66

[116] Adjei I M, Sharma B and Labhasetwar V 2014 Nanoparticles: cellular uptake and cytotoxicity *Adv. Exp. Med. Biol.* **811** 73–91

[117] Oh N and Park J-H 2014 Endocytosis and exocytosis of nanoparticles in mammalian cells *Int. J. Nanomed.* **9** 51–63

[118] Aderem A and Underhill D M 1999 Mechanisms of phagocytosis in macrophages *Annu. Rev. Immunol.* **17** 593–623

[119] Behzadi S, Serpooshan V, Tao W, Hamaly M A, Alkawareek M Y, Dreaden E C, Brown D, Alkilany A M, Farokhzad O C and Mahmoudi M 2017 Cellular uptake of nanoparticles: journey inside the cell *Chem. Soc. Rev.* **46** 4218–44

[120] Hillaireau H and Couvreur P 2009 Nanocarriers' entry into the cell: relevance to drug delivery *Cell. Mol. Life Sci.* **66** 2873–96

[121] Ehrlich M, Boll W, Van Oijen A, Hariharan R, Chandran K, Nibert M L and Kirchhausen T 2004 Endocytosis by random initiation and stabilization of clathrin-coated pits *Cell* **118** 591–605

[122] Nabi I R and Le P U 2003 Caveolae/raft-dependent endocytosis *J. Cell Biol.* **161** 673–7

[123] Stan R V 2002 Structure and function of endothelial caveolae *Microsc. Res. Tech.* **57** 350–64

[124] Chithrani B D, Ghazani A A and Chan W C W 2006 Determining the size and shape dependence of gold nanoparticle uptake into mammalian cells *Nano Lett.* **6** 662–8

[125] Maxfield F R 2002 Plasma membrane microdomains *Curr. Opin. Cell Biol.* **14** 483–7

[126] Falcone S, Cocucci E, Podini P, Kirchhausen T, Clementi E and Meldolesi J 2006 Macropinocytosis: regulated coordination of endocytic and exocytic membrane traffic events *J. Cell Sci.* **119** 4758–69

[127] Geiser M 2010 Update on macrophage clearance of inhaled micro- and nanoparticles *J. Aerosol Med. Pulm. Drug Deliv.* **23** 207–17

[128] Maeda H and Matsumura Y 1989 Tumoritropic and lymphotropic principles of macromolecular drugs *Crit. Rev. Ther. Drug Carrier Syst.* **6** 193–210

[129] Matsumura Y and Maeda H 1986 A new concept for macromolecular therapeutics in cancer chemotherapy: mechanism of tumoritropic accumulation of proteins and the antitumor agent smancs *Cancer Res.* **46** 6387–92

[130] Haley B and Frenkel E 2008 Nanoparticles for drug delivery in cancer treatment *Urol. Oncol.* **26** 57–64

[131] Torchilin V P 2010 Passive and active drug targeting: drug delivery to tumors as an example *Handb. Exp. Pharmacol.* **197** 3–53

[132] Zheng C, Zheng M, Gong P, Jia D, Zhang P, Shi B, Sheng Z, Ma Y and Cai L 2012 Indocyanine green-loaded biodegradable tumor targeting nanoprobes for *in vitro* and *in vivo* imaging *Biomaterials* **33** 5603–9

[133] Singh P, Pandit S, Mokkapati V, Garg A, Ravikumar V and Mijakovic I 2018 Gold nanoparticles in diagnostics and therapeutics for human cancer *Int. J. Mol. Sci.* **19** 1–16

[134] Myer L, Jones D, Tamarkin L and Paciotti G 2008 Nanomedicine-based enhancement of chemotherapy *Cancer Res.* **68** 5718

[135] Libutti S K, Paciotti G F, Byrnes A A, Alexander H R Jr., Gannon W E, Walker M, Seidel G D, Yuldasheva N and Tamarkin L 2010 Phase I and pharmacokinetic studies of CYT-6091, a novel PEGylated colloidal gold-rhTNF nanomedicine *Clin. Cancer Res.* **16** 6139–49

[136] Thiesen B and Jordan A 2008 Clinical applications of magnetic nanoparticles for hyperthermia *Int. J. Hyperthermia* **24** 467–74

[137] Goodman C M, McCusker C D, Yilmaz T and Rotello V M 2004 Toxicity of gold nanoparticles functionalized with cationic and anionic side chains *Bioconjug. Chem.* **15** 897–900

[138] Pan Y, Leifert A, Ruau D, Neuss S, Bornemann J, Schmid G, Brandau W, Simon U and Jahnen-Dechent W 2009 Gold nanoparticles of diameter 1.4 nm trigger necrosis by oxidative stress and mitochondrial damage *Small* **5** 2067–76

[139] Mahmoudi M, Simchi A, Imani M, Shokrgozar M A, Milani A S, Hafeli U O and Stroeve P 2010 A new approach for the *in vitro* identification of the cytotoxicity of super-paramagnetic iron oxide nanoparticles *Colloids Surf. B* **75** 300–9

[140] Wozniak A, Malankowska A, Nowaczyk G, Grzeskowiak B F, Tusnio K, Slomski R, Zaleska-Medynska A and Jurga S 2017 Size and shape-dependent cytotoxicity profile of gold nanoparticles for biomedical applications *J. Mater. Sci., Mater. Med.* **28** 92

[141] Feng Q, Liu Y, Huang J, Chen K, Huang J and Xiao K 2018 Uptake, distribution, clearance, and toxicity of iron oxide nanoparticles with different sizes and coatings *Sci. Rep.* **8** 2082

[142] Hurbankova M, Volkovova K, Hraskova D, Wimmerova S and Moricova S 2017 Respiratory toxicity of Fe_3O_4 nanoparticles: experimental study *Rev. Environ. Health* **32** 207–10

[143] Neha R, Jaiswal A, Bellare J and Sahu N K 2017 Synthesis of surface grafted mesoporous magnetic nanoparticles for cancer therapy *J. Nanosci. Nanotechnol.* **17** 5181–8

[144] Sargent L M *et al* 2014 Promotion of lung adenocarcinoma following inhalation exposure to multi-walled carbon nanotubes *Part. Fibre Toxicol.* **11** 1–18

Chapter 6

Radiation and ultrasound stimulated breast cancer nanomedicine

Abdul K Parchur, Jaidip M Jagtap, Gayatri Sharma and Christopher P Hansen

Radiation and ultrasound therapies are both indispensable therapeutic treatment procedures used in clinics for better healthcare. Remarkable progress in these therapies has increased the overall cancer survival rate. Advancement in nano-medicine enhances the therapy response using newly designed nanomaterials that specifically target tumors compared to conventional cancer therapies. Radiotherapy and ultrasound therapies are the most effectively and commonly used noninvasive treatment plans for treating breast cancer—avoiding the side effects caused by systemic administration of chemotherapeutics and direct surgical removal. Progress in nanomedicine, aimed at improving radiation and ultrasound therapies, have focused mostly on increasing the therapeutic dose specifically to the tumor. The enhancement is twofold: first, greater dose delivered to the tumor increases efficacy and second, more dose delivered to the tumor means less is delivered to off target tissues. In this chapter, we summarize recent developments in radiation and ultrasound therapies using newly designed nanoparticles.

6.1 Introduction

According to 2019 cancer statistics, 1 762 450 new cancer cases (>4800 new cases each day) and 606 880 cancer deaths are expected in the United States; it is the second leading cause of death. Among all cancers, breast cancer will account for 30% (268 600 new cases) and 15% (41 760 deaths) in women in the United Sates in 2019. However, the breast cancer survival rate is 90%. Breast cancer develops in 1 out of 8 women (>12%) creating more than three million survivors in the United States [1, 2]. Outside the US breast cancer is the leading cause of death in 103 countries worldwide, in which Fiji has the highest mortality rate [3]. Radiation and ultrasound therapies are two prominent therapy methods used in treating solid tumors. In radiation therapy, an intense external x-ray irradiation (called external-beam radiation therapy—EBRT) is used to kill the maximum amount of cancer cells, leaving less damage to normal cells and thereby shrinking the tumor. Radiation therapy is an extensively used treatment plan in the clinic to treat up to ~75% of local solid tumors at different stages [4]. Recently Edy *et al* gave a radiation dose of ~40.05–52.5 Gy in 15 hypofractions in three weeks to high risk breast cancer patients with 6–15 MV beam energy [5]. A better understanding of radiobiology is crucial, mostly in recognizing the mechanisms to increase radiation sensitivity and toxicity. So far, numerous approaches for radiation sensitizer and protectors have been proposed or developed in preclinical and clinical studies [6, 7]. The concentration of oxygen in tumor tissues is proportional to cell death, i.e. low oxygen levels in tumors are less effective for radiation therapy. The rapid growth of cancer cells results in hypoxic conditions inside solid tumors, leading to radiation resistant tumors. This represents a serious clinical problem as >50% of tumors undergo radiation therapy [8]. If the tumor is loaded with oxygen enriched nano-particles it can enhance radiation therapy efficacy. Moreover, in 1998, Regulla *et al* confirmed that cells grown on gold (Au) foil are killed 100 times more compared to without the gold foil under x-ray irradiation. Leading to the idea that if the tumors uptake x-ray absorbing nanomaterials (with higher atomic number—Z), it can enhance the radiation dose in the tumors [9]. Gold nanoparticles are capable of (i) absorbing x-ray radiation ~100 times higher than tissue and (ii) releasing extra electrons (Auger effect) in the vicinity of gold nanoparticles [10]. In the clinic, chemotherapeutic drugs are used for the treatment of human cancers. Doxorubicin is the most commonly used chemotherapeutic drug alone or in combination with other drugs (e.g. cisplatin, docetaxel, paclitaxel, tamoxifen, and trastuzumab) in the clinic [11]. Therefore, there is a clinical demand to design, develop, scale up and deliver the next generation radiation enhanced nanomaterials for effective tumor dose enhancement and a need to develop new radiotherapy strategies for effective solid tumor therapy.

Ultrasound radiation is broadly used in nanomedicine for both imaging (diag-nostic; 2–12 MHz range) and therapeutic (treatment; >1 kW cm^{-2}, ~0.5–7 MHz range at the focal spot) purposes. Ultrasound radiation with different frequency/energy can be noninvasively and safely delivered into deep tissue at a high spatiotemporal resolution [10, 12]. Ultrasound began use in medical applications in the 1930s. Ultrasound waves travel in different tissues with different velocities, the resulting echoes from tumor are different than normal tissue, this property can be used in detecting tumors. The Food and Drug Administration (FDA) in the United States approves the therapeutic applications of ultrasound clinically. Nanoparticles' response to ultrasound is designed for both tumor imaging and/or treatment. Also, ultrasound generated much attention as a drug delivery tool due to its deep focus into the tumor tissue and trigger drug release at the tumor sites [12]. High Intensity Focused Ultrasound (HIFU) therapy is a frequently used therapeutic procedure for tumor thermal ablation in clinics, where HIFU is controlled for a quick temperature rise in the tumor tissue at high acoustic intensities to induce coagulative necrosis, i.e. HIFU ablation [13]. In the therapeutic procedure, there are no tumors resistant to ultrasound ablation. During the ablation, proliferating tumor cells and their growing vascularities are killed simultaneously. Ultrasound therapy procedures do not require inserting any instrument/tool into a targeted cancer tumor location during the therapy, making HIFU therapy procedures potentially more attractive than invasive therapies such as, radiofrequency ablation, laser ablation, and micro-wave ablation [14].

This chapter will discuss the advances in radiation and ultrasound therapeutic procedures in treating breast cancer tumors. Also, recent investigations of enhancement in therapeutic response using nanomedicines, and the challenges are discussed. As with any new technology there are safety and toxicity concerns, which will also be discussed.

6.2 Radiation therapy

Radiation therapy and a combination of chemotherapy is a widely used therapeutic treatment procedure in the clinic. Where external high-energy x-ray radiation is used to destroy deep breast tumors. The effectiveness of such therapy procedures depends on the tumor microenvironment (*in situ* tumor oxygenation, tumor vasculature, etc) [16]. Recently, Choi *et al* [15] treated breast cancer cells (MDA-MB-231, 5×10^3 cells/well) that were incubated overnight with sorafenib, paclitaxel, and its combination followed by irradiation of 5 Gy x-ray radiation. Cell viability was determined using an MTT assay (an enzymatic reduction of 3-[4,5-dimethylthiazole-2-yl]-2,5-diphenyltetrazolium bromide). Breast cancer cells treated with a sorafenib, paclitaxel, and radiation therapy cotreatment had a lower half-maximal inhibitory concentration (IC_{50}) than that of paclitaxel or sorafenib without radiation treatment. The mean IC_{50} values for breast cancer cells treated with sorafenib, paclitaxel, sorafenib + RT, paclitaxel + RT, paclitaxel + sorafenib + RT are observed as 7.5 ± 0.4, 4.1 ± 0.1, 5.9 ± 0.3, 3.6 ± 0.4, and ~5.5, respectively. The combined treatment of sorafenib and paclitaxel suppressed breast cancer cell proliferation more effectively than either agent did alone, or with RT (figure 6.1(A)) in a dose-dependent manner (figure 6.1(B)). Furthermore, to understand the therapeutic response ~2.0×10^7 breast cancer cells were injected in female BALB/c mice to grow tumors. Mice were treated with sorafenib (60 mg kg^{-1}, orally), paclitaxel (25 mg kg^{-1}, intraperitoneally), or a combination of both (sorafenib (25 mg kg^{-1}) and paclitaxel (15 mg kg^{-1})) in each group ($n = 10$) in total ~10–12 injections once every two days. Finally, tumors have been irradiated with 3 Gy x-ray radiation. Results show that cotreatment with sorafenib, paclitaxel, and radiation therapy greatly reduces the breast cancer change in tumor volume, body weight, and tumor weight (figures 6.1(C)–(E)) compared to other groups, and no mice death was observed. Also, immunohisto-chemistry study confirms lower B-cell lymphoma (Bcl-2; antiapoptotic gene, which is a breast cancer marker) expression in sorafenib, paclitaxel, and radiation therapy group compared to other groups (figure 6.1(F)) [15]. A recent study by Eom *et al* [17] studied clinicopathological characteristics of 1356 patients diagnosed with breast cancer from November 2006 to November 2011, a total of 605 patients (53.8%) with BCL2-positive expression. In general, BCL2 expression varies due to the molecular subtype, and it is a marker for luminal patients. The effective breast cancer

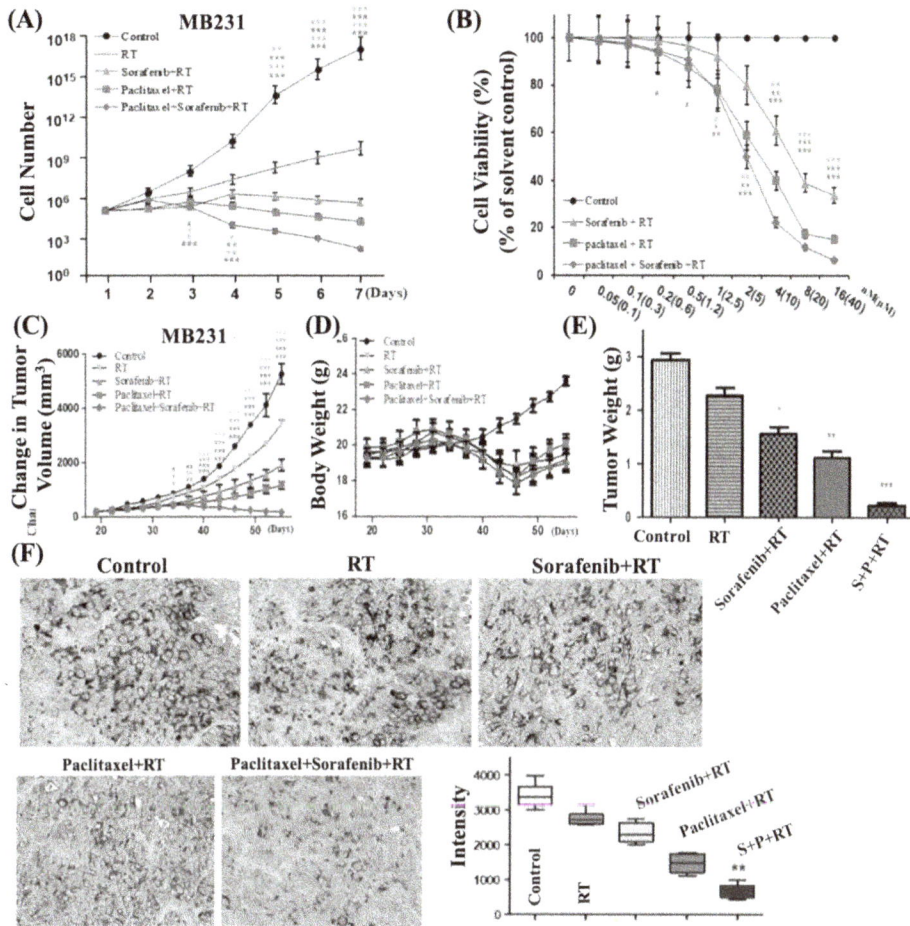

Figure 6.1. Cancer cells (MDA-MB-231) cotreatment with drugs (paclitaxel and sorafenib) and radiation therapy suppressed cell proliferation. The results of various combinations on breast cancer cells using (A) cell proliferation (change in cell number) and (B) change in viability (%). Data indicate mean percentage change in the control group, error bars represent the mean ± SD. Paclitaxel and sorafenib combined with radiation treatment attributes the elevated tumor reduction of breast cancer xenografts *in vivo*. Athymic nude mice with established tumors have been treated with the indicated drugs or radiation treatment. (C) Data represent the mean tumor volume. (D) No substantial effect on body weight in xenografted mice have been attributable to the drugs. (E) Cotreatment with paclitaxel, sorafenib, and radiation therapy showed the highest reduction of the dissected tumor weight in breast cancer xenografts. (F) Anti-apoptosis related proteins in tumor tissues derived from breast cancer cells. Immunohistochemical analysis of B-cell lymphoma (Bcl-2) proteins in the breast cancer tumor tissues following treatment. Assays have been repeated three times, and representative images are shown. MetaMorph 4.6 image-analysis software have been used to quantify the immunostained target protein. $*P < 0.05$, $**P < 0.01$, and $***P < 0.005$ versus control. Reprinted from [15], with permission from Elsevier.

treatment used combined therapy sorafenib, paclitaxel, and radiation therapy with high anticancer activity suggesting it could be a better clinical therapeutic approach for breast cancer patients. It is also essential to recognize toxicity concerns using different drugs in combination with radiotherapy. Recently, FDA approved palbociclib (IBRANCE®, Pfizer Inc.) for the treatment of human epidermal growth factor receptor 2 (HER2) negative and hormone receptor (HR) positive breast cancer patients. Preclinical study results recommend the use of palbociclib (CDK4/6 inhibitor) in combination with radiotherapy. Even though there are potential treatment benefits, clinicians rarely use this combination due to palbociclib toxicity concerns, particularly leukopenia and neutropenia [18]. In radiation therapy, it has some extra benefits such as (i) creating a focal inflammatory response, (ii) exposure to tumor antigens, (iii) inducing maturation of antigen-presenting cells, (iv) enhancing major histocompatibility complex upregulation, (v) leads to danger-associated molecule pattern, and (vi) sensitizes tumor cells to immune-mediated killing to create a potential *in situ* vaccine [19].

6.2.1 Nanoparticle mediated radiation therapy

It is well known that, x-ray radiation kills tumor cells by producing free radicals (reactive oxygen) thereby damaging the tumor cellular DNA [20]. At the same time, a low quantity of oxygen at the center of breast tumors blocks the production of free radicals hindering radiation therapeutic response [21]. The success of radiation therapy depends on the delivery of radiosensitizers to the tumors and reducing hypoxia resistance. Recently, Fan *et al* [21] synthesized sub-50 nm hollow meso-porous silica nanoparticles with multiple framework hybridization. These nano-constructs have been loaded with tert-butyl hydroperoxide (TBHP) and iron pentacarbonyl. A typical transmission electron microscopy (TEM) image of quad-ruple hybridized sub-50 nm mesoporous silica nanoparticle is shown in figure 6.2(A). A schematic illustration of TBHP/Fe(CO)$_5$ loaded mesoporous silica nanoparticle (HMOP-TBHP/Fe(CO)$_5$) used for x-ray-activated synergistic radiodynamic therapy and gas therapy, where Fe(CO)$_5$ act as CO realizing agent and TBHP is •OH realizing agent is shown in figure 6.2(B). During, this radiodynamic therapy process TBHP loaded nanoparticles generate •OH radicals at the center of tumors, which is dramatically enhanced radiation therapy efficacy. Moreover, •OH radicals can attack iron pentacarbonyl and thereby release carbon monoxide molecules for gas therapy. These nanoparticles can be loaded with chemotherapeutic drugs that can further enhance the therapeutic efficacy by slowly releasing the chemotherapeutic drug to the tumor site. These nanoparticles are PEGylated (mPEG-saline, MW = 2000) and show a half-life of ∼49 h, by increasing the size of the nanoparticles a five-fold decrease in half-life has been observed. These nanoparticle mediated radiotherapy procedures show effective therapeutic response against both hypoxic and normoxic cancers [21]. Darfarin *et al* [22] demonstrated that x-ray radiation dose enhancement has been achieved using Au@SiO$_2$ nanoparticles (∼25 nm) treated breast cancer cells. The internalization of amine and thiol functionalized nanoparticles has been found to be 34% and 18%, respectively. Maximum

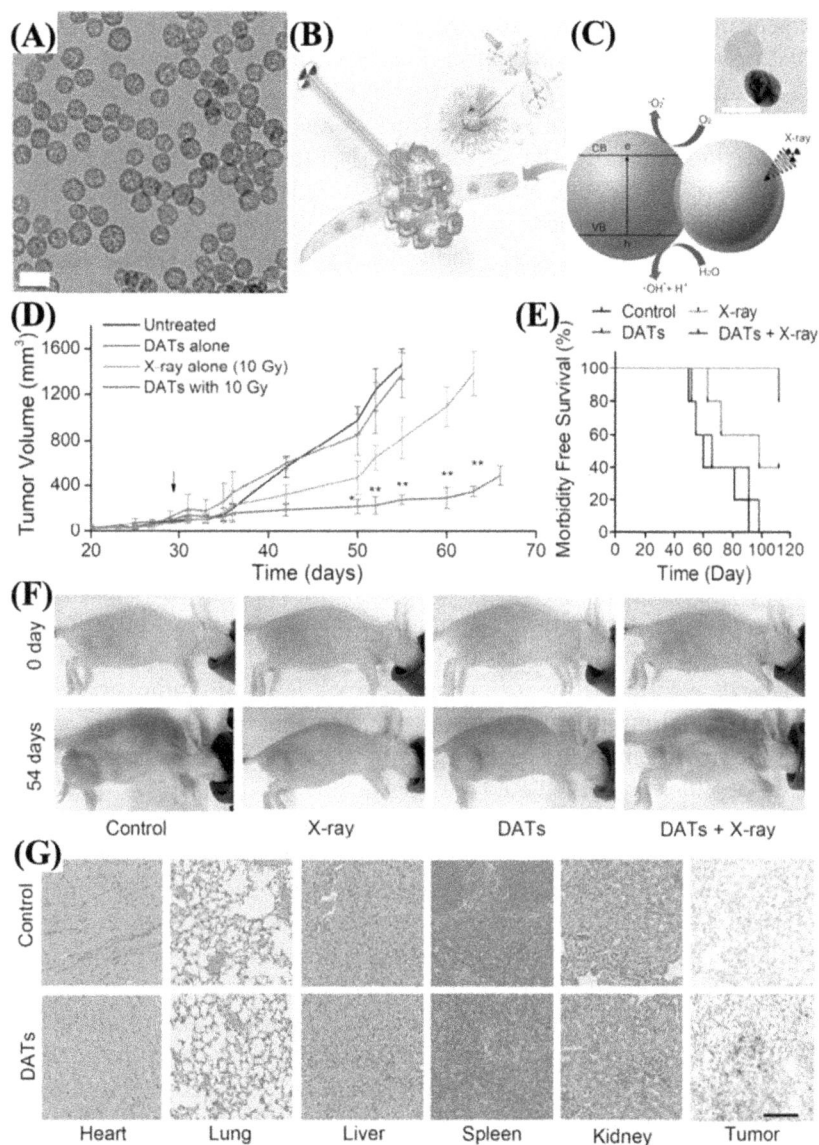

Figure 6.2. (A) Transmission electron microscopy image of sub-50 nm hollow mesoporous silica nanoparticle with multiple framework hybridization, (B) Schematic illustration of HMOP-TBHP/Fe(CO)$_5$ for x-ray activated synergistic radiation therapy and gas therapy. (C) Schematic illustration of the formation of hydroxyl radical (•OH) and superoxide (•O$_2-$) on x-ray radiation induced DAT nanoparticles, inset of figure (C) shows the typical TEM image of DAT nanoparticles. Synergistic enhancement therapeutic response using DAT nanoconstruct and radiation in SUM159 tumor bearing mice compared with control mice, DAT alone, x-ray (10 Gy) alone groups. (D) Change in tumor volume ($n = 5$/ group). (E) Survival curves of SUM159 tumor-bearing mice after various treatments. DAT nanoparticles injected mice after x-ray irradiation shows obvious tumor regression and improvement in survival rate after 60 days of treatments. (F) Digital photographs of the SUM159 tumor bearing mice before and post-54 days' treatment. (G) Representative H&E staining of various tissues from the control and DAT mediated therapy groups. Scale bar = 100 μm. Reprinted from [21, 23], with permission from Nature and American Chemical Society.

enhancement in x-ray radiation dose (8 Gy) has been observed in the breast cancer treated with nanoparticles (~200 ppm). Cheng *et al* [23] synthesized hybrid-nano-structure (dumbbell-like Au–TiO$_2$, called DAT); radiosensitizers can enhance the radiation therapeutic effect on breast tumors at low concentrations while reducing the potential side effects to the surrounding healthy tissue. DAT nanostructures show a synergistic therapeutic effect on x-ray radiation therapy (figure 6.2(C)), due to strong asymmetric electric coupling between the high Au and TiO$_2$ at their interfaces. The inset of figure 6.2(C) shows the TEM image of DAT nanoconstruct. Upon treatment of SUM159 breast cancer cells with DAT nanoconstructs, no significant toxicity was observed, while cells treated with DAT nanoconstructs and irradiated with 10 Gy radiation show ~41% cell viability. This is due to generation of reactive oxygen species (ROS) and secondary electrons from DAT triggered by x-ray irradiation, resulting in higher cancer cell apoptosis. Furthermore, to under-stand the therapeutic response, SUM159 tumor bearing mice (n = 5/group) were injected with DAT and irradiated with x-ray radiation of 10 Gy. Substantial tumor volume growth reduction, and improvement in survival rate has been observed in DAT + x-ray (10 Gy) treated group, compared with untreated, DAT alone, and x-ray alone (10 Gy) (figures 6.2(D)–(F)). Also, therapeutic response has been further confirmed using H&E tumor tissues staining of mice major organs from the control, DAT, x-ray (10 Gy), and DAT + x-ray (10 Gy) therapy group (figure 6.2(G)). Yang *et al* [24] used a combination of chemotherapeutic drug conjugated gold nano-particles and radiation therapy for the enhancement in therapeutic efficacy. A significant enhancement in radiation therapy response (~30%) has been observed using RGD-peptide conjugated gold nanoparticles with cisplatin (435 nM) and x-ray irradiation (2 Gy) compared with breast cancer cells treated with cisplatin and radiation treated group, i.e. even lower x-ray radiation doses can be used to irradiate cancer tumors with better therapeutic efficiency, while reducing the damage to the normal tissues. Gold nanoparticles also show photothermal therapeutic response to NIR irradiation [25–28]. Furthermore, oxygen sensitive microbubbles have been used to enhance the breast cancer radiation therapy efficacy (~3× times) [29].

Thus, the therapeutic efficacy depends on tumor uptake of radio-sensitizer (number of the nanoparticles in the tumor), which also depends on the tumor microenvironment. Flister *et al* [30] show that many aspects of the tumor micro-environment impact on breast cancer tumor growth. Breast cancer cells (MDA-MB-231 (231^{Luc+}) implanted in both SS and SS.BN3 (consomic) rats) had different tumor growth, with higher blood vessels in SS.BN3 group as compared to SS group. This behavior has been further examined by micro-computed tomography, dynamic contrast-enhanced magnetic resonance imaging, NIR fluorescence imaging, and *ex vivo* analysis of primary blood endothelial cells in both SS and SS.BN3 groups [30, 31]. A significantly higher washout rate has been observed in SS rats as compared to SS.BN3 rats with breast tumors. Higher rate of washout results in suboptimum uptake of the nanoparticles in the tumors, thereby minimizing therapeutic response.

6.3 Ultrasound therapy

Ultrasound waves are widely used in both imaging (low intensity ultrasound) and treatment (high intensity ultrasound) of cancer tumors. In recent years focused ultrasound has been used for early detection and noninvasive therapy of breast cancer patients.

High intensity focused ultrasound (HIFU) waves travel through the tissue and accurately focus on breast cancer tumors with minimal damage to normal tissue, thereby converging ultrasound waves produce therapeutic effects. Ultrasound can generate temperature (55 °C–90 °C) at the tumor site depending on the applied frequency, pulse length, focusing, exposure time, pulse intensity, etc, within 10 s and promptly induce cellular death and vascular obliteration in both normal and tumor tissue [32]. Also, ultrasound has been expended for controlled trigger release of chemotherapeutic drugs from nanoparticles or polymers [33]. Using the HIFU therapeutic procedure, both the central tumor zone and tumor periphery attained hyperthermia temperature, even at deep tumor tissues for prolonged time. During this procedure, it is difficult to avoid excessive heating of normal tissue. This can be overcome when ultrasound therapy-HIFU has been performed under magnetic resonance (MR) guidance. MR imaging accurately detects many of the tumors comprising breast cancer tumors. MR imaging technology accomplished of non-invasive tumor temperature measure produced by the ultrasound and several MR imaging parameters independent of temperature. *In vivo* temperature mapping can be used as a real-time control for precisely ablating the tumor tissue, thereby over heating of normal tissue can be avoided for a long time. Furthermore, MR guided HIFU has been used for both tumor ablation and triggered drug delivery hyperthermia [32, 34].

More interestingly, focused ultrasound is the only method used for the disruption of reversible blood–brain barrier for drug delivery to brain cancer tumors without radiation or incision [35]. A noninvasive MR image guided (using 1.5 T MR imaging scanner) procedure for treating human breast cancer tumors using focused ultrasound surgery has been accomplished by Huber *et al* [32] having real-time MR image temperature control in sheep and human breast cancer models. Intense ultrasound waves were focused through the skin and the breast tumor was ablated at ~70 °C with a 2 mm special accuracy in temperature localization. A real-time T2 weighted temperature map has been used for monitoring the therapy zone across the baseline tumor. The histological analysis after post-4 h therapy shows hyperemia, edematous swelling, and mild lymphoplasmacellular infiltration. Post-3 day and 4 week histological analysis confirms a homogeneous-shaped necrosis (up to ~1750 mm^3). These results have shown acute nonperfused region damage during the therapy process. Also, therapy has been performed in human breast cancer having MR visualized tumor size ~2.2 × 2 × 1.4 cm^3. An ultrasound treatment (80 pulses, 9 s each, and 30–50 W power) plan has been adopted by considering targeting tumor volume outlined with a ~2 mm rim for the complete tumor ablation. MR imaging guided temperature monitoring tracks and the related amount of heat delivered to the tumor. Moreover, ultrasound therapy response has been appraised based on cellular proliferation and hormone receptor status in breast cancer tumors. Pre-ultrasound immunohistological stains have shown strong staining for estrogen (ER) and progesterone receptor (PR), B-cell lymphoma 2 (bcl-2) made up of decent proliferative activity, and 30% of p53, whereas post-ultrasound shows negative ER and PR, p53 escalate up to ~90% of tumor cell nuclei [32]. The greatest advantage of MR image guided therapy is its sensitivity to rapid evaluation of ultrasound induced coagulative necrosis in the tumor ablation and adjoining areas. However, MR imaging is expensive compared to other imaging modalities such as ultrasound imaging with poor imaging resolution. Using ultrasound therapy in clinical routine treatment relies on the outcome of the therapy, therapy device, feasibility, efforts and cost. Also, therapy response can be ameliorated by delivering ultrasound sensitizing contrast materials to the targeted tumors, which will be discussed in the following section 6.3.1 [36]. Ultrasound therapy can be favorably used to treat other cancers such as, brain, bone, liver, rectum, pancreas, and pancreas tumors in addition to breast cancer tumors.

6.3.1 Nanoparticles mediated ultrasound therapy

As discussed in section 6.3, ultrasound therapy alone is low cost and can be effectively used for the treatment of deep tumors. In general, conventional chemo-therapy treatment enhances patient survival rates, but they also create toxicity worries to healthy tissues, i.e. adverse side effects. However, these behaviors can be relaxed by precisely delivering chemotherapeutic drugs encapsulated within the ultrasound responsive nanoparticles and externally triggered to release drugs to the targeted tumor. Therapy efficacy can be greatly enhanced and toxicity reduced [37–39]. As discussed in section 6.2.1, the uptake of the nanoparticles in the tumor depends of many factors such as uneven tumor perfusion and biological barriers associated with it. To overcome some of these concerns, Snipstad *et al* [37] designed a multifunctional chemotherapeutic drug delivery system containing microbubbles stabilized by polymeric nanoparticles (NPMBs), capable of ultrasound triggered drug transport to the targeted tumor site. PEGylated poly(2-ethyl-butyl cyanoacry-late)—PEBCA nanoparticles (<200 nm) and surface charge around −2.5 mV and microbubbles (formed using self-assembly of polymeric nanoparticles) with average diameter ~2.6 μm (5×10^8 MB mL^{-1}) have been synthesized. The circulation half-life of polymeric nanoparticles shows ~2.26 h. A schematic illustration of enhanced drug delivery to the targeted tumor site under external ultrasound triggering to NPMBs is shown in figure 6.3(A). Microbubbles boost tumor contrast ~5 min post-injection (figures 6.3(B), (C)) and changes in tumor contrast with time have been shown (figure 6.3(D)). It is disappointing to see that there is ~1% ± 0.35% of nanoparticles (post-6 h injection) in breast tumor and ~69% in liver. In general, for solid tumors, uptake of ~0.7% injected dose has been delivered to the tumor site [40]. However, the advantages of the particles arise due to its PEGylation surface and having half-life more than 2 h. On using lower acoustic pressure (MIs between 0.5–1) there is partial or no uptake achieved, due to unchanged vascular permeability at low pressure. Using high power (~0.5) there is a significant enhancement of

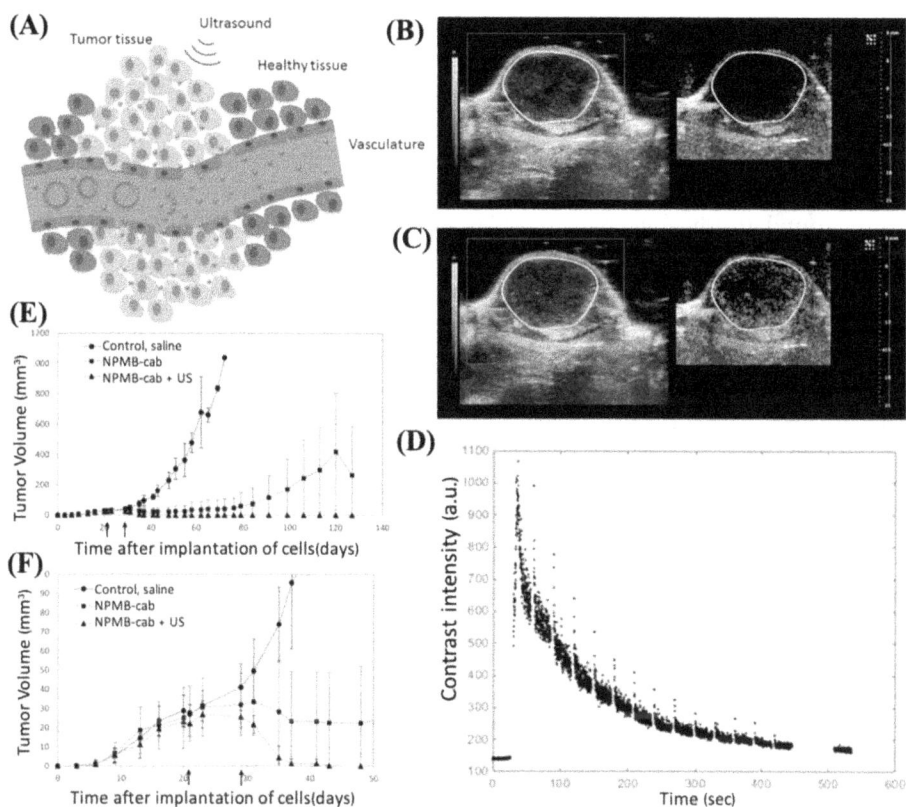

Figure 6.3. (A) Schematic of drug delivery to breast cancer tumor tissue by use of focused ultrasound and nanoparticle-stabilized microbubbles, represented in green. Representative B-mode ultrasound (left) and nonlinear contrast images (right) of a tumor (B) pre-injection and (C) post-injection of nanoparticle-stabilized microbubbles. (D) Dynamic tumor contrast intensity has been quantified, which is shown as a function of time to depicting half-life of microbubbles in blood circulation. (E) Tumor volume as a function of time post-implantation of breast cancer cells. Mice have been treated with saline, nanoparticle-stabilized microbubbles (NPMBs) with cabazitaxel or NPMBs loaded with cabazitaxel and irradiated with ultrasound. Therapy treatments have been performed on day 21 and 29 (indicated with arrows). Data represents mean and standard deviations using $n = 4$ animals/group up to 35 days and $n = 3$ animals/group from 37 days onwards. (F) Zoomed-in of figure (E) from 0–50 days. Reprinted from [37], with permission from Elsevier.

nanoparticle uptake by 2.3 times compared to the untreated group animals with breast tumors. This may be due to an increase in extravasation. Cabazitaxel drug loaded microbubbles release drugs to the targeted tumor, thereby diminishing the tumor size. The change in tumor volume in three groups: control (saline), NPMB-cab (microbubbles loaded with cabazitaxel drug), and NPMB-cab + US (microbubbles loaded with cabazitaxel drug followed by ultrasound) monitored up to 160 days with treatment time 21 and 29 days has been shown in figures 6.3(E), (F). Significant reduction in tumor volume in NPMB-cab + US group confirms efficient ultrasound therapeutic behavior [37].

Recently, Zhao *et al* [41], synthesized siRNA loaded Porphyrin microbubbles capable of photodynamic and gene therapy using a single nanoconstruct with the help of ultrasound. A schematic illustration porphyrin microbubble loaded with siRNA followed by ultrasound-assisted PDT and FOXA1 KD strategy is shown in figures 6.4(A), (B). These microbubbles are ~2 μm in size and zeta potential +25 mV, on loading siRNA surface charge changed to −21 mV confirming successful loading. On ultrasound irradiation microbubbles dissociated into ~103 nm nanoparticles. This breaking behavior of microbubbles under external low frequency ultrasound triggering transforms of microbubbles into nanoparticles for the targeted delivery of photosensitizers and siRNA to the breast tumors. Breast tumor transfection efficiency of siRNA increased by 4× fold and the porphyrin uptake by 8× fold under external ultrasound triggering. The therapeutic response of combined photo-dynamic therapy and siRNA therapy has been studied in a BALB/c mouse model having ~120 mm³ breast tumors. A total of six groups ($n = 6$), including PBS, CpMBs/F, CpMBs/F + US, CpMBs/F + L, CpMBs/NC + US + L and CpMBs/F + US + L (where F = FOXA1-siRNA, NC = scrambled siRNA, US = ultrasound, L = laser) with intravenous injection via the tail vein with microbubbles and irradiated with 650 nm laser at 6 h post-injection (treatment has been performed three times with a time interval of 7 days or tumor volume increases to ~1000 mm³). The change in tumor volume in PBS, CpMBs/F, CpMBs/F + US, CpMBs/F + L, CpMBs/NC + US + L and CpMBs/F + US + L groups has been shown in figure 6.4(C). In the control group, tumor volume increased up to ~18 times, whereas a significant decrease was observed in tumor volume in siRNA, ultrasound, and laser treated group (CpMBs/F + US + L). The digital photograph of PBS; CpMBs/NC + US + L; CpMBs/F + US + L therapy group mice has been shown in figure 6.4(D), and H&E staining of breast tumor slices removed at 72 h post-treatments has been shown in figure 6.4(E). Furthermore, microbubbles based ultrasound therapy can overcome on-target but off-tumor side effect, and improve the accumulation of therapeutic drugs at the targeted tumor site. Improving the selectivity of photo-dynamic treatment can decrease the side effects on normal organs and tissues. Also, siRNA knockdown of FOXA1 has been used to hinder tumor recurrence. Combined ultrasound, siRNA treatment, and photodynamic therapy significantly improve breast cancer therapy efficacy.

6.4 Toxicity concerns

There is no doubt that radiation therapy reduces the risk of breast cancer reappear-ance and death. Also, radiation therapy procedures usually cause ionizing radiation of the heart, to some extent. Recent clinical trials confirm that there is an increase in long-term cardiac side effects (heart disease) associated with post-radiation therapy in women with breast cancer [42, 43]. Cardiac risk estimation needs [15]:
  (i) a relative change in the incidence of heart disease on a per radiation dose basis,
  (ii) projected cardiac radiation exposure based on the therapeutic strategy, and
  (iii) risk of heart disease with no cardiac radiation exposure.

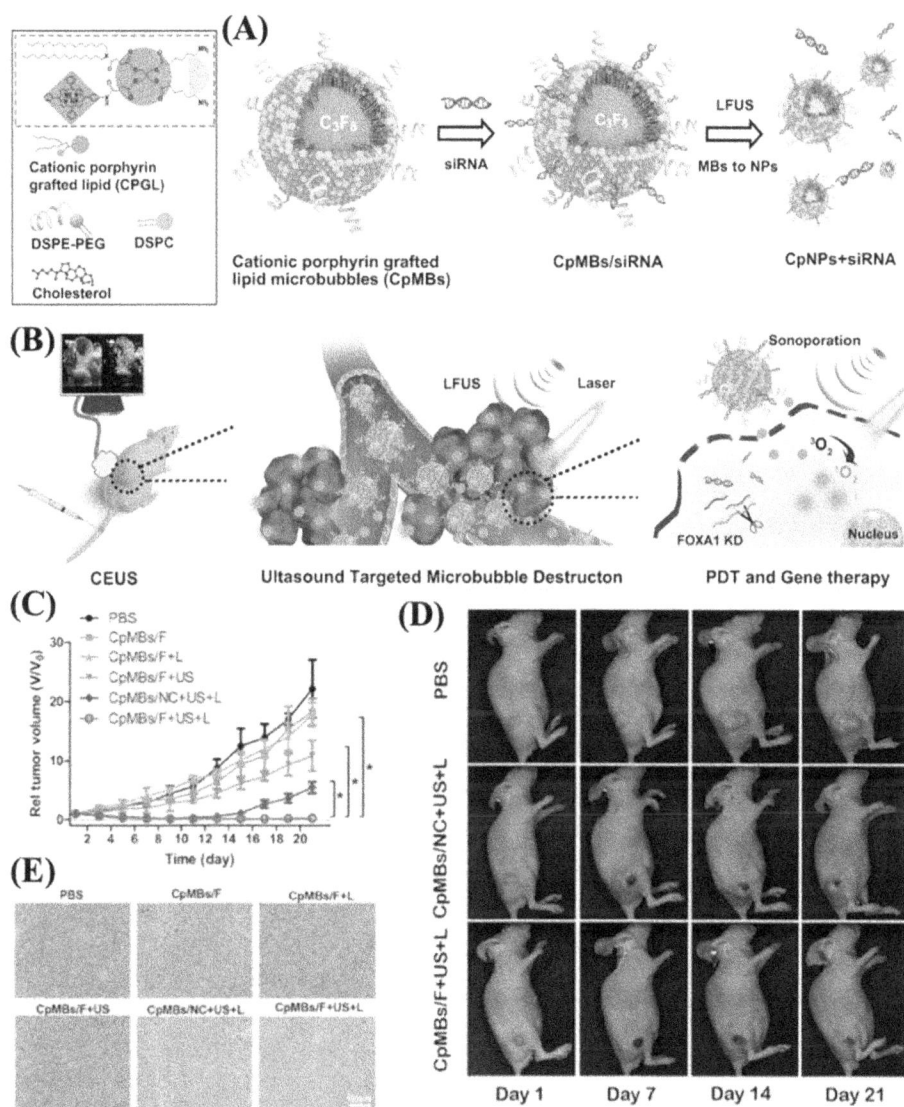

Figure 6.4. (A,B) A schematic illustration of the Porphyrin microbubble loaded with siRNA, ultrasound-assisted photodynamic therapy, and FOXA1 KD treatment strategy. Structure of CpMBs/siRNA and its transformation from microbubbles to nanoparticles using ultrasound irradiation. Enhanced in photodynamic therapy and siRNA transfection effect of CpMBs/siRNA by *in situ* conversion of MBs to NPs with the technology of ultrasound targeted microbubble destruction. *In vivo* therapeutic effects of photodynamic therapy in the breast cancer xenograft bearing mice model. (C) Change in breast cancer tumor volume of different treatment groups PBS, CpMBs/F, CpMBs/F + US, CpMBs/F + L, CpMBs/NC + US + L, CpMBs/F + US + L, having $n = 6$ animals/group, where F = FOXA1-siRNA, NC = scrambled siRNA, US = ultrasound, L = laser, ($*P < 0.05$ vs CpMBs/F + US + L); (D) Representative photographs showing the therapeutic effect of the breast tumor mice after post-treatment in groups, PBS; CpMBs/NC + US + L; CpMBs/F + US + L. (E) H&E staining of breast cancer tumor slices excised at 72 h after different treatments. Reprinted from [41], with permission from Elsevier.

Yun *et al* [14] used meta-analysis involving 1 191 371 cancer patients from 39 studies, to explore the relationship between breast cancer radiation therapy and its consequent risk of cardiovascular disease, 5367 events were included from 137 074 patients (16 studies). There was 1.38 ($P < 0.001$) fold higher risk of cardiac mortality in women treated with x-ray radiation as compared to women who did not receive radiation. Taylor *et al* [15] have observed that cardiac events increased by 7.4% per Gy x-ray radiation dose, with average cardiac radiation dose 3 Gy in right-side and 5 Gy in left-side, and tend to have greater cardiac risk factors than women who received no radiation treatment. In another study, Frederika *et al* [44] confirmed that x-ray dose $\geqslant$30 Gy has been recognized as a marker for increased risk to the heart, with a 2.8–4.7 fold higher cardiac failure rate as a first event compared to no radiation treatment, after compensating for other risk factors. Chowdhary *et al* [18] reported preliminary results using palbociclib (125 mg daily in total 1–21 days association with either letrozole 2.5 mg daily or fulvestrant 500 mg every 28 days) in combination with radiotherapy of breast cancer patients from 2015–8 and toxicity was graded using CT/MR imaging data. These data confirm that palbociclib and radiation therapy results in toxicity Grade 1–2 with no 3+ toxicity concerns, though authors suggest that longer follow-up is necessary to confirm these results [18]. It is commonly observed that left-sided breast cancer radiation treatment greatly influences cardiac toxicity, selected areas of the heart are exposed to about 40–50 Gy radiation dose [45]. Similar to radiation therapy, ultrasound therapy has seen clinical use in recent years and the FDA has approved focused ultrasound therapy for the ablation of breast cancer patients.

6.5 Conclusion

In the clinical therapy process, breast cancer conserving surgery followed by whole breast irradiation of ~50 Gy decreases the cancer recurrence rate up to 88%. Radiation therapy reduces 5.3% overall mortality after 15 years [46]. Using a recent search on the NIH Clinical Trails web page, 'radiation therapy and breast cancer' as key words shows 391 clinical trials are in progress. These are working on early diagnosis and treatment of breast cancer and some other cancer types as well. Whereas 'focused ultrasound and breast cancer' search results in six clinical trials, among them five are cancer trials on breast cancer. These clinical trials use chemotherapeutic drugs, contrast agents in addition to radiation therapy and ultrasound therapy in the treatment procedures. We are still faced with several challenges for the precise detection, treatment, and monitoring of therapy responses in cancer patients. New technologies and treatment strategies have been adopted for better therapeutic outcomes. Furthermore, nanoparticle mediated therapeutic procedures and externally triggered targeted drug delivery systems have been adopted to reduce toxicity and to increase therapeutic effect. Also, MR imaging methods are being used for the accurate assessment of cancer disease. Advanced artificial intelligence algorithms have been used for the quick visualization of therapy response in the operating room, which will help clinicians to make quick assessments during therapy. Even though much development has been made in breast cancer therapeutic procedures, cardiac risk factors and other toxicity concerns cannot be overlooked. There is still a lot of effort needed in developing new nanomedicine based radiation and ultrasound responsive contrast agents, and drugs which show better therapeutic outcome. These efforts will give more room to physicians in the operating room to stop over ablation of normal tissue and decrease cardiac exposure.

References

[1] Caron J and Nohria A 2018 Cardiac toxicity from breast cancer treatment: can we avoid this? *Curr. Oncol. Rep.* **20** 61

[2] Siegel R L, Miller K D and Jemal A 2019 Cancer statistics, 2019 *CA Cancer J. Clin.* **69** 7–34

[3] Bray F, Ferlay J, Soerjomataram I, Siegel R L, Torre L A and Jemal A 2018 Global cancer statistics 2018: GLOBOCAN estimates of incidence and mortality worldwide for 36 cancers in 185 countries *CA Cancer J. Clin.* **68** 394–424

[4] Chen Q, Chen J, Yang Z, Xu J, Xu L, Liang C, Han X and Liu Z 2019 Nanoparticle-enhanced radiotherapy to trigger robust cancer immunotherapy *Adv. Mater.* **31** e1802228

[5] Ippolito E, Rinaldi C G, Silipigni S, Greco C, Fiore M, Sicilia A, Trodella L, D'Angelillo R M and Ramella S 2019 Hypofractionated radiotherapy with concomitant boost for breast cancer: a dose escalation study *Br. J. Radiol.* **92** 20180169

[6] Chen H H W and Kuo M T 2017 Improving radiotherapy in cancer treatment: Promises and challenges *Oncotarget* **8** 62742–58

[7] Maranto C *et al* 2018 STAT5A/B blockade sensitizes prostate cancer to radiation through inhibition of RAD51 and DNA *Repair Clin. Cancer Res.* **24** 1917–31

[8] Benej M *et al* 2018 Papaverine and its derivatives radiosensitize solid tumors by inhibiting mitochondrial metabolism *Proc. Natl. Acad. Sci. U. S. A.* **115** 10756–61

[9] Regulla D F, Hieber L B and Seidenbusch M 1998 Physical and biological interface dose effects in tissue due to x-ray-induced release of secondary radiation from metallic gold surfaces *Radiat. Res.* **150** 92–100

[10] Antosh M P *et al* 2015 Enhancement of radiation effect on cancer cells by gold-pHLIP *Proc. Natl. Acad. Sci. U. S. A.* **112** 5372–6

[11] Liyanage P Y, Hettiarachchi S D, Zhou Y, Ouhtit A, Seven E S, Oztan C Y, Celik E and Leblanc R M 2019 Nanoparticle-mediated targeted drug delivery for breast cancer treatment *Biochim. Biophys. Acta, Rev. Cancer* **1871** 419–33

[12] Miller D L, Smith N B, Bailey M R, Czarnota G J, Hynynen K, Makin I R and Bioeffects Committee of the American Institute of Ultrasound in Medicine 2012 Overview of therapeutic ultrasound applications and safety considerations *J. Ultrasound Med.* **31** 623–34

[13] Adem Y, Nicholas T B and Andrew P G 2019 Colloids, nanoparticles, and materials for imaging, delivery, ablation, and theranostics by focused ultrasound (FUS) *Theranostics* **9** 2572–94

[14] Wu F, Wang Z B, Cao Y D, Xu Z L, Zhou Q, Zhu H and Chen W Z 2006 Heat fixation of cancer cells ablated with high-intensity-focused ultrasound in patients with breast cancer *Am. J. Surg.* **192** 179–84

[15] Choi K H, Jeon J Y, Lee Y E, Kim S W, Kim S Y, Yun Y J and Park K C 2019 Synergistic activity of paclitaxel, sorafenib, and radiation therapy in advanced renal cell carcinoma and breast cancer *Transl. Oncol.* **12** 381–8

[16] Prasad P, Gordijo C R, Abbasi A Z, Maeda A, Ip A, Rauth A M, DaCosta R S and Wu X Y 2014 Multifunctional albumin-MnO(2) nanoparticles modulate solid tumor microenvironment by attenuating hypoxia, acidosis, vascular endothelial growth factor and enhance radiation response *ACS Nano* **8** 3202–12

[17] Eom Y H, Kim H S, Lee A, Song B J and Chae B J 2016 BCL2 as a subtype-specific prognostic marker for breast cancer *J. Breast Cancer* **19** 252–60

[18] Chowdhary M, Sen N, Chowdhary A, Usha L, Cobleigh M, Wang D, Patel K R, Barry P N and Rao R D 2019 Safety and efficacy of palbociclib and radiotherapy in metastatic breast cancer patients: initial results of a novel combination *Adv. Radiat. Oncol.* **4** 453–7

[19] Gunderson A J and Young K H 2018 Exploring optimal sequencing of radiation and immunotherapy combinations *Adv. Radiat. Oncol.* **3** 494–505

[20] Van Houten B, Santa-Gonzalez G A and Camargo M 2018 DNA repair after oxidative stress: current challenges *Curr. Opin. Toxicol.* **7** 9–16

[21] Fan W *et al* 2019 Generic synthesis of small-sized hollow mesoporous organosilica nanoparticles for oxygen-independent X-ray-activated synergistic therapy *Nat. Commun.* **10** 1241

[22] Darfarin G, Salehi R, Alizadeh E, Nasiri Motlagh B, Akbarzadeh A and Farajollahi A 2018 The effect of SiO2/Au core–shell nanoparticles on breast cancer cell's radiotherapy *Artif. Cells Nanomed. Biotechnol.* **46** 836–46

[23] Cheng K *et al* 2018 Synergistically enhancing the therapeutic effect of radiation therapy with radiation activatable and reactive oxygen species-releasing nanostructures *ACS Nano* **12** 4946–58

[24] Yang C, Bromma K, Sung W, Schuemann J and Chithrani D 2018 Determining the radiation enhancement effects of gold nanoparticles in cells in a combined treatment with cisplatin and radiation at therapeutic megavoltage energies *Cancers* **10**

[25] Parchur A K, Sharma G, Jagtap J M, Gogineni V R, LaViolette P S, Flister M J, White S B and Joshi A 2018 Vascular interventional radiology-guided photothermal therapy of colorectal cancer liver metastasis with theranostic gold nanorods *ACS Nano* **12** 6597–611

[26] Parchur A K, Li Q and Zhou A 2016 Near-infrared photothermal therapy of Prussian-blue-functionalized lanthanide-ion-doped inorganic/plasmonic multifunctional nanostructures for the selective targeting of HER2-expressing breast cancer cells *Biomater. Sci.* **4** 1781–91

[27] Li Q, Parchur A K and Zhou A 2016 *In vitro* biomechanical properties, fluorescence imaging, surface-enhanced Raman spectroscopy, and photothermal therapy evaluation of luminescent functionalized CaMoO4:Eu@Au hybrid nanorods on human lung adenocarcinoma epithelial cells *Sci. Technol. Adv. Mater.* **17** 346–60

[28] Urban C, Urban A S, Charron H and Joshi A 2013 Externally modulated theranostic nanoparticles *Transl. Cancer Res.* **2** 292–308

[29] Eisenbrey J R *et al* 2018 Sensitization of hypoxic tumors to radiation therapy using ultrasound-sensitive oxygen microbubbles *Int. J. Radiat. Oncol. Biol. Phys.* **101** 88–96

[30] Flister M J *et al* 2017 Host genetic modifiers of nonproductive angiogenesis inhibit breast cancer *Breast Cancer Res. Treat.* **165** 53–64

[31] Jagtap J, Sharma G, Parchur A K, Gogineni V, Bergom C, White S, Flister M J and Joshi A 2018 Methods for detecting host genetic modifiers of tumor vascular function using dynamic near-infrared fluorescence imaging *Biomed. Opt. Express* **9** 543–56

[32] Huber P E, Jenne J W, Rastert R, Simiantonakis I, Sinn H P, Strittmatter H J, von Fournier D, Wannenmacher M F and Debus J 2001 A new noninvasive approach in breast cancer

therapy using magnetic resonance imaging-guided focused ultrasound surgery *Cancer Res.* **61** 8441–7

[33] Huebsch N, Kearney C J, Zhao X, Kim J, Cezar C A, Suo Z and Mooney D J 2014 Ultrasound-triggered disruption and self-healing of reversibly cross-linked hydrogels for drug delivery and enhanced chemotherapy *Proc. Natl. Acad. Sci. U. S. A.* **111** 9762–7

[34] Hijnen N, Kneepkens E, de Smet M, Langereis S, Heijman E and Grull H 2017 Thermal combination therapies for local drug delivery by magnetic resonance-guided high-intensity focused ultrasound *Proc. Natl. Acad. Sci. U. S. A.* **114** E4802–11

[35] Sun T, Zhang Y, Power C, Alexander P M, Sutton J T, Aryal M, Vykhodtseva N, Miller E L and McDannold N J 2017 Closed-loop control of targeted ultrasound drug delivery across the blood-brain/tumor barriers in a rat glioma model *Proc. Natl. Acad. Sci. U. S. A.* **114** E10281–90

[36] Hsiao Y H, Kuo S J, Tsai H D, Chou M C and Yeh G P 2016 Clinical application of high-intensity focused ultrasound in cancer therapy *J. Cancer* **7** 225–31

[37] Snipstad S *et al* 2017 Ultrasound improves the delivery and therapeutic effect of nano-particle-stabilized microbubbles in breast cancer xenografts *Ultrasound Med. Biol.* **43** 2651–69

[38] Leger P, Limper A H and Maldonado F 2017 Pulmonary toxicities from conventional chemotherapy *Clin. Chest Med.* **38** 209–22

[39] Yang Q, Li P, Ran H, Wan J, Chen H, Chen H, Wang Z and Zhang L 2019 Polypyrrole-coated phase-change liquid perfluorocarbon nanoparticles for the visualized photothermal-chemotherapy of breast cancer *Acta Biomater.* **90** 337–49

[40] Wilhelm S, Tavares A J, Dai Q, Ohta S, Audet J, Dvorak H F and Chan W C W 2016 Analysis of nanoparticle delivery to tumours *Nat. Rev. Mater.* **1** 16014

[41] Zhao R, Liang X, Zhao B, Chen M, Liu R, Sun S, Yue X and Wang S 2018 Ultrasound assisted gene and photodynamic synergistic therapy with multifunctional FOXA1-siRNA loaded porphyrin microbubbles for enhancing therapeutic efficacy for breast cancer *Biomaterials* **173** 58–70

[42] Cheng Y J, Nie X Y, Ji C C, Lin X X, Liu L J, Chen X M, Yao H and Wu S H 2017 Long-term cardiovascular risk after radiotherapy in women with breast cancer *J. Am. Heart Assoc.* **6**

[43] Taylor C W and Kirby A M 2015 Cardiac side-effects from breast cancer radiotherapy *Clin. Oncol. (R. Coll. Radiol.)* **27** 621–9

[44] van Nimwegen F A, Schaapveld M, Janus C P, Krol A D, Petersen E J, Raemaekers J M, Kok W E, Aleman B M and van Leeuwen F E 2015 Cardiovascular disease after Hodgkin lymphoma treatment: 40-year disease risk *JAMA Intern. Med.* **175** 1007–17

[45] Andratschke N, Maurer J, Molls M and Trott K R 2011 Late radiation-induced heart disease after radiotherapy. Clinical importance, radiobiological mechanisms and strategies of prevention *Radiother. Oncol.* **100** 160–6

[46] Piroth M D *et al* 2019 Heart toxicity from breast cancer radiotherapy: Current findings, assessment, and prevention *Strahlenther. Onkol.* **195** 1–12

Chapter 7

Radiotherapy and breast cancer nanomedicine

Madhuri Anuje, Joanna Bauer and Nanasaheb D Thorat

Cancer is the most common cause of death in economically developed countries and the second most common cause of death in developing countries. The global incidence of breast cancer has increased by over 20% since 2008. Breast cancer is the most common and lethal cancer type in women worldwide. Many treatment options for cancer exist; the primary ones include surgery, chemotherapy, radiation therapy, and palliative care. One of the major disadvantages for radiotherapy is destruction of normal tissue adjacent to tumors and in the path of the beam; chemotherapy is a useful treatment for tumor metastasis. Although there is a large library of drugs that can be used in cancer treatment, the problem is selectively killing all the cancer cells while reducing collateral toxicity to healthy cells. Nanoparticles loaded with drugs can be designed to improve efficacy while reducing morbidity. The extremely large surface-area-to-volume ratio of nanocarriers provides an opportunity to manipulate their surface properties for improved treatment. There are several types of drug carriers used for breast cancer commonly available. As of today, only few nanomedicine products have gained US Food and Drug Administration (FDA) approval, and Doxil and Abraxane are the two most successful nanoformulations already widely used for breast cancer treatment.

7.1 Radiotherapy

7.1.1 History of radiotherapy

William Roentgen discovered x-rays in 1895 while studying cathode rays in a gas discharge tube. He observed that another type of radiation was produced that could be detected outside the tube. This radiation could penetrate opaque substances, produce fluorescence, blacken a photographic plate and ionize a gas. He named this unknown radiation x-rays. He also noted that these x-rays could be used to image bones. In fact one of the first known x-ray images ever produced was of his wife Bertha's left hand.

doi:10.1088/2053-2563/ab2907ch7

In 1896, Henri Becquerel was using naturally fluorescent minerals to study the properties of x-rays, which had been discovered in 1895. Using a method similar to that of Roentgen, Becquerel surrounded several photographic plates with black paper and florescent salts like uranium. With the intention of further advancing the study of x-rays, Becquerel intended to place the concealed photographic paper in the sunlight believing that the uranium absorbed the Sun's energy and then emitted it. Unfortunately, he had to delay his experiment because the skies over Paris were overcast. He placed the wrapped plates into a dark desk drawer. After a few days Becquerel returned to his experiment unwrapping the photographic paper and developing it, expecting only a light imprint from the salts. Instead, the salts left very distinct outlines in the photographic paper suggesting that the salts, regardless of lacking an energy source, continually fluoresced. What Becquerel had discovered was radioactivity.

Becquerel's doctoral students Marie Curie with her husband Pierre Curie showed that Becquerel rays could be measured using ionizing techniques and radiation intensity is directly proportional to the amount of uranium in a substance. They also isolated the first known radioactive elements polonium and radium in 1898.

Less than two months after the discovery of x-rays, a medical student in Chicago named Emil Grubbe noted pealing of his hands on exposure to x-rays. He convinced his professor to allow him to assemble his x-ray machine in Chicago in 1896, and that same year used it to treat a woman named Rose Lee with recurrent carcinoma of the breast. By 1960, Grubbe had instructed over 7000 other doctors in the medical use of x-rays i.e. radiotherapy.

During the period, roughly 1920–30, Claude Regaud argued the differential effect of x-rays on cancer and normal tissues could be best obtained by giving the treatment slowly. For cxample, healing was very much better when skin cancer was treated over a period of a week than over a day. This approach, known as fractionation, is one of the most important underlying principles in radiation therapy. To this day, fractionation lies at the heart of many treatment programs currently used in radiation oncology.

An important limitation of the early x-ray machines was their inability to produce high energy, deeply penetrating beams. It was thus difficult to treat deep-seated tumors without excessive skin reactions. Many early advocates of radiation therapy thus relied instead on the placement of radioactive sources in close proximity or even within the tumor, a technique known as brachytherapy. This modality dates back to when Pierre Curie suggested to Danlos that a radioactive source could be inserted into a tumor. It was found that the radiation caused the tumor to shrink.

Ralston Patterson who was a radiologist and who had keen interest in newer advances related to the field was appointed as Director of the Holt Radium Institute in 1931, went on to build a world-recognized center for the treatment of cancer by radiation. Following initial interest in brachytherapy in Europe and the US, its use declined in the middle of the twentieth century due to the problem of radiation exposure to operators from the manual application of the radioactive sources. However, the development of remote after loading systems, which allowed the radiation to be delivered from a shielded safe, in the 1950s and 1960s, reduced the risk of unnecessary radiation exposure to the operator and patients.

In 1949, Dr Harold E Johns, a Canadian medical physicist sent a request to the National Research Council (NRC) asking them to produce Cobalt-60 isotopes for use in a cobalt therapy unit prototype. On October 27, 1951, the world's first cancer treatment with Cobalt-60 radiation took place at Victoria Hospital for a 43-year-old cervical cancer patient. This marked an important milestone in the fight against cancer. Despite advances made in radiation therapy technology, the Cobalt-60 unit remains the world's main radiotherapy machine. Due to its cost effectiveness, reliability and ease of use, it is prevalent in developing countries. Cobalt-60 technology is currently used to treat roughly 70 per cent of the world's cancer cases treated by radiation.

The 1970s and 1980s were characterised by the introduction of innovative devices delivering proton beam. Even if their first clinical use was dated in 1954, it was only by the late seventies that computer-assisted accelerator for protons was successfully applied to treat a different kind of tumor. The major advantage in the use of ion beams is its controllability, which allows providing a superior tool for cancer therapy and difficult-to-treat benign diseases [1].

An exciting development has been the introduction of high energy (megavoltage) treatment machines, known as Linear Accelerators or LINACS. Such machines were capable of producing high energy, deeply penetrating beams, allowing for the very first time treatment of tumors deep inside the body without excessive damage to the overlying skin and other normal tissues. In subsequent years, the field of radiation oncology experienced multiple technologic revolutions. With the advent of computers and newer technological advances, the radiotherapy planning systems underwent a drastic makeover. It was in the 1990s that 3D conformal radiotherapy, a form of radiation therapy where the fields used are designed such that the radiation dose is mostly delivered to the tumor, while the surrounding tissues receive little to no radiation dose. Intensity-modulated radiation therapy (IMRT) is an advanced form of three-dimensional conformal radiotherapy (3D CRT). It uses sophisticated software and hardware to vary the shape and intensity of radiation delivered to different parts of the treatment area. Today, radiation therapy is in the midst of yet another important technologic revolution, namely Image-Guided Radiation Therapy (IGRT).

7.1.2 Mechanism of action of radiotherapy

Radiotherapy mainly acts by killing the tumor cells or by halting their division. This can be the result of direct interaction or indirect interaction as shown in figure 7.1. Several studies have shown that the main pathway of DNA damage, from both x-rays and gamma rays, is through the production of water radicals with 70% of damage caused by free radicals and other reactive oxygen species (ROS) (e.g. OH., NO·, H·, and H_2O_2) and 30% due to secondary electrons and direct fragmentation of the DNA [2]. A **free radical** is an atom or molecule carrying an unpaired orbital electron in the outer shell [3]. An orbital electron not only revolves around the nucleus of an atom but also spins around its own axis. The spin may be clockwise or counterclockwise [3]. In an atom or molecule with an even number of electrons, spins are paired; that is, for every electron spinning clockwise, there is another one spinning counterclockwise. This state is associated with a high degree of chemical

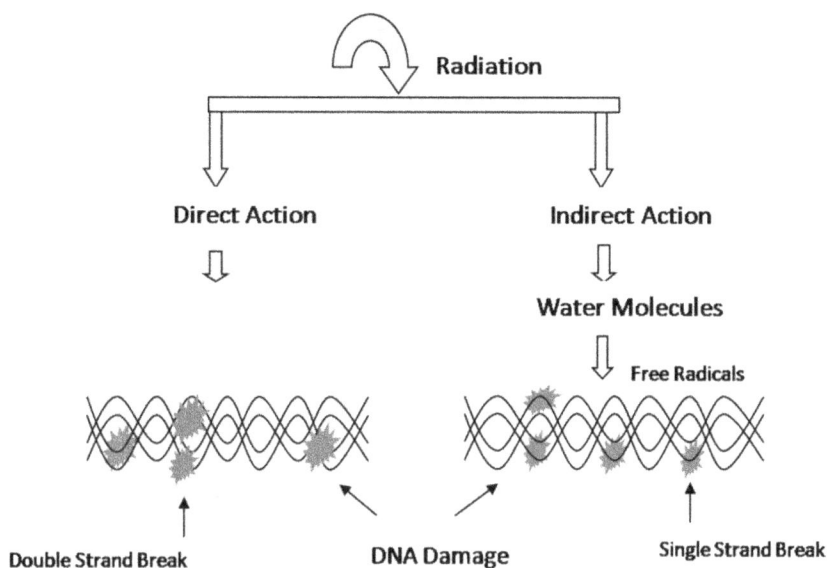

Figure 7.1. Interaction mechanism of radiation with matter.

stability. In an atom or molecule with an odd number of electrons, there is one electron in the outer orbit for which there is no other electron with an opposing spin; this is an unpaired electron [3]. This state is associated with a high degree of chemical reactivity. 80% of a cell is composed of water. When radiation interacts with water ionization occurs as given below,

$$H_2O \rightarrow H_2O^+ + e^-$$

H_2O^+ is an ion radical. An **ion** is an atom or molecule that is electrically charged because it has lost an electron. A free radical contains an unpaired electron in the outer shell, making it highly reactive. H_2O^+ is charged and has an unpaired electron; consequently, it is both an ion and a free radical. The primary ion radicals have an extremely short lifetime, on the order of 10^{-10} s. In the case of water, the ion radical reacts with another water molecule to form the highly reactive hydroxyl radical

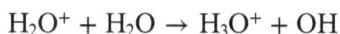

$$H_2O^+ + H_2O \rightarrow H_3O^+ + OH$$

The hydroxyl radical possesses nine electrons; therefore, one of them is unpaired. It is a highly reactive free radical and can diffuse a short distance to reach a critical target in a cell. For example, it is thought that free radicals can diffuse to DNA from within a cylinder with a diameter about twice that of the DNA double helix (4 nm). It is estimated that about two thirds of the x-ray damage to DNA in mammalian cells is caused by the hydroxyl radical [3].

The damage caused to DNA by radiation can result in a range of various lesions, including base damage, single strand breaks (SSBs) or, less frequently, double strand breaks (DSBs). In most cases base damage and SSBs can be effectively repaired by the cell repair mechanism, whereas DSBs, especially when induced at high levels, are difficult to get successfully repaired and therefore more damaging to cells [2].

7.1.3 Classification of radiotherapy

Radiation is delivered in different ways either from outside the body or inside the body. External-beam radiotherapy (teletherapy) or from outside the body (brachytherapy). Teletherapy involves directing high energy radiations from distance (usually source to surface distance 100 cm for LINAC and 80 cm for Telecobalt). External-beam therapy is usually delivered by different techniques such as IMRT, IGRT, SBRT, EBRT etc.

In the case of internal beam therapy (brachytherapy) radiation is usually given within or near the tumor inside the body by using a brachytherapy machine. It involves radioactive material placed directly into the body. A relatively high dose of radiation is given to the tumor whilst healthy surrounding tissue only gets a very small amount of radiation. In some types of cancer these implants may be left in the body permanently. Based on dose rate it is classified as LDR, MDR, HDR and PDR. There are two main forms of brachytherapy—intracavitary treatment and interstitial treatment. With intracavitary treatment, the radioactive sources are put into a space near where the tumor is located, such as the cervix and vagina. With interstitial treatment, the radioactive sources are put directly into the tissues, such as the prostate and breast. Another use of brachytherapy is surface mold brachytherapy, which can be used externally to treat some skin cancers. Classification is shown in figure 7.2. Based on penetration power and ionization they are classified as particulate and photon based radiotherapy.

(A) Photon based radiotherapy

Photons carry less mass and no charge, so their penetration power is less. X-rays and gamma rays are used to treat various cancers. X-rays or gamma rays are sparsely ionizing radiation and considered as low LET (linear energy transfer) electromagnetic radiation. X-rays and gamma rays have exponential dose deposition with tissue depth, therefore, a fraction of total dose is delivered to healthy tissue lying in front of and behind the target. The amount of radiation dose received is being limited by sparing normal tissue in different angles. X-rays are generated by a device

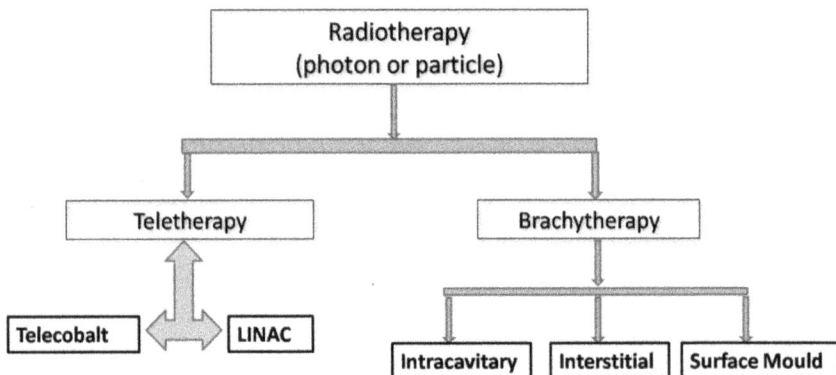

Figure 7.2. Classification of radiotherapy.

that excites electrons (e.g. cathode ray tubes and linear accelerators). In the x-ray therapy, for modern high-energy linear accelerators, that is used to produce photon beams in the range of 4–25 MV, while gamma rays originate from the decay of radioactive substances (e.g. cobalt-60, radium and cesium).

Technological advances in treatment delivery

Technological advances incorporating new imaging modalities, more powerful computers and software, and new delivery systems such as advanced linear accelerators have helped to achieve goal of radiotherapy i.e. much dose to the tumor whilst sparing normal tissue.

(I) Conformal three-dimensional radiation therapy

One of the major advancements in radiation oncology in the early 1990s was the development of conformal three-dimensional (3D CRT) radiation therapy based on CT imaging. In this technique, the prescribed dose volume was made to conform to the target volume and spare more normal tissues. The aim is to deliver radiation to the gross tumor volume (GTV), with a margin for microscopic tumor extension called the clinical target volume (CTV), and a further margin uncertainties from organ motion and setup variations called the planning target volume (PTV) [4].

Although 3D CRT calls for optimal dose distribution, there are many obstacles to achieving these objectives. The most major limitation is the knowledge of the tumor extent [5]. Despite the modern advances in imaging, the clinical target volume (CTV) is often not fully discernible. Depending on the invasive capacity of the disease, what is imaged is usually not the CTV [5]. It may be what is called the gross tumor volume (GTV). Thus, if the CTVs drawn on the cross-sectional images do not fully include the microscopic spread of the disease, the 3D CRT loses its meaning of being conformal [5]. If any part of the diseased tissue is missed or seriously underdosed, it will inevitably result in failure despite all the care and effort expended in treatment planning, treatment delivery, and quality assurance. From the tumor control probability TCP point of view, accuracy in localization of CTV is more critical in 3D CRT [5].

(II) Intensity-modulated radiation therapy

The term intensity-modulated radiation therapy (IMRT) refers to a radiation therapy technique in which nonuniform fluence is delivered to the patient from any given position of the treatment beam to optimize the composite dose distribution [5]. With the development of computer-controlled modulation of radiation intensity and the creation of inverse planning software, radiation oncologists now can create irregularly-shaped radiation beams that precisely conform to the tumor and are dynamic enough to avoid critical organs. In IMRT, little leaflets called multi-leaf collimators (MLC) that move in and out of the radiation beam are used to control the beam pattern during delivery, ensuring that the shape of the beam is more tightly wrapped around the tumor. Through IMRT, radiation oncologists are able to increase the radiation dose to specific tumor sites, deal more effectively with resistant cancer cells and extend the area covered by high-dose radiation to the lymph nodes. As opposed to standard planning techniques, where the dose distribution can only be modified by means of a trial and error approach (changing for instance the field weight, angle and shape), with IMRT, the radiation oncologist designates the doses and dose/volume constraints for the tumor and the surrounding normal organs and the Treatment Planning System (TPS) determines the optimal fluence of each field resulting in a tailored dose distribution (inverse planning). In the past, IMRT was usually delivered using a conventional LINAC with static field geometry. Developments in IMRT techniques have focused on reducing treatment times with arc therapy by converting multiple static fields IMRT into continuously rotating gantry intensity modulation [6]. IMRT is now available in many clinical departments and can be delivered by linear accelerators with static or dynamic multi-leaf collimators or tomotherapy machines. This has allowed improvements in the therapeutic ratio for several tumor sites, such as head and neck cancers, prostate cancers and gynecological cancers [7].

(III) Image-guided radiotherapy
Image-guided radiation therapy (IGRT) may be defined as a radiation therapy procedure that uses image guidance at various stages of its process: patient data acquisition, treatment planning, treatment simulation, patient setup, and target localization before and during treatment [5]. These procedures use imaging technology to identify and correct problems arising from inter and intrafractional variations in patient setup and anatomy, including shapes and volumes of treatment target, organs at risk, and surrounding normal tissues [5]. Imaging systems have been developed that are accessible in the treatment room or mounted directly on the linear accelerator. The accelerator-mounted imaging systems are called on-board imagers (OBIs). One such example is with daily cone-beam CT scans acquired before each treatment. The improved accuracy has made dose escalation feasible, and this has allowed an improvement in the therapeutic ratio for several tumor sites, such as head and neck cancers and prostate cancers [7].

(IV) Stereotactic body radiation therapy
The above technological advancements have enabled SBRT, which precisely delivers very high individual doses of radiation over only a few treatment fractions to ablate small, well-defined primary and oligometastatic tumors anywhere in the body [7]. SBRT systems are capable of producing very conformal treatment plans with a steep dose gradient outside the target. This technique makes possible safe and efficacious treatment across a broad array of anatomic locations, in proximity to critical organs, and even adjacent to or within prior RT fields [6].

(B) Particulate radiotherapy (electron, proton, neutron beams)
Electrons having less penetration power even though it has been widely used for superficial treatments since the early 1950s. The most clinically useful energy range for electrons is 6–20 MeV. At these energies, the electron beams can be used for treating superficial tumors (<5 cm deep) with a characteristically sharp dropoff in dose beyond the tumor. The principal applications are (a) the treatment of skin and lip cancers, (b) chest wall irradiation for breast cancer, (c) administering boost dose to nodes, and (d) the treatment of head and neck cancers [5].

External-beam radiotherapy carried out using heavier particles such as neutrons, protons and heavy ions (helium carbon). The proton is a newly introduced particle for the treatment of cancer. Its dose distribution with depth is unique. For a mono-energetic proton beam, there is a slow increase in dose with depth initially, followed by a sharp increase near the end of range. This sharp increase or peak in dose deposition at the end of particle range is called the Bragg peak [5]. Due to this property of protons or heavy ions there is maximum destructive dose deposition at the tumor site with less radiation to healthy tissue underlying before the tumor. These have particular clinical use in pediatric tumors and in adult's tumors located near critical structures such as spinal cord and skull base tumors, where maximal normal tissue sparing is crucial [7]. Protons can be generated using cyclotrons and synchrotrons while heavy ions are produced using synchrocyclotrons and synchrotrons.

Neutron beams are generated inside neutron generators after proton beams are deflected to a target. They have high LET and can cause more DNA damage than photons. The limitations have been mainly due to difficulty in generating neutron particles as well as the construction of such treatment facilities [7].

7.1.4 Radiation therapy versus radioisotope therapy

Although the intensity, location and timing for external radiation can be well controlled and modulated, its main disadvantages include: (1) the destruction of normal tissue adjacent to tumors and in the path of the beam; (2) the need of high radiation doses for penetrating tissues with a large field or volume; (3) prolonged treatment with the requirement of daily hospital visits for 5–6 weeks; and (4) the use of only selected radiation sources due to the technical requirements and limitations of radiation devices and radiation sources (e.g. high energy x-rays) [8]. Another treatment option available for certain types of cancer is the use of targeted radionuclide therapy as part of nuclear medicine, which is based on administering radioactive substances to patients. Just like chemotherapy, this therapy is a systemic treatment, reaching cells throughout the body by traveling through the bloodstream. However, unlike chemotherapy, these radioactive substances specifically target diseased cells, thus reducing potential side effects. The use of radioisotopes (radionuclide) in clinical practice is well established. Radioisotopes emit energy from the nucleus and generate ionized atoms and free radicals to induce single strand cleavages in DNA [9]. These radioisotopes can be categorized into, α and β particles. Radioisotopes applied in the clinical oncology include beta-emitters, like ^{186}Re, ^{188}Re, ^{166}Ho, ^{89}Sr and ^{90}Y, as well as alpha-emitters, like ^{225}Ac, ^{211}At, and ^{213}Bi, as shown in table 7.1. When used *in vivo*, beta-emitters have profound tissue penetration (20–130 mm) but low linear energy transfer, whereas alpha-emitters have limited penetration (50–80 μm) but a short half-life and the ability to inflict more damage to the cells.

7.1.5 Nanoparticle mediated radionuclide therapy

Targeted radionuclide therapy is often limited by insufficient delivery of radio-nuclide to tumor sites using the currently available targeting strategies. To maximize the therapeutic index and to minimize the outcome of toxicity, it is very important to deliver the radionuclides to the right site at the right concentration and at the right time. Major advantages of nanocarriers are that they can be prepared in sizes <100 nm, and selectively increase the localization of drugs and radionuclides in the tumor tissue while reducing toxicity to normal tissue. There are three generations of nanocarriers: (i) the first generation of nanocarriers (passive targeting) which are rapidly trapped in the reticuloendothelial system (RES) organs (e.g. liver and/or spleen); (ii) the second generation of pegylated nanocarriers (passive targeting), which can evade the RES of the liver and spleen, enjoys a prolonged circulation in the blood and allows for passive targeting through the enhanced permeability and retention (EPR) effect in leaky tumor tissues; (iii) the third generation of nano-carriers (active targeting) has a bioconjugated surface modification using specific antibodies or peptides to actively targeted specific tumor or tissues [10].

Many of the radioisotopes undergo rapid clearance by the kidney. In particular, renal clearance is size dependent, for which size smaller than 5 nm will be excreted rapidly. Radioisotopes as small molecules suffer short circulation time in blood and are unable to achieve therapeutic effect. Another possible elimination process of the radioisotopes is by opsonization, which is an immune process where macromolecules are cleared by the mononuclear phagocyte system (MPS). This biological elimi-nation mechanism can be avoided by loading or attaching nanocarriers to radio-isotope. For example, the physical half-life of ^{89}Sr is 50.5 days, but it is cleared from plasma with an average half-life of 47 h.

Nanoparticles such as liposomes, micelles, or polymeric complex are usually more than 10 nm, which greatly decreases the renal clearance and increases their half-life in blood and the increased size effect. Also, the nanocarriers can prevent opsoniza-tion through PEGylation. The presence of polyethylene glycol (PEG) on the surface of nanoparticles produces steric hindrance, which prevents the adsorption of opsonins. Opsonins are the serum protein present in the blood which adsorb on foreign bodies entering the blood. This particular characteristic of nanocarriers helps prolong the half-life of radiotherapeutic agents in blood. In a tumor-bearing mice model, the half-lives of ^{111}In- and ^{177}Lu-PEGylated liposomes in blood were 10.2 and 11.5 h, respectively, whereas the half-life of ^{111}In-DTPA in blood was extremely short at no longer than 2 h [8]. Also, the retention time of radionuclide within tumors can be increased through improved EPR effect. Abnormal tumor vasculature possesses leaky arterial walls which allow nanoparticles to easily penetrate tumor tissue with enhanced retention time.

7.1.6 Nanoparticles as radiosensitizers

In radiotherapy, tumor cells are killed by high energy x-ray or gamma rays. In the case of deep sited tumors, healthy tissues that lie in the track of photons are exposed to radiation resulting in severe side effects. The main challenge that radiation oncologists and medical physicists face is to minimize normal tissue radiation dose and to increase tumor dose. Radiotherapy needs some improvement in radiation delivery techniques in order to reduce injury to the surrounding tissues. To overcome this problem, radio sensitizers are an appropriate solution. Radio sensitizers are

Table 7.1. Radioisotopes used for therapy [8].

Radionuclide	Particles emitted	Particle's half-life	Range of particles (mm)	Energy of particles (keV)
^{211}At	α	7.2 h	0.08	6000
^{225}Ac	α and β	10 days	0.1	6000–8000
^{212}Bi	α and β	60.6 min	0.09	6000
^{213}Bi	α and β	46 min	<0.1	6000
^{223}Ra	α and β	11.4 days	<0.1	6000–7000
^{212}Pb	α and β	10.6 h	<0.1	7800
^{149}Tb	α	4.2 h	<0.1	400
^{131}I	β and γ	193 h	2	610
$_{90}$Y	β	64 h	12	2280
^{67}Cu	β and γ	62 h	1.8	577
^{186}Re	β and γ	91 h	5	1080
^{177}Lu	β and γ	161 h	1.5	496
^{64}Cu	β	12.7 h	2	1670
^{89}Sr	B	50.5 days	5.5	1460

adjunctive treatments which make tumor cells more susceptible to radiation. They are designed to improve tumor cell killing while having much less effect on normal tissues. Many substances and materials have been reported as radio sensitizers.

Nanoparticles, due to their smaller size, have more cell penetration and fewer adverse effects than conventional radio sensitizers as well as enhanced EPR effect. Among nanomaterials which have this radiosensitizing nature, carbon nanotubes, gold nanoparticles (GNPs) and other metallic nanoparticles should be mentioned [11]. High-Z NP scans intensify the production of secondary electrons and ROS that in turn enhance radiation therapy effects. The most studied NPs are gold-based NPs (GNPs) that have been widely described in particular. The use of lanthanide-based NPs, titanium oxide nanotubes or cadmium selenide quantum dots are also reported. For example, gadolinium-based NPs, besides their high-Z, offer an innovative approach due to their capacity to act as powerful contrast agents in MRI. Interestingly, some authors used silver-based NPs to take advantage of its excellent surface-enhanced Raman scattering and broad-spectrum antimicrobial activities [12]. One method to amplify weak Raman signals is to employ surface-enhanced Raman scattering (SERS). SERS uses nanoscale roughened metal surfaces. In antimicrobial activity small NPs penetrate into cell walls resulting in the damage of bacteria, *Escherichia coli* (*E. coli*).

Metal oxide such as iron oxide is also effective in radiosensitization due to its ROS generation, occurs through the release of iron ions into the cytosol where immediate chelation by citrate or adenosine phosphate will take place. The chelated iron ions can participate in the Haber–Weiss chemistry and thus, catalyze the formation of the highly reactive OH, which is free radical that damages cellular membranes, proteins and DNA [13]. Quantum dots discovered in the early 1980s are

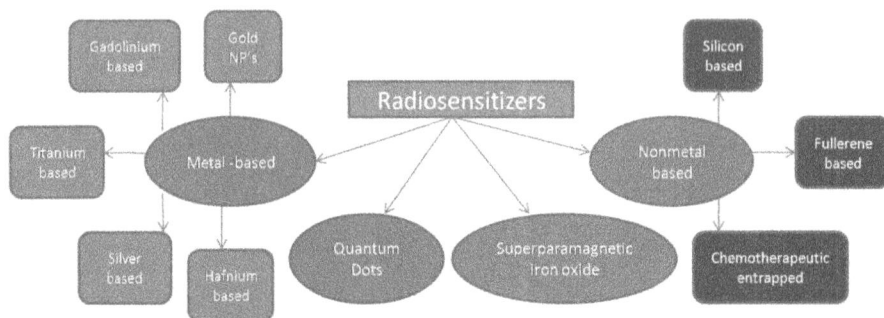

Figure 7.3. Classification of nanoparticles used as a radiation sensitizer [14].

nanocrystals made of semiconductor materials that display quantum mechanical properties due to their small size. Their semiconductor properties are less than those displayed by bulk semiconductors. Quantum dots made from CaF, LaF, ZnS or ZnO have been suggested for use as radio sensitizers. Development of photosensitizing quantum dots has been a very active area of interest. The mechanism of action for these is based on the principle of generation of radicals upon absorption of visible light by the quantum dots [14]. Silica has been used as a carrier or coating material in nanoparticles containing heavy metals for radiosensitization such as gold, FeO_4 or multicomponent cores. Moreover, nanoparticles made of silica alone have also been tested for their potential role in radiosensitization. C_{60} is a fullerene, with unique globular structure consisting of 32 different member rings and containing a total of 60 carbon atoms. Fullerene C_{60} possesses potent anticancer activities and induces certain markers of autophagy in cancer cells. Polymeric nanoparticles have also been formulated using various chemotherapeutic agents either alone or in combination to serve as radio sensitizers. Paclitaxel is a potent chemotherapeutic agent that is also known to be a cell-cycle-specific radiosensitizer. This is because it arrests cell-cycle progression at G2/M, a stage in which the cells are most susceptible to radiation induced damage [14]. Figure 7.3 specifies classifications of nanoparticles for use of radiotherapy sensitizer.

(A) Role of nanoparticles for radiosensitization
In the case of radiosensitizing NPs the core is usually made up of high Z materials such as silver, lanthanide and mostly gold in order to exploit the increased photon absorption [12]. The densely packed metal particles can selectively scatter and/or absorb the high energy gamma/x-ray radiations. This allows for better targeting of cellular components within the tumor tissues allowing for more localized and consolidated damage. When photons interact with nanoparticles they undergo various interactions such as Rayleigh scattering, photoelectric, Compton and pair production depending on energy of interacting photons. In the photoelectric effect the Auger electrons or fluorescent photons are produced when the ejected electrons are replaced with electrons dropping from the higher orbits and energy is released. The fluorescent photons are low energy but have higher coverage range. The Auger electrons have a much shorter range of coverage but can generate much higher

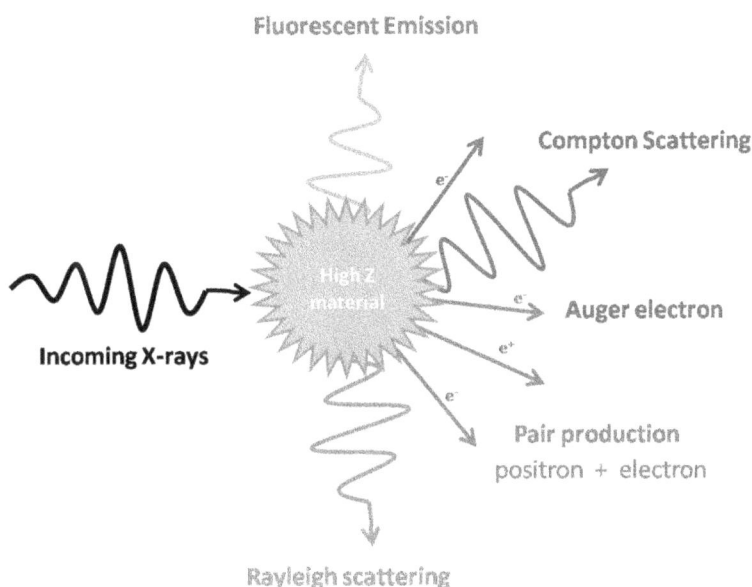

Figure 7.4. Different interaction mechanism of photon with nanoparticle.

ionization density at a localized area [14]. The photoelectric effect is dependent on atomic number, the greater the value of Z, the greater will be the effect of photoelectric interaction. So the resultant secondary species (which is produced from primary photons or electrons) will be higher. Secondary electrons produced by the radiation or by the NPs may also induce further electron emission from NPs; also, it may ionize the water molecule to produce ROS which is highly reactive. All the secondary species may diffuse and damage DNA. Interaction of x-ray with high Z material is shown in figure 7.4.

7.2 Cancers and their staging based treatment modality

As per GLOBOCAN 2018, there will be an estimated 18.1 million new cancer cases (17.0 million excluding nonmelanoma skin cancer) and 9.6 million cancer deaths (9.5 million excluding nonmelanoma skin cancers) in 2018. In both sexes combined, lung cancer is the most commonly diagnosed cancer (11.6% of the total cases) and the leading cause of cancer death (18.4% of the total cancer deaths), closely followed by female breast cancer (11.6%), prostate cancer (7.1%), and colorectal cancer (6.1%) for incidence and colorectal cancer (9.2%), stomach cancer (8.2%), and liver cancer (8.2%) for mortality [15]. Table 7.2 shows new cases and deaths due to cancer estimated in 2018.

For proper cancer treatment it is important to know about tumor extent and growth. Stage refers to the extent of the cancer, such as how large the tumor is, and if it has spread. Staging is important to decide plan treatment, including the type of surgery and/or whether chemotherapy or radiation therapy are needed, e.g. An early-stage cancer may call for surgery while an advanced-stage cancer may need

Table 7.2. New cases and deaths for 36 cancers and all cancers combined in 2018 (Source: [15]).

Cancer site	No. of new cases (% of all sites)	No. of deaths (% of all sites)
Lung	2 093 876 (11.6)	1 761 007 (18.4)
Breast	2 088 849 (11.6)	626 679 (6.6)
Prostate	1 276 106 (7.1)	358 989 (3.8)
Colon	1 096 601 (6.1)	551 269 (5.8)
Nonmelanoma of skin	1 042 056 (5.8)	65 155 (0.7)
Stomach	1033701 (5.7)	782 685 (8.2)
Liver	841 080 (4.7)	781 631 (8.2)
Rectum	704 376 (3.9)	310 394 (3.2)
Esophagus	572 034 (3.2)	508 585 (5.3)
Cervix uteri	569 847 (3.2)	311 365 (3.3)
Thyroid	567 233 (3.1)	41 071 (0.4)
Bladder	549 393 (3.0)	199 922 (2.1)
Non-Hodgkin lymphoma	509 590 (2.8)	248 724 (2.6)
Pancreas	458 918 (2.5)	432 242 (4.5)
Leukemia	437 033 (2.4)	309 006 (3.2)
Kidney	403 262 (2.2)	175 098 (1.8)
Corpus uteri	382 069 (2.1)	89 929 (0.9)
Lip, oral cavity	354 864 (2.0)	177 384 (1.9)
Brain, nervous system	296 851 (1.6)	241 037 (2.5)
Ovary	295 414 (1.6)	184 799 (1.9)
Melanoma of skin	287 723 (1.6)	60 712 (0.6)
Gallbladder	219 420 (1.2)	165 087 (1.7)
Larynx	177 422 (1.0)	94 771 (1.0)
Multiple myeloma	159 985 (0.9)	106 105 (1.1)
Nasopharynx	129 079 (0.7)	72 987 (0.8)
Oropharynx	92 887 (0.5)	51 005 (0.5)
Hypopharynx	80 608(0.4)	34 984 (0.4)
Hodgkin lymphoma	79 990 (0.4)	26 167 (0.3)
Testis	71 105(0.4)	9 507 (0.1)
Salivary glands	52 799 (0.3)	22 176 (0.2)
Anus	48 541 (0.3)	19 129 (0.2)
Vulva	44 235 (0.2)	15 222 (0.2)
Kaposi sarcoma	41 799 (0.2)	19 902 (0.2)
Penis	34 475 (0.2)	15 138 (0.2)
Mesothelioma	30 443 (0.2)	25 576 (0.3)
Vagina	17600 (0.1)	8062 (0.1)
All sites excluding skin	17 036 901	9 489 872
All sites	18 078 957	9 555 021

chemotherapy or radiotherapy. Cancer is often staged twice. The first rating is done before treatment and is called the clinical stage. The second rating is done after treatment, such as surgery, and is called the pathologic stage. The pathologic stage is more precise regarding the extent of the cancer [16].

The TNM staging system is most often used to stage cancer. In this system, the letters T, N and M describe a different area of cancer growth. The T score is a rating of the extent of the primary tumor. The primary tumor is the first mass of cancer cells in the body. If not treated, the primary tumor can grow large. It can also grow through the layers of tissue in which it started. This is called the tumor extension. Once the tumor has grown through the outer edge of a structure, it can grow into other nearby structures. This is called invasion. T scores are based on the presence, size, and extension of the primary tumor. A TX score means that the primary tumor cannot be assessed. A T0 score means there is no primary tumor. It is possible to have cancer but not have a primary tumor. A T score means there are abnormal or cancer cells, but there is no chance for the cells to spread to distant sites. Scores of T1, T2, and so on are based on the primary tumor's size, extension, or both. Higher values mean a greater extent of the cancer [16]. The N category reflects the extent of cancer within nearby lymph nodes. Lymph nodes are small disease-fighting organs that filter lymph. Lymph is a clear fluid within tissue that gives cells water and food. It also collects waste from cells and has white blood cells that fight germs. Lymph drains from tissue into lymph vessels that transport it to the lymph nodes. Cancer cells can invade lymph vessels and travel to lymph nodes. Once in lymph nodes, the cancer cells can multiply and form new tumors [16]. N scores are based on whether there's cancer in nearby lymph nodes and the number or region of nodes with cancer. An NX score means that the lymph nodes can't be assessed. An N0 score means that no cancer was found in the lymph nodes. N1, N2, and N3 scores are based on the number of nodes with cancer or which nodal groups have cancer. Higher values mean a greater extent of the cancer. The M category tells you if the cancer has spread to distant sites. Such sites include distant lymph nodes beyond nearby lymph nodes. Cancer cells can break off the primary tumor and spread to distant sites. This process is called metastasis. Cancer cells can spread to distant sites through lymph or blood. M0 means there is no cancer in distant sites. M1 means there is cancer in distant sites. For each person, stage of cancer can be determined by combining the T, N, M results and other factors specific to the cancer. Stage 0 means there's no cancer, only abnormal cells with the potential to become cancer. This is also called carcinoma *in situ* [16]. Stage I means the cancer is small and only in one area. This is also called early-stage cancer. Stage II and III mean the cancer is larger and has grown into nearby tissues or lymph nodes. Stage IV means the cancer has spread to other parts of your body. It is also called advanced or metastatic cancer.

There are many types of cancer treatment. The types of treatment will depend on the type of cancer and how advanced it is. Treatment can be delivered either single or in combination, such as surgery with chemotherapy and/or radiation therapy. When used to treat cancer, surgery is a procedure in which a surgeon removes cancer from the body. Chemotherapy is treatment with cancer-killing drugs that may be given intravenously or orally. The drugs travel through the bloodstream to stop or slow the growth of cancer cells in most parts of the body. Because chemotherapy kills both fast-growing cancer cells and healthy cells, side effects can include hair loss, mouth sores and nausea. These typically subside upon completion of chemotherapy. Radiation therapy uses high-energy radiation to shrink tumors. It causes the cancer cells to stop dividing or die by damaging their DNA. 'About half of all cancer patients receive some type of radiation therapy sometime during the course of their treatment,' reports the National Cancer Institute. The side effects of radiation include redness on the skin in the area that the radiation passes through, fatigue that does not get better with rest, and sometimes loss of appetite and nausea. Most skin reactions and other side effects improve when treatment ends.

7.3 Cancer nanomedicine

Cancer continues to be one of the most difficult global healthcare problems. Surgery is primarily employed to remove the tumor mass, and plays a pivotal role in the treatment of cancers, but most patients are already in locally advanced or metastatic disease at the time of diagnosis, lacking the opportunity for radical surgery. Radiotherapy, which eliminates cancer cells by DNA damaging, is also mainly applied in treating a tumor confined to a discrete anatomical area. As a systemic therapy, anticancer drugs, either employed in chemotherapy or biotherapy, are able to treat cancer throughout the body, especially for patients with metastases [17].

Although there is a large library of drugs that can be used in cancer treatment, the problem is selectively killing all the cancer cells while reducing collateral toxicity to healthy cells. Nanoparticles loaded with drugs can be designed to improve efficacy while reducing morbidity. Nanomedicine has ushered in a new era for drug delivery by improving the therapeutic indices of the active pharmaceutical ingredients engineered within nanoparticles [18]. In order to obtain optimal therapeutic effects, the right drugs should be delivered to the right location of the right patient at the right time with the right concentration [19]. The justification of cancer nanomedicine relies on enhanced permeation (EP) and retention (R) effect and the capability of intracellular targeting due primarily to size after internalization (endocytosis) into the individual target cells. The EPR effect implies improved efficacy [20]. The tumor vasculature is highly heterogeneous in distribution and also in permeability. Another characteristic of neoplastic tissues is the impaired lymphatic drainage which contributes to increased interstitial fluid pressure [21]. This limits extravasation and transvascular transport of macromolecules, inhibiting the transport of molecules in tumor interstitial space. High tumor cell density and dense tumor stroma are other factors that hamper the movement of active compounds within tumors. These conditions are known as the 'enhanced permeability and retention effect' (EPR) in the tumor microenvironment, which could be favorably used when administering nanoparticles that also exhibit long half-life [21]. Nanocarriers can be used as a passive targeting tool which can exploit EPR effect because they can extravasate into the tumor tissues via the leaky vessels, and then they can localize and accumulate in the tumor microenvironment. The junctions between the cancerous cells range from 100 to 600 nm; therefore, the optimal size of nanoparticles was thought to be between 10 and 100 nm but particle clearance and circulation times should be considered in targeting studies [21].

There are convincing arguments in favor of developing nano-sized therapeutics. First, nanoparticles may help to overcome problems of solubility and chemical stability of anticancer drugs. Uptake and delivery of poorly soluble drugs may be increased by enveloping the compound in a hydrophilic nanocarrier. At the same time, this may increase chemical stability. The PI3K inhibitor and radiosensitizer wortmannin is an example for a drug whose development was stopped because of poor solubility and chemical instability. Using a lipid-based nanocarrier system, the solubility of wortmannin was increased from 4 mg l^{-1} to 20 g l^{-1} while increasing its stability *in vivo* [22]. Second, nanocarrier can protect anticancer compounds from biodegradation or excretion and thus influence the pharmacokinetic profile of a compound. Third, nanotechnology can help to improve distribution and targeting of antitumor medication. Distribution of anticancer drugs is defined by their physicochemical properties and is limited by drug penetration into tumor tissue. However, nanomedicine compounds can be constructed to improve drug penetration and to redirect chemotherapy or targeted compounds selectively to tumor cells or cells of the stromal compartment. Both passive and active targeting strategies are used to redirect anticancer drugs [23]. Fourth, nanocarriers can be designed to release their payload upon a trigger resulting in stimuli-sensitive nanomedicine therapeutics. For example, drugs whose delivery is not primarily pH-dependent, such as doxorubicin, can be conjugated with a pH-sensitive nanoparticle to increase cellular drug uptake and intracellular drug release [24]. Finally, targeted nanomedicine therapeutics may decrease resistance of tumors against anticancer drugs [23].

7.3.1 Physiochemical characteristics of NPs influencing the delivery

The biological performance (including biodistribution, pharmacokinetics, cellular uptake, therapeutic efficacy and side effects) of NPs is controlled by a series of complex and interrelated physicochemical characteristics including size, geometrical shape and surface charge, surface modifications

(A) Size

Size determines the surface-to-volume ratio, and can strongly affect the material uptake and is an important parameter in the design of long-circulating nanoparticles [26]. Smaller NPs still accumulate in tumors due to the enhanced permeability and retention effect (EPR). Smaller NPs will also tend to diffuse further into tumor tissue from the bloodstream, and therefore present a more even distribution in larger tumors than larger NPs [2]. It is generally accepted that particles with a size of less than 10 nm are easily removed from the body by renal excretion, For long-circulation of nanoparticles, the size of particles should be large enough to avoid renal filtration but small enough to minimize opsonization and RES clearance. Based on the current data, nanoparticles of diameter between 10–200 nm appear to fit this description [27].

(B) Shape

Spheres represent the most commonly used shape for designing drug delivery carriers [27]. Shape is now recognized as a crucial parameter in determining the behavior of nano/microparticles in various processes including blood circulation, targeting, cellular uptake and intracellular trafficking. Particle geometry has been shown to have substantial impact on phagocytosis by macrophages. This study [27] demonstrated that unlike spherical particles which are readily phagocytosed by macrophages, the elliptical disk-shaped particles exhibit different interactions with macrophages depending on the local geometry. For elliptical disk-shaped particles, the macrophages were not able to complete phagocytosis when they initially contacted the particle on the flat side of elliptical disk while they exhibited normal phagocytosis when contacted on the pointed end of the particle. When the particles were further stretched to highly elongated worm-like shapes, they exhibited near-complete resistance to phagocytosis by macrophages. In a related study, coated polymer particles with anti-ICAM and injected them intravenously in mice [28]. The elliptical disk-shaped particles exhibited longer circulation times and higher targeting ability compared to spherical particles. Chithrani and colleagues' study demonstrates that Hela cells took up spherical Au NPs with a diameter of 74 and 14 nm at rates 5 and 3.75 fold higher than 74×14 nm rod-shaped Au NPs, respectively [29]. This could be owing to membrane wrapping of cells for entire rod-like NPs takes longer than for spherical particles because of the increase in the surface area [30].

(C) Surface charge

The stronger the charge, the better the colloidal stability of the particles, and hence also a longer shelf life. NPs have surface charges that are either negative or positive and tend to have self–self and self–nonself interactions [17]. Given that the inner side of blood vessels and the cellular surface contains many negatively charged components (including but not limited to glycoprotein, glycocalyx, proteoglycans, etc), positively charged NPs are easily attracted is thought to improve the uptake into cells due to its interaction with the negatively charged lipid membrane [17].

(D) Surface coating strategy for active targeting of drug delivery system

Currently, all the FDA-approved NPs are designed to facilitate tumor drug delivery by passive targeting. However, biological barriers such as high interstitial pressure and vasculature heterogeneity can prevent efficient drug extravasation and distribution inside a tumor [20]. Relying only on the EPR effect, passive targeting alone would, in some cases, not allow sufficient drug loading-NPs to reach the target sites [25]. Also, NP targeted by EPR effect is localized in the interstitial spaces of the tumor, with no specific tumor cell internalizing mechanism. By displaying ligands on their surfaces, NPs can be developed to specifically recognize and bind to cancer cells, being the basis for active targeting [25]. Through this strategy, the intracellular uptake of NPs can be significantly enhanced. Active targeting involves attaching to the surface of NPs other molecules that have specific affinities to interact with cancer tissues. The main motivation is to avoid relying on passive uptake through the EPR effect. This has been achieved, for example, with antibodies, peptides, folates, aptamers, hormones and glucose molecules [26].

Additionally, coating of NPs can help to control the interaction of NPs with the proteins (known as opsonin) of the bloodstream. Proteins present in blood get adsorbed on a foreign body surface to clear it from the body. Modifying the surface of nanoparticles to prevent opsonin adsorption onto particles is the most widely used strategy to prolong particle circulation times. The general objective of surface modification is to decrease particle hydrophobicity and surface charge density, both of which are primary initiators of opsonization [27]. Coating NPs with neutral polymers, such as PEG, polyvinylpyrrolidone (PVP) or erythrocyte membranes can repel opsonization by minimizing protein adsorption to their surface, thus increasing their circulation half-life from a few minutes to several hours [17, 27].

7.3.2 Nanomedicine in clinical cancer care

Their small size, flexible fabrication, and high surface-area-to-volume ratio make them ideal systems for drug delivery. Various types of nanomedicine compounds have been used in clinical cancer care, including viral vectors, drug conjugates, lipid-based nanocarriers, polymer-based nanocarriers, and inorganic nanoparticles [23], which are discussed below. Nanomedicine products in clinical studies or approved for clinical cancer care, are summarized in tables 7.3–7.5.

(I) Viral nanoparticles

Viruses contain genetic materials (either DNA or RNA) which give them the ability to mutate and evolve. Viral nanoparticles (VNPs) are particularly attractive because they are naturally occurring nanomaterials, biocompatible, biodegradable [31], can be produced in large quantities and very cheaply. Poxviruses are large enveloped dsDNA viruses that replicate exclusively in the cytoplasm of infected cells [32]. An elegant way to construct nanoparticles for cancer therapy is the use of tumor-homing viruses engineered to express therapeutic proteins [23]. Myxoma (MYXV) and vaccinia (VACV) viruses have recently emerged as potential oncolytic agents that can infect and kill different human cancer cells [32]. VACV is a potential oncolytic agent because of its rapid life cycle, strong target cell killing activity, inherent ability to preferentially replicate within tumor tissues, large cloning capacity, well-defined molecular biology, and its capacity to infect a variety of human cancer types. A second poxvirus called Myxoma virus (MYXV), from the genus *Leporipoxvirus*, has also emerged as a potential oncolytic agent for treatment of human cancers. In contrast to VACV, which can productively infect a wide range of mammalian hosts, MYXV, can infect only lagomorphs. Although MYXV does not induce any known pathology in humans, or any other non-lagomorph host, this virus can efficiently replicate *in vitro* in a variety of transformed human cancer cells lines [32]. In European rabbits (*Oryctolagus cuniculus*), MYXV causes a lethal disease called myxomatosis [32]. It is a highly infectious and usually fatal viral disease of rabbits, causing swelling of the mucous membranes and inflammation and discharge around the eyes. JX-594 is a poxvirus designed to replicate in tumor cells and destroy them via activation of the EGFR–Ras–MAPK (a chain of proteins in

Table 7.3. Approved drug conjugates and nanocarriers for cancer therapy. Marketed products are listed alphabetically. Alternative names for the products are given in brackets [23].

Product [company]	Material	Drug	Indication
Abraxane® (ABI-007) [Abraxis/ Celgene]	Nanoparticle albumin-bound	Paclitaxel	Breast cancer, pancreatic cancer, non-small-cell lung cancer
DaunoXome® [Galen]	Liposome	Daunorubicin	Kaposi's sarcoma
DepoCyt® [Pacira]	Liposome	Cytosine Arabinoside (cytarabine)	Neoplastic meningitis
Doxil®/Caelyx® [Johnson & Johnson]	Liposome	Doxorubicin	Kaposi's sarcoma, ovarian cancer, breast cancer, multiple myeloma
Genexol-PM® (IG-001) [Samyang Biopharm]	PEG–PLA polymeric micelle	Paclitaxel	Breast cancer, lung cancer, ovarian cancer
Lipo-Dox® [Taiwan Liposome]	Liposome	Doxorubicin	Kaposi's sarcoma, breast and ovarian cancer
Marqibo® [Talon]	Liposome	Vincristine	Acute lymphoid leukemia
Mepact® [Takeda]	Liposome	Mifamurtide MTP-PE	Osteosarcoma
Myocet® [Cephalon]	Liposome	Doxorubicin	Breast cancer (cyclophosphamide)
NanoTherm® [MagforceNanotechnologies]	Iron oxide nanoparticle		Thermal ablation glioblastoma
Oncaspar® [Enzon/Sigma-tau]	PEG protein conjugate	L-Asparaginase	Leukemia
Zinostatin stimalamer® (Zinostatin) [Yamanouchi]	Polymer protein conjugate	Styrene maleic anhydride neocarzinostatin (SMANCS)	Liver cancerRenal cancer

Table 7.4. Polymer–drug conjugates in clinical trials. Products are listed alphabetically. Alternative names for the products are given in brackets [23].

Product [company]	Material	Drug	Indication
ADI-PEG 20 [Polaris]	PEG protein conjugate	Arginine deaminase	Hepatocellular carcinoma, melanoma
AP5280 [Access Pharmaceuticals]	HPMA drug conjugate	Platinum	Solid tumors
CT-2106 [CTI Biopharma] Polyglutamic acid drug	Polyglutamic acid drug conjugate	Camptothecin	Colon cancer, ovarian cancer
DE-310 [Daiichi Pharmaceutical]	Carboxymethyldextran polyalcohol drug conjugate	DX-8951 (camptothecin derivate)	Solid tumors
Delimotecan (Men 4901/T-0128)			CarboxymethyldextranDrug conjugate
T-2513 (camptothecin analog)	Solid tumors		
DOX-OXD (AD-70)	Dextran drug conjugate	Doxorubicin	Solid tumors
MAG-CPT (PNU166148/ Mureletecan) [Pfizer]	HPMA drug conjugate	Camptothecin	Solid tumors
MIX-HAS	HSA drug conjugate	Methotrexate	Kidney cancer
NKTR-102 (Etirinotecan pegol) [Nektar]	PEG drug conjugate	Irinotecan	Breast cancer, ovarian cancer, colorectal cancer
NKTR-105 [Nektar]	PEG drug conjugate	Docetaxel	Solid tumors, ovarian cancer
Pegamotecan (EZ-246) [Enzon]	PEG drug conjugate	Camptothecin	Gastric cancer
PegAsys[Genentech]/PegIntron [Merck]	PEG protein conjugate	IFNα2a/-IFNα2b	Melanoma, leukemia
PEG-SN38(EZN-2208) [BelrosePharma/Enzon]	PEG drug conjugate	SN38 (irinotecan derivate)	Solid tumors, breast cancer lymphoma, colorectal cancer
PK1 (FCE28068) [UK Cancer Research/Pfizer]	HPMA drug conjugate	Doxorubicin	Breast cancer, lung cancer, colorectal
PK2(FCE28069) [UK Cancer Research/Pfizer]	HPMA drug conjugate	Doxorubicin	Hepatocellular carcinoma

PNU166945 [Pfizer]	HPMA drug conjugate	Paclitaxel	Solid tumors
ProLindac (AP5346)	HPMA drug conjugate	DACH–oxaliplatin	Ovarian cancer
Taxoprexin [Protarga]	Docosahexaenoic acid drug conjugate	Paclitaxel	Melanoma, liver cancer, adenocarcinoma, kidney cancer, non-small-cell lung cancer
XMT-1001[Mersana]	Fleximer drug conjugate	Camptothecin	Gastric cancer, lung cancer
Xyotax, Opaxio (CT-2103) [Cell Therapeutics]	Polyglutamic acid (polyglumex) drug conjugate	Paclitaxel	Lung cancer, ovarian cancer

Table 7.5. Lipid-based nanocarriers in clinical trials. Lipid-based nanocarriers are the most widely studied drug delivery systems and several clinical trials are ongoing. Products are listed alphabetically. Alternative names for the products are given in brackets [23].

Product [company]	Material	Drug	Indication
2B3-101 [to-BBB]	Liposome	Doxorubicin	Brain metastases of breast cancer
ALN-VSP [Alnylam]	Lipid nanoparticle	RNAi targeting KSP and VEGF	RNAi targeting KSP and VEGF
Anti-EGFR Immunoliposomes [University of Basel]	Liposome	Doxorubicin	Solid tumors
Aroplatin (L-NDDP) [Aronex]	Liposome	DACHplatin	Colorectal cancer, solid malignancies
ATI-1123 [Azaya therapeutics]	Liposome	Docetaxel	Solid tumors
Atu027 [Silence therapeutics]	Liposome/Lipoplex	siRNA against PKN3	Solid tumors, pancreatic cancer, head and neck cancer
CPX-1 [Celator]	Liposome	Irinotecan HCl/floxuridine	Colorectal cancer
CPX-351 [Celator]	Liposome	Cytarabine/daunorubicin	Acute myeloid leukemia
C-VISA BikDD [MD Anderson Cancer Center]	Lipid conjugate	Pro-apoptotic Bik gene	Pancreatic cancer
DCR-MYC (DCR-M1711) [Dicerna Pharmaceuticals]	Lipid nanoparticle	Dicer substrate RNAi (DsiRNA) targeting MYC oncogene	Solid tumors, multiple myeloma, lymphoma
DPX-0907 (DepoVax) [Immuno Vaccine]	Liposome complex	HLA-A2-restricted peptides, THelper Peptide, polynucleotide adjuvant	Cancer vaccine for ovarian, breast and prostate cancer
E7389 liposome [Eisai]	Liposome	E7389 (Eribulin Mesylate)	Solid tumors
EGFR antisense DNALiposomes [University of Pittsburgh]	Liposome	EGFR antisense DNA	Head and neck cancer

Endo-Tag-1 (MBT-0206) [Medigene]	Paclitaxel	Liposome	Pancreatic cancer, triple negative breast cancer, head and neck cancer
EphA2 Targeting Liposomes [MD Anderson Cancer Center]	siRNA targeting the oncoprotein EphA2	Liposome	Solid tumors
IHL-305[Yakult Honsha]	Irinotecan (CPT-11)	Liposome	Solid tumors
INGN-401 [Introgen/Genprex]	FUS1 gene	Lipid nanoparticle	Lung cancer
Interleukin-2 gene therapy [H. Lee Moffitt Cancer Center]	Interleukin-2 Gene	Liposome	Head and neck cancer
INX-0076 (Brakiva) [Tekmira]	Topotecan	Liposome	Solid tumors
INX-0125 (Alocrest)[Tekmira]	Vinorelbin	Liposome	Solid tumors
L9NC [University of Mexico]	9-Nitro-20(S)-Camptothecin	Liposome (aerosol)	Non-small-cell lung cancer
L-Annamycin [Callisto]	Annamycin	Liposome	ALL, AML, doxorubicin-resistant breast cancer
LE-DT [NeoPharm/Insys]	Docetaxel	Liposome	Solid tumors, pancreas/prostate cancer
LEM-ETU [NeoPharm/Insys]	Mitoxantrone	Liposome	Solid tumors, lymphoma
LEP-ETU(PNU-93914) [NeoPharm/Insys]	Paclitaxel	Liposome	Ovarian cancer, breast cancer, lung cancer
LErafAON [INSYS]	c-Raf antisense oligonucleotide	Liposome	Advanced malignancies
LE-SN38 [NeoPharm/Insys]	SN38 (active metabolite of irinotecan)	Liposome	Colorectal cancer
LiPlaCis [LiPlasome Pharma]	Cisplatin	Liposome	Solid tumors
Lipocurc [SignPath Pharma]	Curcumin	Liposome	Advanced cancer
Lipoplatin [Regulon]	Cisplatin	Liposome	Non-small-cell lung cancer
Liposomal Grb-2 [MD Anderson/Bio-Path]	Grb2 antisense nucleotide	Liposome	Leukemia

(Continued)

Table 7.5. (*Continued*)

Product [company]	Material	Drug	Indication
Liposomal Tretinoin (ATRA-IV) [MD Anderson/NCI]	Liposome	Tretinoin	Refractory Hodgkin disease, kidney cancer, solid tumors
Lipovaxin-MM [Lipotek]	Lipid-based formulationLiposome	Melanoma antigens and IFNγ	Melanoma vaccine
Lipusu [Luye Pharma]	Liposome	Paclitaxel	Solid tumors, gastric cancer, metastatic breast cancer
MBP-426 [Mebiopharm]	Liposome	Oxaliplatin	Gastric cancer
MCC-465 [Mitsubishi Pharma]	Liposome	Doxorubicin	Gastric cancer
MM-302 [Merrimack]	Liposome	Doxorubicin	ErbB2-positive breast cancer
MM-398 (PEP02) [Merrimack]	Liposome	CPT-11 (Irinotecan)	Pancreatic cancer, gastric cancer, glioma
MRX34 [Mirna Therapeutics]	Liposome	miR-RX34	Liver cancer, solid tumors, lymphoma, myeloma
NanoVNB [Taiwan Liposome]	Liposome	Vinorelbine	Solid tumors
NL CPT-11 [UCSF]	Liposome	Irinotecan	Recurrent glioma
Onco-TCS [Inex/Enzon]	Liposome	Vincristin	Non-Hodgkin lymphoma
OSI-211 [OSI Pharmaceuticals]	Liposome	Lurtotecan	Small cell lung cancer, ovarian cancer, head and neck cancer
OSI-7904L [OSI Pharmaceuticals]	Liposome	Thymidylate synthase inhibitor	Colorectal cancer, gastric cancer, head and neck cancer
PNT2258 [ProNAiTherapeutics]	Liposome	DNAi targeting BCL-2	Non-Hodgkin lymphoma, solid tumors
Promitil [Lipomedix]	Liposome	Mitomycin-C Prodrug	Solid tumors
Sarcodoxome [GP Pharma]	Liposome	Doxorubicin	Soft tissue sarcoma
S. CKD602 [Aiza]	Liposome	CKD602 (belotecan)	Solid tumors

Name	Agent	Formulation	Indication
SGT-53 [SynerGene Therapeutics]	p53 plasmid DNA	Liposome	Solid tumors
SGT-94 [SynerGene Therapeutics]	RB94 plasmid DNA	Liposome	Solid tumors
SLIT cisplatin [Transave/Insmed]	Cisplatin	Lipid product (aerosol)	Lung cancer (Aerosol)
SPI-077 [Alza]	Cisplatin	Liposome	Solid tumors, ovarian cancer, head and neck cancer
Stimuvax (BLP25) [Oncothyreon/Merck]	Anti-MUC1 cancer vaccine	Liposome	Non-small-cell lung cancer
STMN-1 [Gradalis]	Pbi-shRNA STMN1	Lipoplex	Solid tumors, metastatic cancer
ThermoDox [Celsion]	Doxorubicin	Thermosensitive liposome	Primary hepatocellular carcinoma, refractory chest wall breast cancer, colorectal liver metastases
TKM-PLK1 (TKM 080301) [Tekmira]	RNAi targeting polo-like kinase 1 (POLO)	Liposome (SNALP)	Liver cancer, neuroendocrine tumors, adrenocortical carcinoma

the cell that communicates a signal from a receptor on the surface of the cell to the DNA in the nucleus of the cell) signaling pathway. In addition, JX-594 expresses granulocyte colony-stimulating factor (G-CSF) (used to stimulate the production of granulocytes in patients undergoing therapy that will cause low white blood cell counts) to potentially increase the immunological antitumor response [23]. This medication is used to prevent infection and neutropenic (low white blood cells) fevers caused by chemotherapy. Several other oncolytic viruses have been tested in clinical trials over recent years, but none of these has reached the market as yet. Their major drawbacks are concerns about biosafety and cytocompatibility [23].

(II) Drug conjugates

Drug conjugates are the most successful nanomedicine therapeutics in clinical cancer care (tables 7.3 and 7.4). Antibody-drug conjugates (ADCs) are monoclonal antibodies (mAbs) attached to biologically active drugs by chemical linkers with labile bonds. By combining the unique targeting of mAbs with the cancer-killing ability of cytotoxic drugs, ADCs allow sensitive discrimination between healthy and diseased tissue. They are defined as nanotherapeutics because of their size scale in the lower nanometer range and their conjugation to active pharmaceutical ingredients [33]. In developing antibody–drug conjugates, an anticancer drug is coupled to an antibody that specifically targets a certain tumor marker (e.g. a protein that, ideally, is only to be found in or on tumor cells). Antibodies track these proteins down in the body and attach themselves to the surface of cancer cells. The biochemical reaction between the antibody and the target protein (antigen) triggers a signal in the tumor cell, which then absorbs or internalizes the antibody together with the cytotoxin. After the ADC is internalized, the cytotoxic drug is released and kills the cancer [34]. Due to this targeting, ideally the drug has lower side effects and gives a wider therapeutic window than other chemotherapeutic agents. Monoclonal antibodies have proved to have an important role in cancer treatment with drugs such as trastuzumab, pertuzumab, cetuximab and rituximab becoming the standard of care in selected solid tumors and lymphomas [35]. The conjugate is usually mono- or oligomeric, intended to improve targeted delivery of the drug without necessarily impacting on drug solubility, stability, or biodegradability [23]. ADCs approved by regulatory authorities include trastuzumab emtansine against HER2-overexpressing breast cancer and brentuximab-vedotin against CD30-positive Hodgkin lymphoma and anaplastic large-cell lymphoma [23]. Trastuzumab has been in clinical use for some time and is effective in both the adjuvant and palliative settings. Its conjugation to emtansine (DM1), a plant-derived inhibitor of microtubules, significantly increases the anti-tumor activity of trastuzumab.

Polymer–drug conjugates are another interesting group of nano-sized (5–20 nm) drug delivery systems which change the pharmacokinetic profile of a drug [36]. More than 15 anticancer conjugates are in clinical development. All accumulate within tumors via the EPR effect. The HPMA copolymer–doxorubicin conjugate PK1 is a novel anticancer agent with a significantly lower frequency of cardiotoxicity and alopecia than free doxorubicin. Study showed promising signs of activity in breast cancer and non-small-cell lung cancer [23]. Radio peptides represent a specific class of drug conjugates because their size (around 1 nm) is at the limit of the conventional definition of a nanomedicine agent. The most widely used therapeutic radio peptides are DOTATOC [37] and DOTATATE. Radiopeptides are built from a peptide component determining the specificity of the compound, and a chelator binding the radioisotope (e.g. 90Y or 177Lu) [23].

(III) Lipid-based nanocarriers
There are different forms of lipid-based nanocarriers [38] (table 7.5), the most frequently investigated are liposomes (closed phospholipid bilayers) and micelles (normal phase, oil-in-water micelles). Lipid nanocarriers provide a carrying capacity three to four orders of magnitude greater than ADCs, which typically comprise 1–6 drug molecules per monoclonal antibody. Pegylated liposomal doxorubicin (Doxil® or Caelyx®) was the first nanocarrier approved by the US Food and Drug Administration (FDA) in 1995 [39, 40]. To date, five more lipid nanocarriers have been approved for clinical use: non-pegylated liposomal doxorubicin (Myocet®) [41], non-pegylated liposomal daunorubicin (DaunoXome®) [42], non-pegylated liposomal cytarabine (DepoCyt®) [43], vincristine sulfate liposomes (Marqibo®) [44], and liposomal mifamurtide (Mepact®) [45, 46].

However, the nanoformulations of doxorubicin, daunorubicin, and vincristine prolong the half-life of the cytotoxic compounds and profoundly improve their toxicity profiles. Lipid nanocarriers, and in particular pegylated liposomes, tend to accumulate in tumor tissue due to the enhanced permeability of intra- and peritumoral vessels and somewhat prolonged retention in cancerous tissue [23]. In several preclinical experiments, the surface of liposomes was coated with anti-HER2, anti-EGFR, anti-VEGFR2, or other antibodies [47–49]. Cellular uptake depended on the interaction of the target antigen on the cell surface and the antibody coating the surface of the liposome.

(IV) Natural polymers: protein and peptide nanocarriers
Biological protein-based nanoparticles such as albumin, silk, keratin, collagen, elastin, corn zein, and soy protein-based nanoparticles are advantageous in having biodegradability, bioavailability, and relatively low cost. Many protein nanoparticles are easy to process and can be modified to achieve desired specifications such as size, morphology, and weight. Protein nanoparticles are used in a variety of settings and are replacing many materials that are not biocompatible and have a negative impact on the environment [50]. Albumin-nanoparticle-bound paclitaxel (nab-paclitaxel; Abraxane®) is used for the therapy of breast cancer, non-small-cell lung cancer, and pancreatic cancer. Paclitaxel is a cytotoxic compound isolated from the bark of the pacific yew tree (*Taxus brevifolia*). Since paclitaxel itself is poorly water-soluble, it is dissolved in Cremophor EL, a polyethoxylated castor oil. However, this solvent is the reason for the frequent allergic reactions to the drug. Therefore, alternative formulations were investigated. By incorporating paclitaxel into an albumin-nanoparticle, drug solubility is improved and the use of castor oil can be avoided [23]. In clinical trials, nab-paclitaxel improved response rates compared to conventional paclitaxel in breast cancer patients [51]. In combination

with gemcitabine, nab-paclitaxel increased survival compared to gemcitabine monotherapy in patients with pancreatic cancer [52].

(V) Natural polymers: glycan nanocarriers

The terms glycan meaning compounds consisting of a large number of monosaccharides (also called simple sugars, are the simplest form of sugar and the most basic units of carbohydrates) linked glycosidically. Glyco-nanomaterials take advantage of the unique physical properties of the nanoscale such as catalytic, photonic, electronic, or magnetic properties that are not seen in the bulk as well as from the properties of glycans such as water solubility, biocompatibility, structural diversity, and targeting properties [53]. Currently, there are no approved glycan nanocarriers for systemic cancer therapy [23]. However, glycan nanoparticles have been tested in proof-of-concept phase 1 and in phase 2 trials. In a phase 2 clinical trial, cyclodextrin nanoparticles were used as carriers for camptothecin in patients with advanced solid tumors. CRLX101 (formerly IT-101) showed fewer side effects than free camptothecin and the overall response rate was 64% [54]. CALAA-01 is another polymer nanoparticle (50–70 nm) based on cyclodextrin and PEG in clinical development for siRNA delivery. It is targeted to human transferrin receptors overexpressed on cancer cells to silence the expression of the M2 subunit of ribonucleotide reductase. Intracellular release of siRNA is triggered by an acidic endosomal pH [55]. Preclinically, chitosan-based glycan nanoparticles have been used to develop an oral formulation of the cytotoxic compound gemcitabine. Gemcitabine-loaded chitosan nanoparticles were about 95 nm in diameter. Intestinal uptake increased 3–5 fold compared to free oral gemcitabine [56].

(VI) Synthetic polymer-based nanocarriers

Because of their chemical versatility, synthetic polymer nanocarriers are a promising tool for nanomedicine therapeutics [57]. Polymer nanoparticles such as PLA (polylactic acid) have been functionalized with aptamers to impart targeting capabilities to develop therapeutic delivery systems for the treatment of cancer systemically. For example, nanoparticles of PEGylated PLA with a terminal carboxylic acid functional group were used to encapsulate rhodamine-labeled dextran. The carboxylic acid group was used to conjugate an RNA aptamer that binds to the prostate-specific membrane antigen (PSMA), a well-known transmembrane protein that is overexpressed on prostate cancer epithelial cells. Prostate LNCaP epithelial cells, which express PSMA, showed a preferred uptake of over 75 fold in comparison to the control cell line *in vitro*. When tested *in vivo*, anti-PSMA aptamer-functionalized PLGA nanoparticles showed a near-fourfold increase in tumor accumulation in comparison to nonfunctionalized nanoparticles at 24 h when injected systemically. The strategy provided for creating targeting drug delivery nanoparticles is amenable to encapsulating different drug molecules (e.g. paclitaxel, cisplatin, docetaxel, or small interfering RNA) or attaching different aptamers [58].

As an example, the PEG–PGA polymeric micelle NC-6004(Nanoplatin®) for the delivery of cisplatin was tested in a phase 1 trial. Compared to the free drug, NC-6004 was associated with less oto- and neurotoxicity as well as nausea, and the disease control rate was encouraging [59]. The first targeted polymeric nanoparticle in clinical trials is BIND-014. It is made up of docetaxel-encapsulated PLGA–PEG nanoparticles targeted to prostate-specific membrane antigen (PSMA). In various animal models, BIND-014 has delivered up to 10 times more docetaxel to tumors than free docetaxel. The nanoparticle system exhibited prolonged residence time in circulation, and more extensive retention in the tumor. In addition, tumor shrinkage in animal models was the same at a fifth of the dose of free docetaxel [23].

(VII) Inorganic nanoparticles for cancer therapy
Inorganic nanoparticles are used for a variety of applications, including tumor imaging, enhancement of radiotherapy, or drug delivery. Iron oxide nanoparticles are mainly used for diagnostic purposes [60] and some are tested in clinical studies for magnetic resonance imaging of tumors. NanoTherm® is an aqueous colloidal dispersion of iron oxide nanoparticle. Upon instillation in the tumor, thermal ablation is performed with an alternating magnetic field applicator (magnetic hyperthermia). The most robust data are available for glioblastoma [61]. NanoTherm® has gained marketing approval in several European countries. Recently, the first clinical trial with hafnium oxide nanoparticles has started. The nanoparticles are investigated in patients with soft tissue sarcoma as a radio-sensitizer [62]. No inorganic nanoparticle for drug delivery has reached marketing approval as yet. Some of them are in early clinical testing, for example pegylated colloidal gold–TNFα particles for cancer therapy [63] and silicon nanocarriers for parenteral peptide delivery [64].

7.4 Breast cancer

Breast cancer is the most common and lethal cancer type in women worldwide [65]. The global incidence of breast cancer has increased by over 20% since 2008 [66]. Close to 1.5 million new cases of breast cancer are reported each year, accounting for 25% of all cancer cases. Overall, breast cancer is the second leading cause of mortality just behind lung cancer, and among females, it has long been the top cause of cancer death (15% of all female cancer patients) [66]. As per the National Cancer Institute report, estimated deaths due to breast cancer in 2018 were about 40 920 [67]. Discovering new effective and safe forms of treatment for this prevalent and deadly malignant disease is, therefore, critical.

7.4.1 Breast anatomy and cancer

Before learning about breast cancer it is necessary to know about breast anatomy. The breast is highly complex. It goes through more changes than any other part of the human body—from birth, puberty, pregnancy and breastfeeding, right through to menopause. Breast tissue extends from the collarbone, to lower ribs, sternum (breastbone) and armpit. Each breast contains 15–20 glands called lobes, where milk is produced in women who are breastfeeding. These lobes are connected to the nipple by 6–8 tubes called ducts which carry milk to the nipple. The ring of darker breast skin is called the areola and the raised tip in the areola is called the nipple

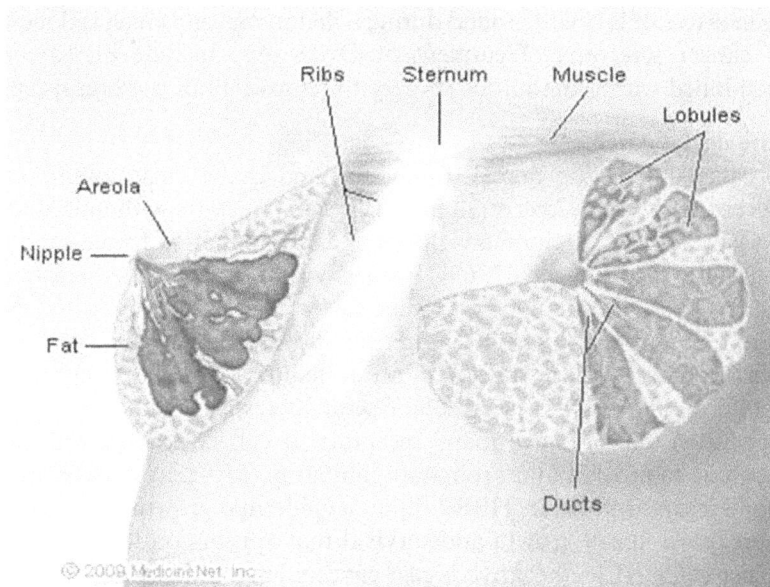

Figure 7.5. Illustration of breast anatomy, Image credit www.medicinenet.com [68].

(figure 7.5). The breast and armpit contain lymph nodes and vessels carrying lymph fluid and white blood cells. Much of the rest of the breast is fatty tissue.

The breast, like any other part of the body, consists of billions of microscopic cells. These cells usually multiply in an orderly fashion—new healthy cells continue to divide and replace the ones that have died. However, sometimes cells develop abnormalities (mutations). This occurs when the genes that usually check that cells are replicating correctly fail to detect mutations. When this happens, abnormal cells continue to divide and multiply, sometimes growing quite rapidly. At this stage, a growth may not be cancerous. It could be a 'noninvasive tumor' which remains contained in the duct or the lobe. A tumor is considered cancerous once it is able to invade surrounding tissue. These cancers require treatment because if they continue to grow and spread, they could become life-threatening. Breast cancer usually begins in the ducts that carry milk to the nipple or within the smaller structures of the lobes. A small number of cancers start in other tissues of the breast; these are called sarcomas and lymphomas. Although many types of breast cancer can cause a lump in the breast, not all do. Breast cancer can spread when cells break away from the main tumor and are transported to other parts of the body. This can happen via the lymphatic system or the bloodstream.

7.4.2 Classification of breast cancer—noninvasive and invasive

(A) Noninvasive ductal carcinoma

Ductal carcinoma *in situ* (DCIS) is the presence of abnormal cells inside a milk duct in the breast. DCIS is considered the earliest form of breast cancer. DCIS is noninvasive, meaning it hasn't spread out of the milk duct and has low risk of becoming invasive. It is usually found during a mammogram, which is done as a part of breast cancer screening. Treatment of DCIS may include breast-conserving surgery combined with radiation or surgery to remove all of the breast tissue.

(B) Invasive ductal carcinoma
It is also called infiltrating ductal carcinoma, and is the most common invasive breast cancer (about 50%–75% of all breast cancers), it starts with milk ducts of the breast. Invasive lobular carcinoma is the second most common breast cancer (about 5%–15% of all breast cancer) [67, 69]. It starts with lobules of the breast. Tubular, mucinous (colloid) and papillary carcinomas and carcinomas with medullary features are less common invasive features, as shown in table 7.6.

All invasive breast cancer and DCIS are tested for hormone receptor (estrogen and progesterone). A hormone receptor-positive (estrogen and/or progesterone receptor-positive), tumors have many receptors; it can be treated with hormonal therapy such as tamoxifen and aromatase inhibitors. Most invasive breast cancers are hormone receptor-positive. HER2 (human epidermal growth factor receptor 2), a protein involved in cell growth and survival that appears on the surface of some breast cancer cells. HER2-negative breast cancers have little or no HER2 protein. HER2-positive breast cancers have a lot of HER2 protein. HER2-positive tumors can be treated with the targeted therapy drug trastuzumab (Herceptin). About 10%–15% newly diagnosed breast cancers are HER2-positive.

7.4.3 Conventional modalities for treating breast cancer

The main types of treatment for breast cancer are surgery, radiation therapy (RT), chemotherapy (CT), endocrine (hormone) therapy (ET), and targeted therapy [70]. Breast conservation surgery is the trending approach in the treatment of localized breast cancer. The surgery is preceded by neoadjuvant therapy to shrink tumor bulk. Surgery is usually followed by adjuvant therapy to ensure full recovery and minimize the risk of metastases [70]. Cancer cells that may not be seen during surgery can be killed by radiation to reduce the risk of local recurrence of cancer. RT is a process in which cancer cells are exposed to high levels of radiation directly. RT after surgery shrinks the tumor in combination with CT. But there are some side effects of RT, such as decreased sensation in the breast tissue or under the arm, skin problems in the treated area, for example, soreness, itching, peeling, and/or redness, and at the end of treatment the skin may become moist and weepy [68]. The purpose of hormonal therapy is either adding or blocking hormones. The female hormones estrogen and progesterone can promote the growth of some breast cancer cells. Therefore, hormone therapy is required to block or lower the levels of estrogen and progesterone to prevent growth of cancer cells. In chemotherapy cytotoxic drugs are administered to kill cancer cells. Chemotherapy may be recommended as adjuvant chemotherapy or neoadjuvant chemotherapy. Adjuvant chemotherapy is the systemic therapy given to patients after surgery to treat undetected breast cancer cells. Neoadjuvant chemotherapy is given before surgery to shrink large cancers so that they can easily be removed by lumpectomy [70].

Table 7.6. Prevalence and tumor characteristics of different types of invasive breast cancer [69].

Types of invasive breast cancer	Proportion of all invasive breast cancers	Tumor characteristics	Prognosis
Invasive ductal carcinoma (IOC)	50%–75%	Hard tumor texture, tumor is irregular, star-shaped, cell features vary, DCIS often present	Prognosis varies with stage and grade of tumor
Invasive lobular carcinoma (ILC)	5%–15%	Normal, slightly firm or hard tumor texture, cells most often appear in single file order, tumors are most often ER-positive and HER2-negative	Prognosis varies with stage and grade of tumor, For any given stage or grade, prognosis is similar to that of IDC, Pattern of metastases is slightly different from IDC
Tubular carcinoma.	1%–5%	Tumors are most often small, often no palpable tumor, cells form tube-like structures, tumors are almost always ER-positive and HER2-negative	Prognosis is usually better than for IDC, rare for cancer to spread to lymph nodes or other parts of the body
Mucinous (colloid) carcinoma	1%–5%	Soft tumor, often no palpable tumor, cells are surrounded by excess mucous (mucin), tumors are most often ER-positive and HER2-negative	More common among older women, tends to have a good prognosis, less common for cancer to spread to lymph nodes
Carcinomas with medullary features	Less than 1%	Soft tumor, cells have a sheet-like appearance, tumors are typically ER-negative, PR-negative and HER2-negative (often called triple negative)	More common among younger women and women with a BRCA1 gene mutation
Invasive papillary carcinoma	Less than l%	Soft tumor, Cells appear as fingerlike branches	More common in women after menopause, tends to have a good prognosis

* Percentage is higher in cancers found through mammography screening.
ER-positive = estrogen receptor-positive.
ER-negative = estrogen receptor-negative.
PR-negative = progesterone receptor-negative.
HER2-negative = HER2 receptor-negative.

(A) Surgery

Surgery is the removal of the tumor and some surrounding healthy tissue during an operation. Surgery is also used to examine the nearby axillary lymph nodes, which are under the arm. The types of surgery include the following.

(I) Lumpectomy

This is the removal of the tumor and a small, cancer-free margin of healthy tissue around the tumor [71]. Most of the breast remains. For invasive cancer, radiation therapy to the remaining breast tissue is generally recommended after surgery known as breast-conserving surgery. For DCIS, radiation therapy after surgery may be an option depending on the patient and the tumor. The first treatment goal is to extend life as much as a mastectomy would. The second goal is to reduce the chance of the cancer coming back. Also, enough tissue should be spared so that the breast will look as similar as possible after treatment to the way it looked before treatment. A lumpectomy may also be called, a partial mastectomy, quadrantectomy, or a segmental mastectomy.

(II) Mastectomy

This is the surgical removal of the entire breast. There are several types of mastectomies for which an oval shaped cut is often first made around the areola. The breast tissue is then detached and removed. Mastectomy causes a scar [72].

(III) Lymph node removal and analysis

Invasive breast cancer can spread outside the breast through lymph. Most of the lymph in breast drains to axillary lymph nodes. The remaining lymph drains to nodes near the breastbone or collarbone [72]. There are two types of surgery to remove lymph nodes, sentinel lymph node biopsy spreads and axillary lymph node dissection.

A sentinel lymph node biopsy finds and removes a small number of lymph nodes from under the arm that receive lymph drainage from the breast. This procedure helps avoid removing multiple lymph nodes with an axillary lymph node dissection for patients whose sentinel lymph nodes are mostly free of cancer. The smaller lymph node procedure helps lower the risk of several possible side effects. Those side effects include swelling of the arm called lymphedema, the risk of numbness, as well as arm movement and range-of-motion problems with the shoulder, which are long-lasting issues that can severely affect a person's quality of life. By injecting a radioactive tracer or dye behind or around the nipple, sentinel lymph nodes can be detected. The dye or tracer travels to the lymph nodes, arriving at the sentinel node first. If a dye is used it changes color or if a tracer is used it gives off radiation at the lymph node region [71].

An axillary lymph node dissection removes many lymph nodes from under the arm. These are then examined for cancer cells. The actual number of lymph nodes removed varies from person to person. An axillary lymph node dissection may not be needed for all women with early-stage breast cancer with small amounts of cancer in the sentinel lymph nodes [71]. Women having a lumpectomy and radiation therapy who have a smaller tumor and no more than two sentinel lymph nodes with cancer may avoid a full axillary lymph node dissection [71]. This helps to reduce the risk of side effects and does not decrease survival. If cancer is found in the sentinel lymph node, whether more surgery is needed to remove more lymph nodes depends on the specific situation [71].

(B) Radiation therapy

A radiation therapy regimen, or schedule (see below), usually consists of a specific number of treatments given over a set period of time. Radiation therapy often helps lower the risk of recurrence in the breast. In fact, with modern surgery and radiation therapy, recurrence rates in the breast are now less than 5% in the 10 years after treatment, and survival is the same with lumpectomy or mastectomy. If there is cancer in the lymph nodes under the arm, radiation therapy may also be given to the same side of the neck or underarm near the breast or chest wall.

Radiation therapy may be given after or before surgery:

- Adjuvant radiation therapy is given after surgery. Most commonly, it is given after a lumpectomy, and sometimes, chemotherapy. Patients who have a mastectomy may not need radiation therapy, depending on the features of the tumor. Radiation therapy may be recommended after mastectomy if you have a larger tumor, cancer in the lymph nodes, cancer cells outside of the capsule of the lymph node, or cancer that has grown into the skin or chest wall, as well as for other reasons.
- Neoadjuvant radiation therapy is radiation therapy given before surgery to shrink a large tumor, which makes it easier to remove. This approach is uncommon and is only considered when a tumor cannot be removed with surgery [71].

Radiation therapy can cause side effects, including fatigue, swelling of the breast, redness and/or skin discoloration or hyperpigmentation, and pain or burning in the skin where the radiation was directed, sometimes with blistering or peeling.

Radiation therapy schedule

Radiation therapy is usually given daily for a set number of weeks.

(I) After a lumpectomy

Radiation therapy after a lumpectomy is external-beam radiation therapy given for 3–4 weeks if the cancer is not in the lymph nodes. If the cancer is in the lymph nodes, radiation therapy is given for 5–6 weeks. This often starts with radiation therapy to the whole breast, followed by a more focused treatment to where the tumor was located in the breast for the remaining treatments. This focused part of the treatment, called a boost, is standard for women with invasive breast cancer to reduce the risk of a recurrence in the breast. Women with DCIS may also receive the boost. For women with a low risk of recurrence, the boost may be optional. For some women only the lumpectomy site is treated by APBI (accelerated partial breast irradiation) [72]. Guidelines for radiation therapy after lumpectomy with and without Neoadjuvant therapy are given in table 7.7

Table 7.7. Radiation therapy after lumpectomy. Neoadjuvant (preoperative) therapy was not given. (Source: NCCN Guidelines for Patients, Invasive Breast Cancer, 2018 [72]).

T Stage	N stage	Where radiation needed
T1 T1	Clinical and pathologic N stages are N0 Clinical and pathologic N stages are N0 Or Pathologic N stage is N0	• No radiation for very low risk cancers • Part of the breast for certain low risk cancers • Whole breast ± boost • Regional lymph nodes may be treated when cancer is at high risk
T0 or T1	Pathologic N stage is N1	• Whole breast ± boost • Regional lymph nodes + at risk axillary tissue may be treated
T0 or T1	Pathologic N stage is N2 or N3	• Whole breast ± boost to surgery site • Infraclavicular lymph node • Supraclavicular lymph node • Internal mammary lymph nodes • Axillary tissue at risk for cancer

Neoadjuvant (preoperative) therapy was given

T Stage	N stage	Where radiation needed
T2 or T3	Clinical and pathologic N stages are N0	• Whole breast ± boost • Regional lymph nodes may be treated when cancer is at high risk
T0–T3	Clinical N1 downstaged to pathologic N0	• Whole breast ± boost • Regional lymph nodes + at risk axillary tissue may be treated
T0–T3	Clinical N stage is N2 or N3 Or Pathologic N stage is N1–N3	• Whole breast ± boost • Infraclavicular lymph node • Supraclavicular lymph node
T4	Any N stage	• Internal mammary lymph nodes • Axillary tissue at risk for cancer

(II) After a mastectomy

For those who need radiation therapy after a mastectomy, it is usually given five days a week for 5–6 weeks. When radiation is given, the chest wall and sometimes drain sites will be treated. The mastectomy scar will get more radiation called a boost. Regional lymph nodes often get treated too. If axillary lymph nodes were removed, other tissue in the area is treated instead. Radiation is often done when a recurrence is likely. The risk is higher for T4 tumors and N2 or N3 disease. Radiation is also advised when breast cancer remains after preoperative therapy [72] (table 7.8).

(III) Partial breast irradiation

Partial breast irradiation (PBI) is radiation therapy that is given directly to the tumor area instead of the entire breast. It is more common after a lumpectomy. Targeting radiation directly to the tumor area usually shortens the amount of time that patients need to receive radiation therapy. However, only some patients may be able to have PBI. PBI can be done with standard external-beam radiation therapy that is focused on the area where tumor was removed and not on the entire breast.

(IV) Brachytherapy

PBI may also be done with interstitial brachytherapy by using plastic catheters or a metal, placed temporarily in the breast. These require the placement under anesthetic of a number of needles or tubes across the tumor bed, either with a template or freehand. The treatment volume is generally the tumor cavity plus a 1–2 cm margin. The dose is custom-shaped to this treatment volume. The dose can be delivered using LDR brachytherapy, typically over 4–5 days, or fractionated HDR brachytherapy, typically treating twice a day for 4–5 days. The large number of catheters in an interstitial implant allows more control over skin and chest wall doses, especially with HDR dose optimization. The dose within the tumor is more homogeneous with lots of smaller hot-spots, as compared with the one large hot-spot obtained with single-catheter techniques [73]. Breast HDR brachytherapy can involve short treatment times, ranging from 1 dose to 1 week. It can also be given as 1 dose in the operating room immediately after the tumor is removed. These forms of focused radiation therapy are currently used only for patients with a smaller, less-aggressive, and lymph node-negative tumor [71].

(V) Intensity-modulated radiation therapy

Intensity-modulated radiation therapy (IMRT) is a more advanced way to give external-beam radiation therapy to the breast. The intensity of the radiation directed at the breast is varied to better target the tumor, spreading the radiation more evenly throughout the breast. The use of IMRT lessens the radiation dose and may decrease possible damage to nearby organs, such as the heart and lung, and the risks of some immediate side effects, such as peeling of the skin and burns, compared with women with smaller breasts. IMRT may also help to lessen the long-term effects on the breast tissue, such as hardness, swelling, or discoloration, that were common with older radiation techniques. IMRT is not recommended for everyone.

Table 7.8. Radiation therapy after mastectomy. Neoadjuvant (preoperative) therapy was not given. (Source: NCCN Guidelines for Patients, Invasive Breast Cancer, 2018 [72]).

T Stage	N stage	Where radiation needed
T1 or T2 tumor removed with large and cancer-free margins	Pathologic N stage is N0	• No radiation therapy
T1 or T2 tumor removed with small and cancer-free margins	Pathologic N stage is N0	• Chest wall may be treated • Regional lymph nodes may be treated when cancer is at high risk
T1 or T2 with cancer in margins	Pathologic N stage is N0	If a second surgery is not an option • Chest wall + Regional nodes + at high risk axillary tissue may be treated
T3	Pathologic N stage is N0	• Chest wall + regional lymph nodes + at high risk axillary tissue may be treated
T0–T3	Pathologic N stage is N1	• Chest wall
T0–T3	Pathologic N stage is N2 or N3	• Infraclavicular lymph node • Supraclavicular lymph node • Internal mammary lymph nodes • Axillary tissue at risk for cancer

Neoadjuvant (preoperative) therapy was given

T Stage	N stage	Where radiation needed
T0–T3	Clinical N1 downstaged to N0	• Chest wall + regional lymph nodes + at high risk axillary tissue may be treated
T0–T3	Clinical N stage is N2 or N3 Or Pathologic N stage is N1–N3	• Chest wall • Infraclavicular lymph node • Supraclavicular lymph node
T4	Any N stage	• Internal mammary lymph nodes • Axillary tissue at risk for cancer

(C) Therapies using medication

Systemic therapy is the use of medication to destroy cancer cells. This type of medication is given through the bloodstream to reach cancer cells throughout the body. Common ways to give systemic therapies include an intravenous (IV) tube placed into a vein using a needle, an injection into a muscle or under the skin, or in a pill or capsule that is swallowed (orally).

The types of systemic therapies used for breast cancer include:

- chemotherapy;
- hormonal therapy;
- targeted therapy;
- immunotherapy.

They can also be given as part of a treatment plan that includes surgery and/or radiation therapy.

(I) Chemotherapy

Chemotherapy is the use of drugs to destroy cancer cells, usually by ending the cancer cells' ability to grow and divide. It may be given before surgery to shrink a large tumor, make surgery easier, and reduce the risk of recurrence, called neo-adjuvant chemotherapy. It may also be given after surgery to reduce the risk of recurrence, called adjuvant chemotherapy. A chemotherapy regimen, or schedule, usually consists of a combination of drugs given in a specific number of cycles over a set period of time. Chemotherapy may be given on many different schedules depending on what worked best in clinical trials for that specific type of regimen. It may be given once a week, once every two weeks (also called dose-dense), once every three weeks, or even once every four weeks. There are many types of chemotherapy used to treat breast cancer [71]. Common drugs include:

- Capecitabine (Xeloda);
- Carboplatin (available as a generic drug);
- Cisplatin (available as a generic drug);
- Cyclophosphamide (available as a generic drug);
- Docetaxel (Taxotere);
- Doxorubicin (available as a generic drug);
- Pegylated liposomal doxorubicin (Doxil);
- Epirubicin (Ellence);
- Eribulin (Halaven);
- Fluorouracil (5-FU, Efudex);
- Gemcitabine (Gemzar);
- Ixabepilone (Ixempra);
- Methotrexate (Rheumatrex, Trexall);
- Paclitaxel (Taxol);
- Protein-bound paclitaxel (Abraxane);
- Vinorelbine (Navelbine).

A patient may receive one drug at a time or a combination of different drugs given at the same time. Research has shown that combinations of certain drugs are sometimes more effective than single drugs for adjuvant treatment. The following drugs or combinations of drugs may be used as adjuvant therapy for early-stage and locally advanced breast cancer [71]:

- AC (doxorubicin and cyclophosphamide);
- EC (epirubicin, cyclophosphamide);
- AC or EC (epirubicin and cyclophosphamide) followed by T (doxorubicin and cyclophosphamide, followed by paclitaxel or docetaxel, or the reverse);
- CAF (cyclophosphamide, doxorubicin, and 5-FU);
- CEF (cyclophosphamide, epirubicin, and 5-FU);
- CMF (cyclophosphamide, methotrexate, and 5-FU);
- TAC (docetaxel, doxorubicin, and cyclophosphamide);
- TC (docetaxel and cyclophosphamide).

Therapies that target the HER2 receptor may be given with chemotherapy for HER2-positive breast cancer (see targeted therapy, below). An example is the antibody trastuzumab. Combination regimens for early-stage HER2-positive breast cancer may include [71]:

- AC-TH (doxorubicin, cyclophosphamide, paclitaxel, trastuzumab);
- AC-THP (doxorubicin, cyclophosphamide, paclitaxel, trastuzumab, pertuzumab);
- TCHP (docetaxel, carboplatin, trastuzumab, pertuzumab);
- TCH (docetaxel, carboplatin, trastuzumab);
- TH (paclitaxel, trastuzumab).

The side effects of chemotherapy depend on the individual, the drug(s) used, and the schedule and dose used. These side effects can include fatigue, risk of infection, nausea and vomiting, hair loss, loss of appetite, diarrhea, constipation, early menopause, weight gain, and chemo-brain. These side effects can often be very successfully prevented or managed during treatment with supportive medications, and they usually go away after treatment is finished. Rarely, long-term side effects may occur, such as heart damage, nerve damage, or secondary cancers such as leukemia and lymphoma. Many patients feel well during chemotherapy and are actively taking care of their families, working, and exercising during treatment, although each person's experience can be different [71].

(II) Hormonal therapy
Hormonal therapy, also called endocrine therapy, is an effective treatment for most tumors that test positive for either estrogen or progesterone receptors (called ER-positive or PR-positive). This type of tumor uses hormones to fuel its growth. Blocking the hormones can help prevent a cancer recurrence and death from breast cancer when used either by itself or after chemotherapy. Hormonal therapy may be given before surgery to shrink a tumor, make surgery easier, and lower the risk of recurrence. This is called neoadjuvant hormonal therapy. It may also be given after surgery to reduce the risk of recurrence. This is called adjuvant hormonal therapy.

Types of hormonal therapy
(i) Tamoxifen
Tamoxifen is a drug that blocks estrogen from binding to breast cancer cells. It is effective for lowering the risk of recurrence in the breast that had cancer, the risk of developing cancer in the other breast, and the risk of distant recurrence. Tamoxifen works well in women who have been through menopause and those who have not. Tamoxifen is a pill that is taken daily by mouth. Common side effects of tamoxifen include hot flushes as well as vaginal dryness, discharge or bleeding. Very rare risks include a cancer of the lining of the uterus, cataracts, and blood clots. However, tamoxifen may improve bone health and cholesterol levels [71].

(ii) Aromatase inhibitors (AIs)
As breast and ovarian cancers require estrogen to grow, AIs are taken to either block the production of estrogen or block the action of estrogen on receptors. AIs decrease the amount of estrogen made in tissues other than the ovaries in postmenopausal women by blocking the aromatase enzyme [71]. This enzyme changes weak male hormones called androgens into estrogen when the ovaries have stopped making estrogen during menopause. In contrast to premenopausal women, in whom most of the estrogen is produced in the ovaries, in postmenopausal women estrogen is mainly produced in peripheral tissues of the body. Because some breast cancers respond to estrogen, lowering estrogen production at the site of the cancer (i.e. the adipose tissue of the breast) with aromatase inhibitors has been proven to be an effective treatment for hormone-sensitive breast cancer in postmenopausal women. These drugs include anastrozole (Arimidex), exemestane (Aromasin), and letrozole (Femara). All of the AIs are pills taken daily by mouth. Only women who have gone through menopause or who have had medicines to stop the ovaries from making estrogen can take AIs. The side effects of AIs may include muscle and joint pain, hot flushes, vaginal dryness, an increased risk of osteoporosis and broken bones, and increased cholesterol levels. Research shows that all AIs work equally well and have similar side effects [71].

(iii) Ovarian suppression
Ovarian suppression is the use of drugs or surgery to stop the ovaries from producing estrogen. It may be used in addition to another type of hormonal therapy for women who have not been through menopause. There are two methods used for ovarian suppression.

One is use of gonadotropin or luteinizing releasing hormone (GnRH or LHRH) drugs to stop the ovaries from making estrogen, causing temporary menopause. Goserelin (Zoladex) and leuprolide (Eligard, Lupron) are types of these drugs [71]. They are given by injection and stop the ovaries from making estrogen for 1–3 months. The effects of GnRH drugs go away if treatment is stopped [71].

Another is surgery to remove the ovaries, which also stops estrogen production. While this is permanent, it can be a good option for women who have finished having children, especially since the cost is typically lower over the long term [71].

(III) Targeted therapy

Breast cancer targeted therapies are used to treat patients whose breast cancer cells overexpress certain characteristic proteins on their surface allowing an abnormal growth pattern [74]. Targeted therapy is a treatment that targets the cancer's specific genes, proteins, or the tissue environment that contributes to cancer growth and survival. These treatments are very focused and work differently than chemotherapy. This type of treatment blocks the growth and spread of cancer cells while limiting damage to healthy cells. The first approved targeted therapies for breast cancer were hormonal therapies. Then, HER2-targeted therapies were approved to treat HER2-positive breast cancer.

HER2-targeted therapy

The HER2 protein represents the most common overexpressed receptor signature in breast cancer and is considered a relevant biomarker for treatment [74].

(i) Trastuzumab

The first recombinant antibody approved by the FDA to target HER2-positive breast cancers was trastuzumab or Herceptin followed by other agents like pertuzumab and lapatinib [74]. Currently, patients with stage I to stage III breast cancer should receive a trastuzumab-based regimen often including a combination of trastuzumab with chemotherapy, followed by completion of one year of adjuvant trastuzumab. Patients receiving trastuzumab have a small (2%–5%) risk of heart problems. This risk is increased if a patient has other risk factors for heart disease or receives chemotherapy that also increases the risk of heart problems at the same time [71]. These heart problems may go away and can be treated with medication.

(ii) Pertuzumab (Perjeta)

This drug is approved as part of neoadjuvant treatment for stage II and stage III breast cancer in combination with trastuzumab and chemotherapy [71].

(iii) Neratinib (Nerlynx)

This oral drug is approved as a treatment for higher-risk HER2-positive, early-stage breast cancer. It is taken for a year, starting after patients have finished one year of trastuzumab [71].

Many of the following drugs are used for advanced or metastatic breast cancer.

(iv) Drugs that target the CDK4/6 protein

In breast cancer cells, CDK4/6 proteins stimulate cancer cell growth. Drugs including abemaciclib (Verzenio), palbociclib (Ibrance), and ribociclib (Kisqali) are approved for women with ER-positive, HER2-negative advanced or metastatic breast cancer and may be combined with some types of hormonal therapy [71].

(v) Lapatinib (Tykerb)

This is for women with HER2-positive advanced or metastatic breast cancer when other medications are no longer effective at controlling the cancer's growth. It may be combined with the chemotherapy capecitabine, the hormonal therapy letrozole, or the HER2-targeted therapy trastuzumab [71].

(vi) Larotrectinib (Vitrakvi)

This drug is for breast cancer with an *NTRK* fusion that is metastatic or cannot be removed with surgery and has worsened with other treatments [71].

(vii) Talazoparib (Talzenna)

Talazoparib is used for women with locally advanced or metastatic HER2-negative breast cancer and a BRCA1 or BRCA2 gene mutation.

(IV) Immunotherapy

The immune system is made up of a number of organs, tissues, and cells that work together to protect from foreign invaders that can cause disease. When a disease- or infection-causing agent, such as a bacterium, virus, or fungus, gets into your body, your immune system reacts and works to kill the invaders. This self-defense system works to keep you from getting sick. Immunotherapy uses substances—either made naturally by your body or man-made in a lab—to boost the immune system to stop or slow cancer cell growth, stop cancer cells from spreading to other parts of the body, to be better at killing cancer cells.

Immunotherapy, also called biologic therapy, is designed to boost the body's natural defenses to fight the cancer. It uses materials made either by the body or in a laboratory to improve, target, or restore immune system function.

(i) Atezolizumab (Tecentriq)

In 2019, the US Food and Drug Administration (FDA) approved a combination of atezolizumab plus protein-bound paclitaxel for locally advanced triple negative breast cancer that cannot be removed with surgery and metastatic triple negative breast cancer. In addition, it is only approved for breast cancers that test positive for PD-L1 [71]. PD-1 is a type of checkpoint protein found on T cells. PD-L1 is another checkpoint protein found on many healthy cells in the body. When PD-1 binds to PD-L1, it stops T cells from killing a cell. Still, some cancer cells have a lot of PD-L1 on their surface, which stops T cells from killing these cancer cells. An immune checkpoint inhibitor medicine such as atezolizumab, stops PD-1 from binding to PD-L1 allows T cells to attack the cancer cells [75].

(ii) Pembrolizumab (Keytruda)

This is a type of immunotherapy that is approved by the FDA to treat metastatic cancer or cancer that cannot be treated with surgery. These tumors must also have a molecular alteration called microsatellite instability-high (MSI-H) or DNA mismatch repair deficiency (dMMR) [71]. MSI-H describes cancer cells that have a greater than normal number of genetic markers called microsatellites. Microsatellites are short, repeated, sequences of DNA. Cancer cells that have large numbers of microsatellites may have defects in the ability to correct mistakes that occur when DNA is copied in the cell [76]. Different types of immunotherapy can cause different side effects. Common side effects include skin reactions, flu-like symptoms, diarrhea, and weight changes [71].

7.4.4 Challenges in radiotherapy and drug delivery

Surgery and radiation therapy modalities are used mainly for eradicating the primary tumor and locoregional cancerous tissues. Their value tends to decline as the cancer progresses and metastasizes [77]. The efficiency of radiotherapy relies on the irreversible damage that the ionizing radiation provokes to the DNA of injured cells, which eventually kills them or avoids their reproductive cycle, controlling in this way the progress of the tumor. During the treatment of tumoral tissues with radiotherapy special care must be taken to avoid radiation exposure of surrounding healthy tissues in order to provide the patient with an improved quality of life in the medium and long term future [78]. So exposure to healthy tissue is an important challenge for the radiation oncologist and physicist. Chemoradiotherapy, the concurrent administration of chemotherapy and radiotherapy, is part of the standard of care and curative treatment for many cancers. However, the combination treatment also significantly increases toxicity. For example, chemoradiotherapy in lung cancer can carry a mortality risk of approximately 5%, which is higher than either chemotherapy or radiotherapy alone [8]. Many cancers, such as pancreatic cancer and glioblastoma, are also relatively resistant to radiotherapy. There is a need to further improve the therapeutic efficacy of radiotherapy in these less radio-responsive tumors. Radiotherapy treatment is performed at tissue or organ size scales, not at cellular level, so it is important to delimit as precisely as possible the volume of the injured region to be irradiated. This area delimitation also includes certain normal tissue that could be somehow affected by tumor cells and which is recommended to be removed in order to achieve a better control of the tumor. Therefore, it is today a challenge to kill and control tumors at cellular level, for example using targets or markers able to identify and selectively attach to tumor cells, allowing a more localized treatment and eradication of the malign cells, whereas the harmful secondary effects induced on the healthy ones would be significantly reduced [78].

Although drug therapy can achieve systemic treatment, their current success rates are typically suboptimal. Conventional drug delivery approaches suffer from several limitations like lack of selectivity and cytotoxicity by non-targeted cells. Therefore, successful targeting strategies must be determined to overcome non-specific uptake by non-targeted cells [70]. There are several obstacles that limit their effectiveness. Table 7.9 summarizes these challenges. These challenges actually fall into three major categories. Items 1–4 are issues derived from suboptimal biodistribution of the drug in body, that is, too little drug in tumor tissues (so, suboptimal efficacy) and too much in healthy tissues (so, high toxicity). Items 5–7 are related to the poor response to the drug even though it reaches the tumor, while the last two are related to the inherent properties of the drug or drug combination itself [77].

Nanomedicine has the potential to overcome at least some of these limitations. Nanomaterials have an interesting impact on various fields of application due to their desirable properties, such as the large surface area-to-volume ratio, thus allowing a wide range of molecules to be coated around their surface [79], which provides an opportunity to manipulate their surface properties for improved treatment, for example, cancer targeting, extended circulation, increased endocytosis and transcytosis, in order to gain more efficient access into tumor sites, metastatic sites and cancer cells [77]. Moreover, by entrapping in or binding onto nanocarriers, the therapeutic agents can also gain better stability, increased solubility and

Table 7.9. A summary of the key challenges to breast cancer drug therapy and the ways nanomedicine can be used to tackle these challenges [77].

Challenges to breast cancer drug therapy	How nanomedicine can help
1. Insufficient specificity for breast cancer	Passive targeting and active targeting by nanomedicine to increase tumor drug level and decrease noncancer drug levels
2. Inefficient access of drugs to metastatic sites such as brain and bone	Many nanomedicine formulations inherently may improve brain and bone penetration
3. Undesirable pharmacokinetics such as quick clearance and short half-life	Use of strategies such as PEGlyation to extend the circulation time
4. Dose-limiting toxicity of the anticancer drugs or the excipients, for example, surfactants and organic co-solvents	Increased tumor specificity as above; controlled drug release from nanocarrier; solvent-, surfactant-free nanoformulation
5. Drug resistance at cellular level, for example, increased drug efflux transport	Passive and active targeting both may enhance endocytosis; some nanoformulations may inhibit drug efflux mechanisms; co-delivery of agents that target drug resistance mechanisms
6. Drug resistance at tumor microenvironment level, for example, lower pH, hypoxia, cancer microenvironment crosstalk and so on	Targeting tumor microenvironment; use of stimulus-responsive nanoformulations such as pH-responsive devices
7. Difficulty in eradicating cancer stem cells	Targeting cancer stem cells
8. Undesirable pharmaceutical properties of the drugs, for example, low aqueous solubility, poor *in vivo* stability	Many nanocarriers can achieve drug solubilization and can protect unstable drugs
9. Suboptimal dosing schedule and sequence, especially when combinations of multiple drugs are involved	Careful optimization of dosing schedule and sequence; use of nanocarrier to co-deliver multiple drugs

controlled release kinetics [77]. Drug combinations may also be co-delivered for increased synergistic or additive anticancer effects [78]. The use of these features to tackle the limitations of breast cancer drug therapy is summarized in table 7.9.

7.5 Breast cancer nanomedicine

Radiation therapy after surgery shrinks the tumor in combination with chemotherapy. But there are some side effects of radiation therapy, such as decreased sensation in the breast tissue or under the arm, skin problems in the treated area (including soreness, itching, peeling, and/or redness) and at the end of treatment the skin may become moist and weepy [6]. Systemic anticancer pharmaceuticals are usually hydrophobic, untargeted and toxic compounds, which can cause severe side effects [84]. Anthracyclines and taxanes are two mainstay chemotherapeutics used in advanced breast cancer chemotherapy over the last three decades [84, 85]. The family of anthracyclines, especially doxorubicin formulations, has been reported to induce cardiotoxicity [85]. Also, taxanes (paclitaxel and docetaxel) have several side effects including bone marrow suppression, hypersensitivity reactions, cutaneous reactions and dose-limiting neurotoxicity [85]. Because cytotoxic drugs typically carry numerous dose-limiting normal tissue side-effects, it is generally impractical to overcome this form of drug resistance simply by increasing the drug dose. To improve the therapeutic ratio of cancer chemotherapy, it is therefore critical to establish alternative approaches that may improve accumulation and prolong retention of cytotoxic drugs in drug-resistant cancer cells without causing additional normal tissue side-effects [86].

Nanocarrier-based drug delivery systems for chemotherapeutic drugs act efficiently on multiple malignant sites. There are several types of drug carriers commonly available such as polymeric dendrimers, micelles, microspheres, liposomes, quantum dots (QDs), nanoemulsions, gold nanoparticles (GNPs), and hydrogels, which require various methods of drug attachment including encapsulation, covalent binding, and adsorption. The various drug delivery platforms such as systemic (organic and inorganic), localized, and receptor-based are used for the model of breast cancer therapy as well as nanotherapy for breast cancer stem cell and for tumor microenvironment as shown in figure 7.6.

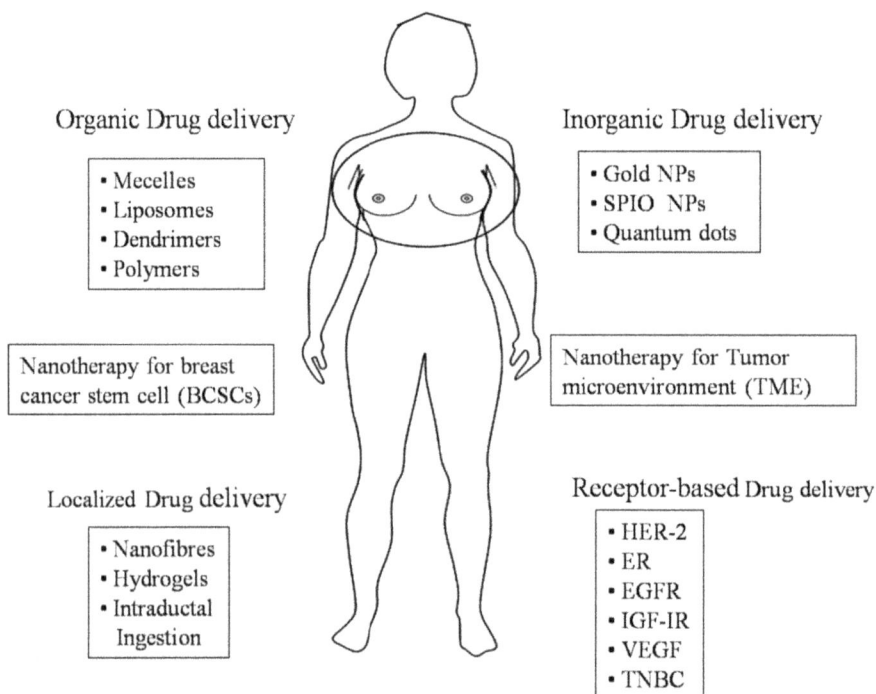

Figure 7.6. Drug delivery system for breast cancer. Reproduced with permission from reference [87], under the terms of the Creative Commons Attribution Non-Commercial License. Copyright © 2017 Singh *et al* published and licensed by Dove Medical Press Limited.

7.5.1 Latest trend for the development of nanomedicine based breast cancer treatment

Targeting of nanocarriers is of two types: active targeting and passive targeting. Passive targeting is achieved by incorporating the therapeutic agent into a nano-carrier that passively reaches the target organ. It is the accumulation of active ingredients at a specific site due to various physicochemical, pathophysiological and anatomical attributes [70]. Active targeting is usually achieved by conjugating a specific targeting ligand to the surface of nanoparticles that provide accumulation of nanoparticles at the desirable targeting site [70].

(A) Passive cancer targeting

Materials of small size can preferentially accumulate in tumors over normal tissues because of the leaky vasculature and poorly developed lymphatic drainage in tumor tissues. This passive targeting effect is known as EPR. EPR can enhance the drug delivery specificity for solid malignant diseases such as breast cancer. In addition to EPR, nanocarriers may further improve the anticancer effect of their loaded drug at a cellular level. Even without involving specific receptor-mediated activities, a nanocarrier can enter cancer cells by passive endocytosis mechanisms such as macropinocytosis to potentiate the efficacy of drugs that act on intracellular targets (e.g. RNA drugs, paclitaxel, and doxorubicin). Overall, even though passive targeting is generally not a highly specific and efficient drug delivery approach, it should not be ignored in the discussion of the latest nanomedicine research because of its simplicity. In fact, quite a number of recently developed breast cancer nanotherapies still rely on passive targeting.

(B) Active targeting directly on breast cancer cells

Active targeting, or ligand-mediated targeting, consists of binding ligands to the surface of the nanocarriers so that it can bind specifically to surface molecules or receptors overexpressed in the targeted tissue. It not only allows an accumulation of the nanocarrier at the targeted site, but also might induce receptor-mediated endocytosis, improving the intracellular delivery of the carried content [88].

7.5.2 Drug delivery systems for breast cancer

Systemic drug delivery approaches for breast cancer—organic and inorganic

The most common drug delivery approaches are based on organic and inorganic particles. The organic particles used for drug delivery application are micelles, liposomes, polymers, dendrimers, and nanogels [82, 87]. They have versatile surface building blocks for efficient endocytosis and loading. NPs have a multifunctional surface-modifying property that directs the cell to the tumor vasculature. The technology of encapsulating chemotherapeutic drugs using a nanoscale device is the best approach with regard to decreased side effects and improved bioavailability of drugs for breast cancer [87]. The surface modifier may be soluble polymer, pH or temperature-sensitive lipids or polymers, specific ligands, such as antibody, peptide, folate, transferring, etc [83]. Such surface modifications provide functionality, which in turn makes nanocarriers an efficient delivery system. Different surface modifiers can be used, as shown in table 7.10. For conjugating monoclonal antibody to the surface of nanocarriers, covalent coupling is a widely used technique. A monoclonal antibody attached on the surface of the nanocarrier provides improved specificity for targeting.

(A) Organic drug delivery approaches
(I) Micelles

Polymeric micelles (PMs) are colloidal particles prepared from conjugates of water-soluble polymers with phospholipids or long-chain fatty acids and other surfactants. Micelles are used for the delivery of water-insoluble chemotherapeutic drugs. They were first proposed by Paul Ehrlich for targeted drug delivery to diseased cells. Micelles accumulate at poorly vascularized tumors and enhance permeability and retention, and increase the half-life of anticancer agents [87]. Amphiphilic methoxypoly(ethylene glycol) grafted polyphosphazene with glycine ethyl ester side groups (PPP-g-PEG/GlyEt) has been synthesized and utilized for the preparation of doxorubicin loaded polymeric miceller systems. Doxorubicin loaded polymeric micelles have demonstrated better loading, sustained release with significant higher uptake in adriamycin-resistant human breast cancer MCF-7 cell line [89]. Moreover, fabricated immune micelles (antibodies bound to the surface of micelles) were also used in breast

Table 7.10. Surface modifiers for nanocarriers [70].

Surface modifiers	Example	Characteristics
Polymer	PEG	Hydrophilic, provide long circulation time
Endogenous ligand	Albumin A	protein that acts as an important transporter of nutrients and water-insoluble molecules and specifically accumulates in tumor tissues
Antibodies	Trastuzumab	Anh-HER-2
	Pertuzumab	Anti-HER-2
	Tykerb	Anti-HER-2
	Cetuximab	Anti-EGFR
	Bevacizumab	Anti-VEGF
ER-positive and lines *in vivo*	IR-3	Inhibits IGF-I-mediated growth of ERnegative cell
Receptor agonist/antagonist	Folate binding protein	Bind with folate receptor
	Transferrin	Bind with Transferrin receptor
Aptamer (phosphorprotein)	OPN-R3	Bind with Osteopontin
Peptide	RGD	Bind with $\alpha v \beta_3$ integrin receptor
Others	Chondroitin sulfate	Bind with Sle^x oligosaccharide glycosaminoglycans

adenocarcinomas. Thermosensitive docetaxel loaded injectable miceller systems have also been prepared using Pluronic F127 and Tween 80. Studies show that docetaxel loaded injectable miceller systems remained in the tumor mass for more than six days after post intratumoral injection. Results suggested that the mixed miceller system provided a promising locally delivered vehicle for a long period of time with improved efficacy [90]. Vasoactive intestinal peptide grafted sterically stabilized nanomicellar paclitaxel has also prepared for multidrug resistant breast cancer cells. Paclitaxel loaded both nanomicellar formulations have significantly inhibited cell growth in a dose-dependent fashion in drug sensitive MCF-7 cells. Both formulations were approximately 7-fold more potent than paclitaxel dissolved in DMSO [91]. Amphiphilic 4-armed star-shaped chlorin-core diblock co-polymers based on methoxypoly(ethylene glycol) and poly (epsilon-caprolactone) have been synthesized and miceller systems were prepared with paclitaxel. The chlorin-core micelle with paclitaxel loading exhibited synergistic therapeutic effects on MCF-7 breast cancer cells and improve the cytotoxicity of paclitaxel significantly in MCF-7 cells after irradiation [92]. In another study of the use of paclitaxel PM formulation in metastatic breast cancer patients, the Genexol-PM response rate was observed to be 58.5% compared to plain drugs that are in clinical trial Phases I and II. However, in SK-BR-3 cells, antibody-decorated NPs have shown 53.4% and 38.6% higher cellular uptake than the plain micelles in Phases I and II, respectively [93].

(II) Liposomes

Liposomes are spherical vesicle microparticles that contain single or multiple bilayered membrane structures and were first described in 1965. Their size varies from 50–200 nm and they have a tendency to accumulate in tumor cells with an enhanced permeability and retention [87]. Both lipophilic and hydrophilic drugs can be entrapped into liposomes because of their biphasic character. Lipophilic drugs are very poorly soluble in water, hence entrapped in the lipid bilayers of liposomes [70]. Hydrophilic drugs may be entrapped inside the aqueous core of liposomes or located in the external water phase. Moreover, conventional liposomes have major limitations as they are rapidly cleared by RES due to adsorption of opsonin proteins on the phospholipid membrane of liposomes [70]. Liposomal drug delivery to tumor is affected by long circulation time, stability (drug retention), and small vesicle size of liposome. The prolonged circulation achieved by pegylated doxorubicin liposomes (Doxil®, Caelyx ®; Alza Pharmaceuticals, San Bruno, CA, US) of size less than 100 nm proved that surface coating of a hydrophilic polymer PEG provides favorable *in vivo* pharmacokinetic properties. Pegylated liposomes, also known as sterically stabilized or 'stealth' liposomes, prevent opsonization, hence engulfment by mononuclear phagocytes is avoided and prolonged circulation is achieved. Anthracyclines have been encapsulated in different kinds of liposomes and the cardio-toxic effects of anthracyclines can be reduced by liposomal biodistribution [94]. Myocet ® (Cephalon Europe, Maison Alfort, France) is a non-pegylated liposomal formulation of doxorubicin and is approved for the treatment of metastatic breast cancer in Europe in combination with Cyclophosphamide. Caelyx ® is a pegylated liposomal formulation of Doxorubicin and is approved for the treatment of metastatic breast cancer, Kaposi's sarcoma, and refractory ovarian cancer in Europe and the US. Recently, an antibody (trastuzumab) conjugated thermosensitive liposomal system has been prepared using pore-forming protein listeriolysin O. Developed systems have demonstrated highly specific binding and internalization into the mammary epithelial cells that overexpress the human epidermal growth factor receptor 2 (HER2). Antibody conjugated liposomes delivered 22-fold greater concentration of calcein to mammary epithelial cells compared to the conventional approach [95].

Moreover, liposomes are reported to play a role in direct inhibition of P-gp by anionic membrane lipids. A previous study on Rhodamine retention using P-gp and BCRP substrate in breast cancer cell line MCF-7 showed that liposome encapsulation was increased in MCF-7/P-gp cells compared to MCF-7/wild-type cells [87]. Liposomes are reported as an effective delivery system for siRNA- or oligonucleotide-based therapy, and liposome-based drug formulations are currently used in clinical protocols. The encapsulation of drugs in liposomes reduces the toxicity through biodistribution. The therapeutic application of liposomes as a drug carrier for the delivery of paclitaxel has also been evaluated in human ovarian cancer [87].

(III) Polymers

Polymeric NPs (size 3–200 nm) that are formulated by binding a copolymer to a polymer matrix are widely used as drug delivery carriers. Polymeric nanoparticles have been developed with the need of nanocarriers for hydrophobic chemotherapeutics and hormone regulators, such as taxanes, camptothecin, cisplatin and tamoxifen [85]. Polymers are classified as natural, synthetic, biodegradable, and nonbiodegradable forms. The most commonly used natural polymers are cellulose, chitosan, alginate, and gelatin, which are mildly immunogenic in nature. The most important property of polymeric nanoparticles is the controlled drug delivery not only via the diffusion and swelling of nanostructures but also as the response to changes in environments such as pH and temperature or to the external stimulus such as light and ultrasound [85].

Moreover, synthetic polymers such as poly-ε-caprolactone (PCL), poly-(lactic-*co*-glycolide), and polylactide (PLA) have a high rate of solubility and permeability. Such polymers are biocompatible and biodegradable with slow degradation rate, with good drug stability and release. Numerous polymer–drug conjugates, such as poly (D,L-lactide-*co*-glycolide) (PLGA), PEG, dextran, and *N*-(2-hydroxypropyl) methacrylamide (HPMA), have been tested in drug delivery research [96]. Chemotherapeutic drugs like paclitaxel, doxorubicin, camptothecins, and platinates have been clinically tested in drug conjugates for multiple cancers. Genexol®-PM, paclitaxel-doped poly(D, L-lactic acid) is one of the polymeric nanomedicine in clinical trials for breast cancer treatment. Paclitaxel is an antimitotic chemotherapeutic which inhibits the depolymerization of the microtubule and results in cell-cycle arrest and inhibition of mitosis. It has been demonstrated clinically that Genexol-PM is able to deliver higher dose of paclitaxel (suggested dosage, 300 mg m^{-2}) than the conventional chemotherapy (175 mg m^{-2}) without additional dose-limiting toxicity. The half-life of Genexol-PM is 18.8 ± 3.1 h, which is 1.8-times longer than 6.8 ± 1.4 h for free paclitaxel [85]. It has been shown that polymeric NPs have a higher loading capacity for poorly water-soluble drugs, more stability, and more physicochemical properties (solubility, stability) compared to liposomes. The hybrid PM, developed by coating a PEG–phospholipid copolymer envelope on a nuclear PLGA NP, has improved therapeutic index with reduced toxicity [97]. A cisplatin-modified Pt(IV)-based PLGA–PEG NP was also reported with a significantly improved efficacy in breast cancer patients [98]. Lee and Nan proposed a novel combination of the drug delivery system for HER2-overexpressing metastatic breast cancer via HER2-targeted HPMA copolymer conjugates in combination with a tyrosine kinase inhibitor (PKI-166). Their study on targeting HER2 receptors via extracellular (via TRZ binding) and intracellular (via PKI-166 binding) kinase domains suggested the synergistic effect from a drug conjugate delivery system for anticancer activity [96]. Therefore, a novel drug delivery system using a polymer with different mechanisms of action can bring forth a promising targeted therapy to overcome the limitations of the individual drug. Although polymeric nanoparticles are effective drug carriers for hydrophobic chemotherapeutics or hormone regulators, there are still some shortcomings. The first concern is the stability of polymeric nanoparticles under harsh conditions such as high temperature and radiation during the sterilization, an essential step in nanomedicine manufacture [85]. Moreover, the presence of toxic residual solvents introduced during the manufacture of polymers and the exocytosis caused by undissolved polymeric structures cannot be ignored [85].

(IV) Dendrimers
Dendrimers are highly branched molecules of size 1–15 nm with a well-defined core, an interior region, and a large number of terminal groups. Dendrimers contain three different regions: core, branches, and surface. The monodispersity, water solubility, encapsulation ability, and large number of functionalizable peripheral groups of dendrimers make these an ideal carrier for drug delivery [70]. In 1978, Vogtle first described the nanotechnology platforms for drug delivery using dendrimers. Like other nanocarriers, the biocompatibility and pharmacokinetics of dendrimers are easy to predict and can be controlled [87]. There are various drug platforms that have been synthesized as delivery vehicles such as polyetherhydroxylamine (PEHAM), poly-amidoamine (PAMAM), polyesteramine, polypropyleneimine, and polyglycerol [87]. There are three architectural domains in dendrimers: (i) the core of the dendrimer, (ii) that is attached by branch cell layers, i.e. dendrons, (iii) having the multivalent surface with a larger number of reactive sites. Because of the presence of unimolecular structures, dendrimer–drug conjugates are more stable, easy to formulate and sterilize as compared to liposomes and micelles [70]. Polypropylenimine based dendrimers have been developed for delivering of 31 nt triplex forming oligonucleotide (ODN) in breast, prostate and ovarian cancer cell lines. Dendrimers enhanced the uptake of ODN by approximately 14 fold in MDA MB-231 breast cancer cells. Dendrimers exerted their effect in a concentration- and molecular weight-dependent manner. A biocompatible polyester dendrimer composed of the natural metabolites, glycerol and succinic acid has been developed for the encapsulation of camptothecins, and their derivatives. Cellular uptake and efflux measurements in MCF-7 cells showed an

increase of 16-fold for cellular uptake and an increase in drug retention within the cell when using the dendrimer vehicle [70]. The water solubility and size of dendrimers increased by PEGylation which helped to improve the retention and biodistribution characteristics. Several groups have shown that cell toxicity strongly correlates with the dendrimers end. The surface functional groups of dendrimers are amines that are decorated with protons, benzyloxycarbonyl- or tert-butoxycarbonyl-protecting group's ethylenediamine ligands, or dansyl fluorescence labels. The doxorubicin-containing poly ion complex micelle accumulates in the nucleus of drug-resistant MCF-7 cells and is also considered to have a potent antiproliferative effect on targeted tumor. The cytotoxicity of MCF-7 breast cancer cells was examined *in vitro* using low-generation (G0, G1, and G2) PAMAM-like polymers. However, dendrimers–drug conjugate has an antineoplastic agent and is covalently attached to the peripheral groups of the dendrimers, and has distinct advantages over drug-encapsulated systems. For local delivery in breast cancer, doxorubicin–G4–PAMAM complexes were encapsulated into the liposomes. These were formulated with HEPC and showed enhanced activity towards the MDA-MB435 breast cells compared to the individual dendrimers. Thus, the methods for delivering the dendrimers-based NPs for transport of drugs into the specific area of malignant cells could be the best approach for delivery of NPs and to treat cancer cells [87].

(B) Inorganic drug delivery approaches
(I) Gold nanoparticles
GNPs are used in chemotherapy for several cancers. Due to their small size (approximately 130 nm) and specificity, they circulate throughout the tumor cells. GNP coating acts as a biomarker for the cancer diagnosis and is used as a probe for transmission electron microscopy and antimicrobial agents. Gold nanoparticles have also exhibited unique chemical and physical properties for transporting and unloading the pharmaceuticals. They are essentially inert, non-toxic, and in the size range of 1 nm to 150 nm. Moreover, they can be functionalized and modified to improve the versatility of gold nanoparticles. Significant efforts have been devoted over the past years to the fabrication of gold nanoparticles for the delivery of anticancer drugs. Tamoxifen–poly(ethylene glycol)–thiol gold nanoparticle conjugates have been recently developed to selectively target and deliver plasmonic gold nanoparticles to estrogen receptor-positive breast cancer cells. Developed systems demonstrated 2.7 fold enhanced drug potency *in vitro* compared to conventional ones [70].

The conjugation of GNPs to transferrin molecules was tested in breast cancer cells, and the results showed higher cellular uptake of transferrin molecules bound to GNPs in comparison to unbound molecules. PEG-conjugated liposomes were used for anticancer drug delivery [87]. The gold nanorod–bombesin (GNR–BBN) conjugates showed extraordinary *in vitro* stabilities and high binding efficacy toward prostate and breast tumor cells. Dumbbelllike Au–Fe(3)O(4) nanoparticles (NPs) have been made and coupled with Herceptin and a platin complex. The platin–Au–Fe(3)O(4)–Herceptin NPs act as a target-specific nanocarrier for delivery of platin into HER2-positive breast cancer cells (Sk-Br3) with strong therapeutic effects [70]. Balakrishnan *et al* targeted the breast cancer EGFR/VEGFR-2 signaling pathway using AuNPs–Qu-5, and reported its role in inhibition of migration, invasion, angiogenesis, and metastasis of breast cancer cells. This group has studied significant inhibition of multiple proteins such as p-PI3K, Akt, Snail, Slug, vimentin, N-cadherin, and p-GSK3β with treatment with AuNPs–Qu-5 [99]. Another report showed that triple negative breast cancer MDA-MB-231 cells were inhibited by phytochemical compounds such as gallic acid capped with GNPs [87].

(II) SPIO-NPs
Superparamagnetic iron oxides (SPIOs) nanoparticles have the ability to control the physical and chemical properties of particles such as their shape, size, and surface chemistry. SPIO-NPs have several applications in detection of inflammatory diseases and targeting of surface markers on tumors. SPIO consists of two components, an iron oxide core and a hydrophilic coating of the magnetic particle biomolecule, which allow it to deliver nano-derived biomolecules in a targeted area. Biopolymers such as PEG, polyacrylic acid, dextran, alginate, polyethyleneimine, and poly (vinyl alcohol) (PVA) are used as coating reagents for the surface stabilization of SPIOs [87]. They bind to tumor sites for delivery of antibodies, enzymes, proteins, drugs, or nucleotides. The uptake of SPIO-loaded PLA–tocopheryl PEG succinate (SPIO–PNPs) by MCF-7 breast cancer cells was confirmed through TEM in several experiments [100]. SPIO-targeted biomarkers have been developed for tumor cell imaging and detection. SPIO–Herceptin detects over-expression of HER2/neu (c-erbB-2) tyrosine kinase receptor in the metastatic breast cancer [101]. SPIOs are most efficiently used in magnetic resonance imaging and macrophage processing. However, knowledge concerning breast cancer and metastatic lymph nodes injection of SPIOs is lacking and needs to be explored. In future, SPIO-NPs could be applied as an effective treatment agent in breast cancer therapy.

(III) Quantum dots (QD)
Quantum dots (QDs) are tiny semiconductor particles a few nanometers in size, having optical and electronic properties that differ from larger LED particles. QD nanocrystals have a tunable wavelength, high brightness, anti-photo bleach, and optical properties, and are used as probes for many biological and biomedical applications. The conjugation of surface-modified QDs with antibodies, peptides, or other biomolecules enables their application in clinical oncology targeting. Bae *et al* synthesized a bimodal imaging nanoprobe by conjugating monoclonal antibodies and perfluorocarbon (PFC)/QD nanoemulsions for the detection of surface antigens on breast cancer cells (SK-BR-3, MCF-7, MDA-MB468) and also proposed that PFC/QD nanoemulsion had a great capacity for imaging therapy of tumor cells [102].

(C) Localized drug delivery approaches
Localized drug delivery has more impact as a therapeutic option for early-stage cancers compared to the systemic drug. There are natural (dextran, chitosan, hyaluronic acid, gelatin, collagen polypeptides) and synthetic polymers that are used intratumorally in the cancerous tissue for drug delivery to cure breast cancer [87]. Furthermore, hydrogel formation of the NPs or polymers system, nanofiber with versatile morphology and tensile strength, and intraductal injection using

microcatheter enhance the performance of ongoing smart drug delivery therapy [87]. A nanofiber is a cross-linked polymer characterized by tensile strength and chemical nature. Biodegradable polymers such as PEG, PLGA, chitosan, PVA, PLA, poly-ethylene oxide, and PCL are used for preparing nanofibers for drug delivery applications [87]. Generally, nanofibers that carry drugs follow several basic designs —nanofibers with homogenous structures in which the drug is dispersed throughout the polymer matrix, core–shell nanofibers for which the matrix carrying the drug is covered by pure polymer and nanofibers with the pharmacologically active compounds immobilized on their surface [103]. Another type of nanofiber, electrospun, was found to be bioactive and biocompatible similar to a human extracellular matrix, which supports diverse cells to grow into fabricated tissues. A nanofiber-based platform has been prepared to evaluate migration of metastatic breast cancer cells. Curcumin-loaded PCL nanofibers were tested in breast cancer cell line MCF-7, and exhibited 15% more cytotoxicity compared to the commercial drug [104].

Hydrogels are 3D, cross-linked networks of water-soluble polymers. Hydrogels can be made from virtually any water-soluble polymer, encompassing a wide range of chemical compositions and bulk physical properties. Furthermore, hydrogels can be formulated in a variety of physical forms, including slabs, microparticles, nanoparticles, coatings, and films. Hydrogels are water-insoluble molecules, chemically or physically linked into a polymer chain. The unique physical properties of hydrogels have sparked particular interest for their use in drug delivery applications. Their highly porous structure can easily be tuned by controlling the density of cross-links in the gel matrix and the affinity of the hydrogels for the aqueous environment in which they are swollen. Their porosity also permits loading of drugs into the gel matrix and subsequent drug release at a rate dependent on the diffusion coefficient of the small molecule or macromolecule through the gel network [105]. Chitosan is particularly attractive for clinical and biological applications due to its low toxicity, biocompatibility and biodegradability. Thus, a chitosan based hydrogel system should lead to enhanced concentrations of therapeutic payloads at tumor sites, minimize concerns about side effects, and ultimately raise the therapeutic index [106]. Hydrogels for tumor cell therapy are developed in the form of microspheres or NPs. They are fabricated using protein and glycosaminoglycan components of breast tissue, which stimulate the growth of human breast cells. The most common endothermal hydrogel was based on chitosan, which is formed in tumor tissue after intratumor injection [107]. Chitosan hydrogels based on temperature-responsive hydroxyl butyl, poly (vinyl alcohol), thermosensitive poly (ethylene glycol)-grafted, chitosan chloride/glycerophosphate and chitosan/bifunctional aldehyde have been investigated, but not tested in the preclinical trial for breast cancer application. However, only a few of the hydrogels have *in situ* gelling properties. Another platform for local and sustained delivery with high efficiencies in *in vitro* and *in vivo* breast cancer mice model was reported via siRNA encapsulation in oligopeptide-terminated poly (β-amino ester) NPs [87].

Ductal carcinoma *in situ* is a noninvasive early cancer, occurring in the lining of the breast milk duct, and represents 80% of breast cancers diagnosed. The use of a microcatheter for collecting ductal cells can improve the ductal epithelium cells detection in abnormal breast cells. Ductal lavage procedure using microcatheter has been reported for cytological analysis with high efficiency for collecting breast epithelial cells. A study of 507 women who had a high risk of breast cancer was conducted to evaluate nipple aspirate fluid and ductal lavage and proved that ductal lavage was more sensitive and safer than nipple aspiration [107]. Moreover, chemotherapy through localized drug delivery was achieved by intraductal injection of chemotherapeutic drugs 5-fluorouracil and estradiol into mammary papilloma for the improvement of the immune response [108]. Detection of ductal cellular abnormalities can provide additional information to reduce the risk of breast cancer and also help in ongoing drug therapy.

(D) Receptor-based active targeting
Breast cancer growths are regulated by multiple receptors, and inhibition at the receptor provides a new avenue for cancer therapy. Studies on receptor targeting are being used in clinical trials in patients with metastatic breast cancer. Several studies have focused on receptors HER2, EGFR, IGF-IR, and VEGFR, which revealed specific targets for breast cancer cells [87]. Surface modification of nanoparticles using specific ligand conjugation characterizes active targeting. For example, specific cell surface receptors are overexpressed on breast cancer cells. Nanoparticle surface modification with peptides, aptamers, monoclonal antibodies and small molecules which bind to the overexpressed receptor may increase cell-specific uptake via receptor-mediated endocytosis (RME), whereby the drug delivery system DDS accumulates inside the target cell and delivers the drug payload. Upon encountering the acidic environment of the endosome, a portion of the cell membrane which envelopes the DDS, transfers the DDS from the extracellular to intracellular domain, and finally, ligand-receptor complexes dissociate, releasing free receptors, which are recycled to the cellular plasma membrane. The drug delivery approach is shown in figure 7.7

(I) HER2 (human epidermal growth factor receptor-2)
The human epidermal growth factor receptor (HER) family of receptors plays a central role in the pathogenesis of several human cancers. They regulate cell growth, survival, and differentiation via multiple signal transduction pathways and participate in cellular proliferation and differentiation. The family is made up of four main members: HER1, HER2, HER3, and HER4, also called ErbB1, ErbB2, ErbB3, and ErbB4, respectively [109]. Unlike other members of the HER family (HER1, HER3 and HER4), there is no identifiable ligand for HER2, but HER2 can dimerize with other HER receptors. The heterodimerization leads to the phosphorylation of the HER kinase domains. This activates several signal pathways related to transcription, translation and protein stability, including the MAPK, PI3K/Akt and phospholipase Cγ. Therefore, HER2 is important in the regulation of gene expression, cell mobility, growth, proliferation, apoptosis and other cellular responses, especially for cancerous cells [85]. HER2/HER3 is the most active and tumor promoting dimerized combination. HER2 is overexpressed by ~20% – 30% of human breast tumors, which promotes the growth of cancer cells. In about one of

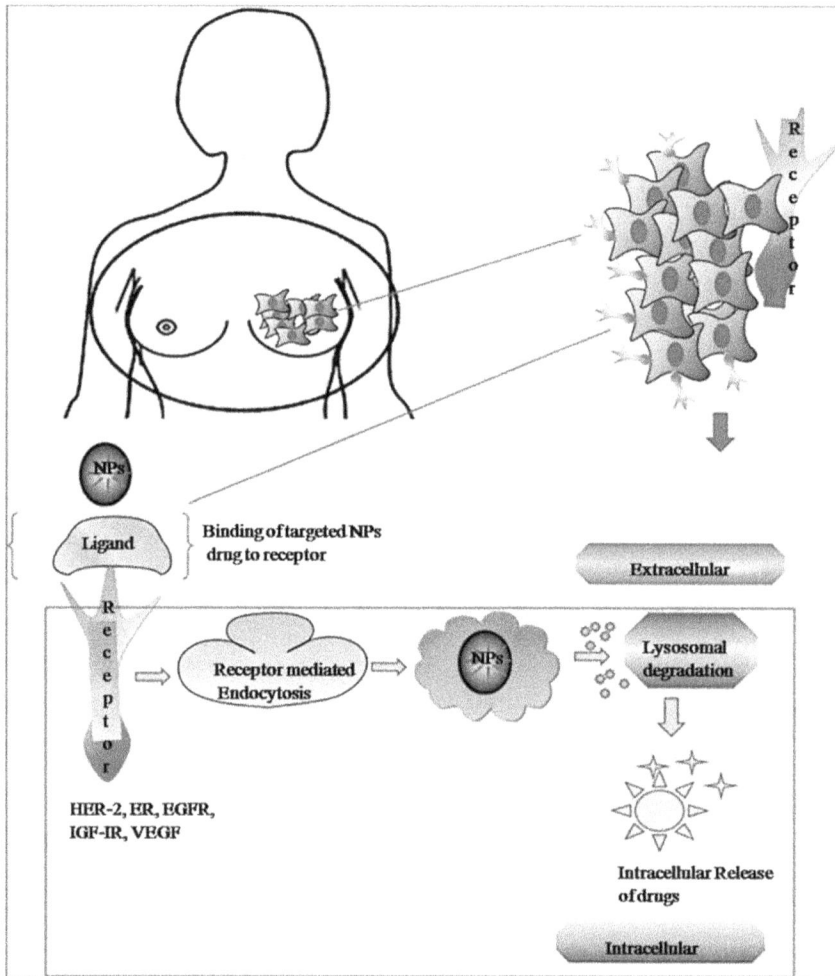

Figure 7.7. Receptor-mediated drug delivery to metastatic breast cancer cells. Nanocarrier-based drug targeting using receptor-mediated pathways governs the major therapeutic approach for the active sites in tumor cells. Ligand nanoparticle conjugate binds to the receptors (EGFR, ER,VEGFR, HER2, IGF-IR) on the membrane, mediates internalization of nanoparticles through endocytosis, and releases the drugs by lysosomal degradation to the active sites of tumor cells. Reproduced with permission from [87] under the terms of the Creative Commons Attribution Non-Commercial License. Copyright © 2017 Singh *et al* published and licensed by Dove Medical Press Limited.

every five breast cancers, the cancer cells have a gene mutation that makes an excess of the HER2 protein. HER2-positive breast cancers tend to be more aggressive than other types of breast cancer [110]. HER2 has been reported to be overexpressed in breast cells. However, the overexpression of HER2 can be utilized in the development of targeting nanomedicine to treat HER2-positive breast cancer. The first clinical HER2 targeting agent is the humanized monoclonal antibody, trastuzumab (Herceptin®). There are two antigen-specific sites in trastuzumab, which can bind to HER2, preventing the activation of tyrosine kinase and HER2 dimerization. In addition, trastuzumab activates the immune effector cells for antibody-dependent cell mediated cytotoxicity to promote cancer cell death. The presence of trastuzumab also enhances the therapeutic effects of other chemotherapeutics. Therefore, trastuzumab is widely used as a single agent or combined with other chemotherapeutics in adjuvant therapy to reduce breast cancer recurrence rate and improve the overall survival of HER2-positive breast cancer patient's clinically [85]. Table 7.11 summarizes a number of recent studies on HER2-targeted nanomedicine.

Table 7.11. Nanomedicine for treatment of triple negative breast cancer target [77].

Target	Nanocarrier	Therapeutic agent(s)	Key outcomes
EGFR	pH/redox dual-sensitive cationic unimolecular NP	siRNA	GE11 peptide, an anti-EGFR peptide, was found to significantly enhance the cellular uptake of NPs in MDA-MB-468 TNBC cells
	RNA-NPs decorated with EGFR-targeting aptamer	Anti-miRNA	Strong accumulation of the NPs in orthotopic TNBC tumor model with reduced renal and liver clearance was observed
	Immunoliposomes decorated with anti-EGFR antibody	Doxorubicin	Phase I study indicated good tolerability and recorded clinical activity
Folate receptor	Micelles of copolymer functionalized with folate	Orlistat	*In vitro* and *in vivo* anticancer activities through PARP inhibition reported
	Folate-conjugated liposomes	Benzoporphyrin derivative	Monolayer and 3D MDA-MB-231 cell model was more responsive to the targeted formulation
CXCR4	pH-sensitive immunoliposome conjugated with anti-CXCR4 antibody	siRNA silencing oncoprotein lipocalin-2	Significant lipocalin-2 knockdown and reduction in cell migration reported
	Nanostructured lipid carriers (AMD3100 coated as targeting ligands)	CXCR4 antagonist AMD3100 and photothermal therapy agent IR780	Able to reduce TNBC metastasis and achieve improved photothermal therapy at the same time

(II) ER (estrogen receptor)

Estrogen plays a critical role in the development of breast cancer, because it can stimulate the growth of breast tissues by expressing growth factors via the function of estrogen receptor (ER) as a transcription activator. Anti-estrogen therapy is successful in ER-positive breast cancer therapy targeting ER receptors [111]. Two receptors for estrogen have been identified: ER α and ER β [112]. ER α becomes a transcription repressor upon binding of certain antagonists such as tamoxifen. But the patients with advanced breast cancer will develop resistance to tamoxifen after some time [113]. By selective estrogen receptor modulators(SERM) or selective estrogen receptor down-regulators (SERD), the targeting effects of estrogen on ER receptor may be blocked at the cellular level [114]. The efficacy of anti-estrogens and aromatase inhibitors is limited because only 60%–70% of the breast cancers are ER-positive [113]. As ER-negative breast cancers are more aggressive [115], other targeted therapies are required. Rochefort *et al* (2003) have reviewed two general approaches of targeted therapy in the case of ER-negative hormone-independent breast cancers. First is to transform ER-negative into ER-positive cancer cells by gene therapy or ER gene reexpression. Second is the targeting of genes or proteins which are actively involved in the progression of the corresponding breast cancer subset.

(III) EGFR (epidermal growth factor receptor)

Overexpression of EGFR has been reported in poorly differentiated triple negative and inflammatory breast cancer cells. EGFR gene was identified in the early 1980s, and the clinical interest in the gene began in the late 1990s with the development of inhibitors. There are several members of the EGFR family reported, including EGFR (also known as ErbB1 and HER1), HER2 (also known as HER2/neu and ErbB2), ErbB3 (HER3), and ErbB4 (HER4). Out of these, HER2 was overexpressed in breast cancer. It has been proven that the EGFR expression was correlated with an increased copy number of the gene and protein overexpression in breast cancer. The increased EGFR gene copy number and protein over-expression were observed in ER-negative, PR-negative, HER2 negative (triple negative) breast cancer patients [87]. Although drugs including cetuximab, lapatinib, gefitinib, and others have been developed to target the EGFR, the overall clinical outcome is poor. EGFR signaling and the relationship between triple negative and inflammatory breast cancer-targeted therapies are the current topic of interest in the field of breast tumor therapy. Inhibitors of the PI3K/AKT/mTOR pathway for deregulation in triple negative breast cancer are in early-phase clinical trials [87].

(IV) VEGF (vascular endothelial growth factor)

VEGF was recently reported as a potential target in breast cancer. VEGF was first identified by Ferrara in the 1980s [83]. The VEGF family and their corresponding tyrosine kinase receptors are known to play an important role in angiogenesis when upregulated by various hormones, cytokines, and transforming growth factors [70]. It is associated with the development, progression, and metastasis of breast cancer, through receptors such as VEGFR-2 (also known as flk/kdr), VEGFR-1 (also known as flt), and VEGF-C (homolog of VEGF gene family). VEGF-2 (flk) and VEGF-1 (flt) are expressed on vascular endothelial or non-endothelial cell-specific receptor [87]. In breast cancer patients, significantly higher plasma levels of VEGF and lower plasma levels of sVEGFR-1 are found as compared to healthy controls. As VEGF has a critical role in tumor growth and metastasis, the anti-VEGF blocking antibody bevacizumab showed remarkable results in the treatment of metastatic colorectal cancer and it has recently been approved by the FDA. Recently, bevacizumab (directed against VEGF) has been evaluated in Phase III trials of metastatic breast cancer and renal cell carcinoma [116]. In patients with metastatic breast cancer during a phase I/II study of bevacizumab combination with trastuzumab, 54% response rate was reported but at least six patients experienced some cardiotoxicity. In this, HER2 amplification is associated with an increase in VEGF gene expression, demonstrated by preclinical models [117]. Other targeted therapy includes small-molecule VEGF receptor inhibitors such as PTK787, SU11248, etc [70].

(V) IGF-IR (insulin-like growth factor)

Breast cancer growth is regulated by receptor tyrosine kinases (RTKs), and the inhibition of the receptors, thus, could be the targets for anticancer therapy. The growth and differentiation of normal breast cells are mediated by IGF-IR signaling. Additionally, it stimulates mitogenesis and apoptosis of tumor cells. RTKs contain two domains—intracellular tyrosine kinase domain and extracellular ligand-binding domain. The ligands binding to the IGF-IR activate tyrosine kinases and induce conformational changes [118]. After activation, antiapoptotic effects of the IGF-IR are mediated via the Akt/PI3K pathways and IGF-IR is overexpressed in many cell types [119]. Inhibition of these RTKs helps in reduction of cell growth and drug development [120]. Several reports demonstrate the evidence for overexpression and hyperactivation of the IGF-IR in the early stages of breast cancer [121–123]. Nordihydroguaiaretic acid is a phenolic compound and is reported as a direct inhibitor of both IGF-IR and the HER2/neu receptor in breast tumor cells and induces apoptosis. Thus, negative expression of IGF-IR with a potential inhibitor can play an important role in breast cancer therapy.

(VI) Targeting triple negative breast cancer

Breast cancer that is estrogen receptor-negative, progesterone receptor-negative and HER2-negative is known as targeting triple negative breast cancer (TNBC). About 15%–20% of breast cancer cases are TNBC [77]. This fairly common breast cancer subtype is particularly difficult to treat and more deadly for several reasons. It is considered the most aggressive form of breast cancer and has higher incidence of BRCA-1 mutations. Its aggressiveness means that when it is diagnosed, it is more often at an advanced stage [77]. Without the expression of those three receptors, TNBC is not sensitive to hormone therapy and anti-HER2 therapy. Chemotherapy becomes the only useful drug therapy, but about 60% of TNBC cases were found to be chemoresistant. Besides HER2, several other molecular targets have been studied for active targeting of TNBC. EGFR is overexpressed in up to half of the TNBC cases and has a high density on the cell surface. Table 7.11 shows studies of nanodelivery to TNBC, including the use of anti-EGFR peptide, aptamer and monoclonal antibody. All of them indicate good cellular uptake by the TNBC cells and good efficacy *in vitro* and *in vivo*. Overall, EGFR is a promising target for nanotherapy of TNBC [77].

Folate receptor is also commonly targeted for nanodelivery because folate receptor is expressed in 50%–86% of metastatic TNBC patients and these patients generally have poorer prognosis. Other than folate receptor, C-X-C chemokine receptor type 4 is a potentially useful TNBC cell target, as it is often highly expressed in TNBC cells. The biggest appeal of this target is that its expression is associated with high risk of TNBC metastasis [77].

(E) Nanotherapy of breast cancer stem cells

It is suggested that a small fraction of cancer cells have the capacity for self-renewal and differentiation to multiple cancer cell types. They are often termed as cancer stem cells or tumor initiating cells. As long as these cells survive the anticancer treatment, they may serve as the 'seeds' to enable the formation of a full tumor (i.e. tumorigenic) [77]. In addition, accumulating evidence indicates that these cells are highly aggressive, have strong metastatic potential and are generally more drug resistant. It is, therefore, promising to prevent cancer relapse and metastasis by targeting these cells with nanomedicine [77]. Breast cancer stem cells (BCSCs) were initially discovered in 2003 by Al-Hajj *et al*. BCSCs actually do not have many well-characterized biomarkers for targeting. CD44 receptor is by far the most popular because it is highly expressed on BCSCs and also serves as a crucial signaling platform for the integration of the cues (e.g. growth factors and cytokines) from the tumor microenvironment (TME) [124]. Besides anti-CD44 monoclonal antibodies, hyaluronic acid is also a popular ligand for CD44 targeting, as CD44 is a receptor of hyaluronic acid. Other valuable targets of BCSCs include CD133. Swaminathan *et al* developed PLGA nanoparticles decorated with anti-CD133 antibody for pacli-taxel delivery [125]. Interestingly, this study showed that while free paclitaxel effectively inhibited tumor growth at the beginning, the tumors rebounded quickly once the treatment was discontinued, whereas no such problem was encountered with the CD133-targeted nanoparticles. The study demonstrated that the key benefit of cancer stem cell nanodelivery may actually be preventing breast cancer recur-rence, which is critical in breast cancer treatment. One key factor that makes BCSC targeting difficult is that BCSCs often enter a dormant state and their population is low [126]. One recent strategy proposed to achieve improved BCSC targeting is through dual targeting. Qiao *et al* showed that the hyaluronic acid and DCLK1 antibody dual-decorated nanoparticles target the mammospheres better than single-targeted system [127]. Table 7.12 lists nanotherapy for BCSCs.

(F) Tumor microenvironment-based nanotherapy

Tumor microenvironment (TME) provides the necessary conditions for the growth, invasion, and metastasis of cancer cells, akin to Stephen Paget's hypothesis of 'seed and soil.' Hence, the TME is a significant target for cancer therapy via

Table 7.12. Nanomedicine for breast cancer stem cell therapy [77].

Target	Nanocarrier	Therapeutic agent(s)	Key outcomes
CD44	PLGA-*co*-PEG micelles decorated with anti-CD44 antibodies	Paclitaxel	Using a new fluorescent cancer cell model, they were able to demonstrate improved sensitivity of cancer stem cells to paclitaxel
	PLGA nanoparticles coated with hyaluronic acid	Salinomycin and paclitaxel	Surface coating of hyaluronic acid led to a 1.5 fold increase in uptake into the CD44+ MDA-MB-231 cells and highest *in vitro* activity
	Chitosan-decorated Pluronic F127 nanoparticles	Doxorubicin	Significant improvement in doxorubicin delivery to CD44+ cells was reported with strong cytotoxicity
	Nanoparticles consist of four polymers PLGA Pluronic F127, chitosan and hyaluronic acid	Doxorubicin and irinotecan	Nanoparticles effective in cancer stem cells *in vitro* and *in vivo*, with up to ~500 times of enhancement versus simple mixture of two drugs
CD133	PLGA nanoparticles decorated with anti-CD133 antibody	Paclitaxel	Effective in decreasing the number of MDA-MB-231 mammospheres and colonies that are refractory to paclitaxel
CD44+DCLK1	PLGA–PEG nanoparticles dual grafted with hyaluronic acid and DCLK1 antibody	No drug	Dual-grafted nanoparticles exhibited a targeting effect toward CSCs *in vitro* and *in vivo* stronger than when only one targeting moiety was used

nanoparticles. From the nanodelivery perspective, the TME is moderately more acidic (pH 6.5–6.9, as low as 5) than the physiological pH of normal tissue (7.2–7.5) and more hypoxic, and these properties can be exploited for nanodelivery. Targeting the TME may interfere in its interactions with the cancer cells to achieve effective treatment. To date, several strategies have been studied for nanodelivery to breast cancer TME. These include pH-responsive delivery, targeting hypoxia, targeting TAMs and targeting other TME components [77].

(I) pH-responsive drug delivery

This acidic TME is due to the higher glycolysis rate of cancer cells to obtain the energy required for survival by converting glucose into lactic acid. The pH variation in tumor cells may play an important role in designing a pH-responsive cancer targeting system [128]. When these nanoparticles reach tumors where the micro-environmental pH is slightly acidic, a pH-dependent structural transformation occurs. The acidic environment at the tumor site triggers the protonation of pH-sensitive moieties, thereby disrupting the hydrophilic–hydrophobic equilibrium within the nanoparticle, in turn causing structural transformation and the release of therapeutic cargo loaded inside [128]. Poly(histidine) (pHis) is an attractive candidate that has been extensively used for the fabrication of apH-sensitive drug delivery system. Several pH-responsive nanodelivery systems were, therefore, developed for breast cancer treatment (table 7.13(A)). In general, drug release was all faster at lower pH due to acid hydrolysis of the linkages in the nanocarrier.

(II) Nanotherapy of hypoxia-related events

Another unique characteristic is hypoxia, wherein cells residing deep in the tumor mass are deprived of oxygen due to irregular vasculature networks inside the solid tumor such as breast cancer. The cells in these hypoxic regions proliferate more slowly than well-oxygenated cells, and these slow-growing cells are less susceptible to conventional anti-proliferating drugs [128]. Due to the central role of hypoxia in enhancing tumor angiogenesis, metastasis, epithelial to mesenchymal transition, tumor invasiveness and suppression of immune reactivity, there has arisen great interest in the development of nanoparticles that can target the hypoxic regions within the tumor [128]. This microenvironment may also lead to the so-called 'pan-chemoresistance' to a broad range of anticancer drugs [77]. Intervention of these complex, unfavorable events is very challenging. Drug compounds may hardly access and accumulate in the poorly vascularized TME in an efficient manner. The passive and active targeting effects of nanomedicine may improve the tumoral drug concentrations to better tackle selected hypoxia-related events for cancer treatment [77].

For instance, liposome of disulfiram was developed to reverse pan-chemoresist-ance caused by hypoxia-induced nuclear factor-κB [129]. The study showed that liposomal disulfiram was effective in disrupting the nuclear factor-κB pathway in spheroid cultured breast cancer cell model manifesting cancer stem cell character-istics and pan-chemoresistance, and this translated into significant *in vitro* and *in vivo* efficacy. In another series of studies of CRLX101 [130], an investigational

Table 7.13. Tumor microenvironment-based nanotherapy: pH-responsive nanocarriers [77].

Nanocarrier	Therapeutic agent(s)	Key outcomes
(A) pH-responsive nanocarriers		
Micelles of amphiphilic copolymer joined by β-thiopropionate linkage	Doxorubicin	Linkage can undergo acid hydrolysis. Drug release 80% at pH 5.2% versus 35% at pH 7.4 after 100 h
Chitosan-based glycolipid-like nanocarrier	Doxorubicin	More cytotoxic to MCF-7 breast cancer cells than to SKOV3 ovarian cancer cells because the former cell line has more acidic extracellular environment
pH-responsive liposomes	Paclitaxel	Faster paclitaxel release in acidic pH and more effective both *in vitro* and *in vivo* on breast cancer models
pH-responsive triblock copolymeric micelles with cell-penetrating peptides	Doxorubicin	Doxorubicin release was pH-dependent, about 65% released at pH 5.0 and 32% at pH 7.4. More cytotoxic than free doxorubicin on breast cancer cells
Acidity-sensitive linkage-bridged block copolymer nanoparticles	siRNA	PEG surface layer detached in response to tumor acidity to facilitate cellular uptake, and siRNA rapidly released within tumor cells due to the hydrophobic PLGA layer
(B) Targeting of tumor-associated macrophages		
Abraxane	Paclitaxel	In addition to EPR and gp60 targeting, Abraxane may increase the CD80+ CD86+ M1 macrophage subpopulation and work against M2 cells to provide additional anticancer effects
Legumain-targeting liposomal nanoparticles	Hydrazinocurcumin	By inhibiting the STAT3 activity of TAM, TAM got 're-educated' and switched to M1-like macrophages, leading to inhibition of 4T1 cell migration and invasion *in vitro* and suppression of tumor growth, angiogenesis and metastasis *in vivo*
PLGA nanoparticles with mannose	Doxorubicin	Significantly improve the anticancer effect of the nanoparticles in triple negative breast cancer, indicating depletion of TAM

(C) Targeting other targets in tumor microenvironment			
Target	Nanocarrier	Therapeutic agent(s)	Key outcomes
Stromal cells	Liposomes decorated with lipidated cathepsin B inhibitor	NS629	By targeting cathepsin B, selective targeting and internalization of liposomes observed, leading to enhanced delivery *ex vivo* and *in vivo* in an orthotopic breast cancer model
	Cellax® (nanoparticles of acetylated carhoxymethylcellulose linked with PEG)	Docetaxel	Reported higher MTD and lower tumor growth and metastasis than Abraxane in multiple xenograft models; also, decreased α-smooth muscle actin content in 4T1 and MDA-MB-231 model by 82% and 70%, respectively
MMP-9	Liposome with degradable lipopeptides	Carboxyfluorescein as fluorescent dye	Lipopeptide degraded by MMP-9, substantial increase in release rate in the presence of MMP-9
MMP-2	Liposome modified with chlorotoxin	Doxorubicin	Chlorotoxin-modified liposomes exhibited higher *in vitro* toxicity and *in vivo* targeting efficiency to 4T1 tumors than nonmodified liposomes, and could prevent lung metastasis with low systemic toxicity
FSH receptor on tumor vasculature	Nano-graphene oxide with FSH antibody	Doxorubicin	Vasculature accumulation of GO FSHR-rnAb conjugates in tumor at early time points; enhanced drug delivery efficiency in MDA-MB-231 metastatic sites

nanoparticle–drug conjugate with a camptothecin payload, the nanoformulation was evaluated alone or in combination with antiangiogenic drugs such as bevacizumab in murine breast cancer model. Tumors tend to develop resistance to antiangiogenic drugs by upregulation of hypoxia-inducible factor-1α. This can be blocked with camptothecin, but this compound is very poorly soluble and causes strong systemic toxicity. These issues can be addressed by delivering the drug as nanoformulation CRLX101. It was found effective in durably blocking the hypoxia-inducible factor-1α, restoring the cancer sensitivity to bevacizumab, improving tumor perfusion and reducing hypoxia [131].

He *et al* reported the fabrication of dual-sensitive nanoparticles with hypoxia and photo-triggered release of the anticancer drug [132]. The authors developed dual stimuli nanoparticles through the self-assembly of polyethyleneimine–nitroimidazole micelles (PEI–NI) further co-assembled with Ce6-linked hyaluronic acid (HC). Hypoxia-mediated activation was achieved by the incorporation of nitroimidazole (NI), a hypoxia-responsive electron acceptor. Hydrophobic NI segments would be converted to hydrophilic 2-aminoimidazole under hypoxic conditions, thereby aiding in the release of the anticancer drug (doxorubicin, DOX) loaded inside the nanoparticles.

(III) Nanotherapy of tumor-associated macrophages
There are several subtypes of macrophages in tumor-associated macrophages (TME) and they are promising targets. M1 macrophages are known to be involved in inflammatory processes and may have significant antitumor effects. In contrast, activated M2 macrophages, which are generally involved in the wound-healing events in tumors, may behave antagonistically to M1 as TAM. TAMs play instrumental roles in various processes such as matrix modeling, neoangiogenesis and local immunity suppression that facilitate cancer growth, invasion and spreading. They have been targeted with nanodelivery systems to achieve TAM reprogramming, suppression, depletion and recruitment prevention.

Table 7.13(B) lists the nanodelivery systems that may achieve anticancer effects via macrophages. Interestingly, it was found that a part of Abraxane's activity may be derived from its activity in increasing the CD80+ CD86+ M1 macrophage subpopulation that works in an antagonistic manner to the M2 subpopulation.

(IV) Other promising targets in TME
Stromal cells are connective tissue cells such as fibroblasts and pericytes. They were long shown to play crucial roles in mammary gland development and breast cancer progression [133]. Their activities can be interfered with for breast cancer treatment with nanotherapeutics [77,134] (table 7.13(C)). Cellax is a nanoparticle made of acetylated carboxymethylcellulose–PEG for docetaxel delivery [135]. It has been clinically evaluated for breast cancer treatment, and is claimed to be superior to Abraxane in many aspects including reduced tumor growth and metastasis. What is appealing is that Cellax was found to decrease α-smooth muscle actin content by 82% and 70%, respectively, versus no significant change in free docetaxel and Abraxane groups. The reduction in smooth muscle actin by Cellax contributed to a

substantial increase in tumor perfusion and tumor vascular permeability and reduction in tumor matrix and interstitial pressure versus control. It will be interesting to see if these outcomes are translatable in a clinical setting.

TME also has upregulated levels of enzymes such as matrix metalloproteinase (MMP), which is predominantly envolved in tumor development and proliferation. The upregulated levels of MMP enzymes in TME makes them the most common target for enzyme-based TME nanoparticles [128]. There are other valuable targets in the TME such as MMPs. MMP-9 is a MMP subtype highly expressed in metastatic breast tumors. The protease activity of MMP-9 was heavily involved in extracellular matrix remodeling and angigogenesis in TME, and can be exploited for triggering drug release. MMP-2 is another potential MMP target associated with advanced breast cancer. In addition, it is also possible to target tumor vasculature to enhance tumoral drug delivery [77].

7.5.3 Updated status of nanomedicine application for breast cancer treatment

As of today, only a few nanomedicine products have gained US Food and Drug Administration (FDA) approval, and Doxil and Abraxane are the two most successful nanoformulations already widely used for breast cancer treatment in clinical settings. Their development and the ways they are incorporated into the standard drug therapy for cancer treatment also provide good lessons for the nanomedicine researchers and clinicians. In addition, some promising nanoformulations that already entered clinical phase are also introduced.

(A) Doxil

This product was formulated to improve the balance between the efficacy and toxicity of doxorubicin therapy. It contains doxorubicin (Adriamycin), a member of the anthracycline group, enclosed in an 80–90 nm size unilamellar liposome coated with PEG that allows the drug to stay in the bloodstream longer so that more of the drug reaches the cancer cells [136–138]. Doxorubicin is believed to act on cancer cells by two different mechanisms: intercalation into DNA and disruption of topoisomerase II-mediated DNA repair and generation of free radicals that result in damage to cellular membranes, DNA and proteins [139].

Doxil is the first FDA-approved nanodrug (1995). Doxorubicin used to be the most important chemotherapy drug for breast cancer treatment; however, it is also notorious for causing congestive heart failure. These problems may occur during therapy or sometimes months to years after receiving doxorubicin. In some cases the heart problems are irreversible. Myocardial damage may lead to congestive heart failure and may occur as the total cumulative dose of doxorubicin HCl approaches 550 mg m^{-2}. The risk may be increased in a patient using certain medicines that may affect heart function or who has a history of heart problems, receiving radiation treatment to the chest area, or previous therapy with other anthracyclines (e.g. epirubicin) or cyclophosphamide. This cardiotoxicity is dependent on its cumulative dose (36% incidence when the total dose is $>600 \text{ mg m}^{-2}$) [77]. Doxil was, therefore, designed to reduce the systemic toxicity without compromising the anticancer effects of doxorubicin. It turns out to be the most successful product that demonstrates the capability of PEGylation to avoid premature elimination of nanocarriers by the reticuloendothelial system, so that extended circulation time can be achieved. It also confirms that the nanoformulation can achieve good intratumoral drug level by EPR passive targeting effect [77]. Liposomal doxorubicin is also called Doxil (Johnson & Johnson, USA), Caelyx (Janssen-Cilag, Europe), Evacet (Liposome company INC.) and lipodox (Sun Pharma) [140].

Doxil was shown to markedly suppress tumor growth rates and improve survival [140]. The liposomes alter pharmacologic and pharmacokinetic parameters of conventional doxorubicin so that drug delivery to the tumor is enhanced while toxicity normally associated with conventional doxorubicin is decreased [141]. Moreover, the use of Doxil led to a major (~3-fold) risk reduction of cardiotoxicity versus free doxorubicin [142]. In a Phase III trial, 22 509 women with metastatic breast cancer were treated with Doxil 50 mg m^{-2} or doxorubicin 60 mg m^{-2}. Doxil and doxorubicin were comparable with respect to progression-free survival and overall survival. The overall risk of cardiotoxicity, however, was significantly higher with doxorubicin than Doxil [77].

With reduced cardiotoxicity, Doxil helps relieve the cumulative dose cap and enables lower risk, extended doxorubicin treatment and, thus, substantially increases the versatility of this drug. Doxil has already been combined with several other chemotherapy drugs (e.g. cyclophosphamide and 5-fluorouracil, cisplatin and infusional fluorouracil, cyclophosphamide followed by paclitaxel) and targeted therapy such as trastuzumab for advanced breast cancer treatment in clinical trials [143–146]. All of the studies indicated good efficacy and low toxicity including cardiotoxicity, even in elderly or cardiotoxicity-prone patients. It should be noted that a recent meta-analysis showed that although Doxil reduces cardiotoxicity substantially, it also leads to new side effects such as skin toxicity and mucositis, but these side effects are clinically much less serious than cardiotoxicity [147]. Overall, these studies provide good precedents that a nanomedicine can work effectively and safely in combination with standard drug therapy.

The FDA approved the first generic version of Doxil (doxorubicin hydrochloride liposome injection, February 2013) made by Sun Pharma Global FZE, a subsidiary of India's Sun Pharmaceutical Industries Ltd, Lipodox is the second generation of pegylated liposomal doxorubicin composed of distearoyl phosphatidyl choline (DSPC) and cholesterol with surface coating of PEG [140]. Lipodox has a circulation half-life of 65 h. However, due to the long circulation time of the pegylated drug, stomatitis (inflammation of mucus lining) became the new dose-limiting toxicity. Doxil and lipodox (both pegylated) accumulate at tumor site by passive targeting mechanism. Unfortunately, both have more side effects than Myocet (non-pegylated doxorubicin). Moreover, liposomal formulations are more expensive than the non-liposomal drugs [140]. Another concern is the toxicity of liposomal formulations especially that of the pegylated liposomes, such as various skin reactions and hypersensitivity reactions [140].

Myocet is a non-pegylated liposomal doxorubicin citrate made by Enzon Pharmaceuticals for Cephalon in Europe and for Sopherion Therapeutics in the United States and Canada. Myocet received Canadian regulatory approval on

December 21, 2001 and is indicated for the first-line treatment of metastatic breast cancer in combination with cyclophosphamide [140]. Myocet has a different pharmacokinetic profile from doxorubicin, resulting in an improved therapeutic index (less cardiotoxicity and equal anticancer activity) [148].

(B) Abraxane (Albumin-bound Paclitaxel, Paclitaxel protein bound)

Paclitaxel (Taxol, Bristol-Myers Squibb) belongs to an important class of antitumor agents called taxanes. Taxanes are cell-cycle-specific agents and bind with high affinity to microtubules resulting in inhibition of mitosis and cell death [140]. Taxanes are highly hydrophobic molecules and therefore have very low solubility in water. In order to solubilize paclitaxel, a derivative of castor oil, called polyethoxylated castor oil (cremophore EL, a non-ionic surfactant) and ethanol (50:50) is used [140]. However, the use of the solvent-based formulation is associated with serious and dose-limiting toxicities. Cremophore itself causes hypersensitivity reactions (premedication with a steroid (oral dexamethasone) and antihistamine (diphenhydramine) was required) and is known to leach out plasticizers from standard intravenous injection tubing, necessitating special infusion set [140]. The albumin-bound paclitaxel,(or Nab-paclitaxel) Abraxane, however, do not contain cremophore. Therefore, no premedication or special tubing is required to administer abraxane.

Abraxane (Abraxis Bioscience/Celgene) was approved by the FDA in January 2005 for treatment of metastatic breast cancer [140]. Paclitaxel is noncovalently bound onto 130 nm nanoparticle processed from human serum albumin. It turned out that not only the solvent/surfactant-related adverse effects have been avoided and no premedication is required, by exploiting the natural interactions between albumin and the gp60/caveolin-1 receptor pathway, but also Abraxane is associated with rapid and preferential delivery and accumulation of paclitaxel at the tumor site [77]. When compared with solvent-based paclitaxel, Abraxane is associated with a 9-fold greater penetration of paclitaxel into tissues via transporter-mediated pathways, a 33% higher intratumoral drug concentration, a 10-fold higher mean maximal concentration of free paclitaxel and a 4-fold lower elimination rate [149–151]. The clinical performance is also excellent. In GeparSepto trial, which involved 1229 women with previously untreated unilateral or bilateral primary invasive breast cancer, it was found that substituting solvent-based paclitaxel with nab-paclitaxel significantly increased the proportion of patients achieving a pathologic complete response rate after anthracycline-based chemotherapy, and suggested that these results might lead to replacement of solvent-based paclitaxel by nab-paclitaxel for primary breast cancer treatment [152].

Clinicians have also explored the combination of Abraxane with other standard anticancer drugs. In a study in HER2+ breast cancer patients, the combination of Abraxane with carboplatin and trastuzumab was shown to be efficacious and generally safe [153]. In another trial, it was shown that the addition of bevacizumab to Abraxane prior to dose-dense doxorubicin/cyclophosphamide significantly improved the pathologic complete response rate compared to chemotherapy alone in patients with triple negative, locally advanced breast cancer or inflammatory

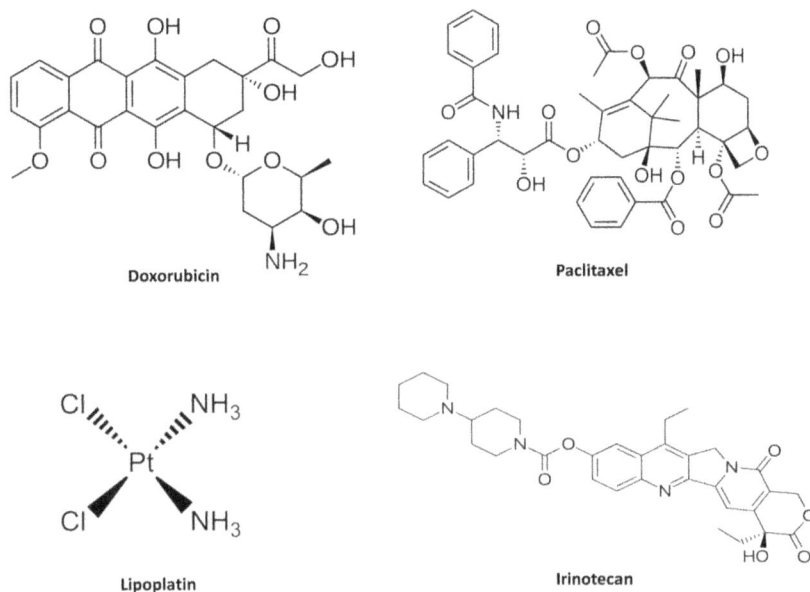

Figure 7.8. Chemical Structures of Doxil, Abraxane, Lipoplatin, Onivyde.

breast cancer [154]. In general, Abraxane has demonstrated comparable or superior efficacy over solvent-based paclitaxel for breast cancer treatment, and like Doxil, it can be safely combined with standard chemotherapy or targeted therapy.

(C) Other investigational nanoformulations
This section will introduce some investigational nanoformulations that showed promise in clinical trials. Lipoplatin is a liposomally-encapsulated (average diameter, 110 nm) drug product of the FDA-approved cytotoxic agent cisplatin. The major concern about the use of cisplatin and other platinum compounds is nephrotoxicity. Patients receiving these agents need to be hydrated to prevent renal damage. In the Lipoplatin product, cisplatin (cis-diamino dichloro-platinum (figure 7.8)) is encapsulated in a liposome shell composed of dipalmitoyl phosphatidyl glycerol, soy phosphatidyl choline, cholesterol and methoxy-polyethylene glycol–distearoyl phosphatidylethanolamine lipid conjugate. The ratio of cisplatin to lipids is 8.9%:91.1% (w/w) [155, 156]. Lipoplatin is used against pancreatic cancer in combination with gemcitabine as first-line treatment. Lipoplatin accumulates in the cancer tissues by the extravasation of the nanoparticles through the defective vasculature of the tumor tissue during neoangiogenesis. Lipoplatin had mild hematological and gastrointestinal toxicity and did not show nephron, neuro, ototoxicity or any other side effects characteristic of cisplatin.

With the promising early data, Lipoplatin has been extensively evaluated and has successfully completed a number of clinical studies from Phase I to III trials [157]. Its official indication is for non-small cell lung cancer. In a number of clinical trials, Lipoplatin demonstrated enhanced cisplatin retention in tumor tissue and substantially reduced renal toxicity, peripheral neuropathy, ototoxicity and myelo-toxicity [158–160]. This product also has potential to be included in the treatment of breast cancer. In a Phase II study of Lipoplatin/vinorelbine combination in HER2/neu-negative metastatic breast cancer, complete response and partial response were achieved in the majority of patients (9.4% and 43.8%, respectively), with only 9.4% showing disease progression. No grade 3/4 nephrotoxicity or neuropathy, both key toxicities of cisplatin, were noted.

Onivyde® is an FDA-approved (2015) nanoliposomal formulation of irinotecan [161]. Its official indication is for metastatic pancreatic cancer. In a Phase III trial on patients with gemcitabine-based chemotherapy-resistant metastatic pancreatic cancer, Onivyde plus 5-fluorouracil/leucovorin was significantly better than 5-fluo-rouracil/leucovorin only in terms of several clinical parameters [162]. Irinotecan is not a commonly used drug for breast cancer; however, in a Phase I study with advanced refractory solid tumors that include breast cancer, the disease control rate with Onivyde was 45.5% [163]. All studies showed that it is well tolerated.

Besides these two products, there are other promising anticancer nanoformula-tions that have entered clinical phase. One example is Genexol-PM that is being developed by Samyang Biopharm in Korea [164]. Genexol-PM is a solvent-free formulation of paclitaxel delivered by polymeric micelles made of their proprietary poly (ethylene glycol)–poly (lactic acid) block co-polymers. In the Phase I and II trials on metastatic or recurrent breast cancer, it was reported that a higher maximal tolerable dose can be used, with 12.2% complete response and 46.3% partial response [165]. An ongoing Phase III trial is being conducted in Korea now.

7.6 Conclusion

With modern surgery and radiation therapy, recurrence rates of cancer in the breast are now less than 5% in the 10 years after treatment. In radiotherapy normal cells lying in the track of the photon beam are exposed. Radionuclide therapy like chemotherapy is an alternative option in which radioisotope travels through the bloodstream; these radionuclide substances specifically targets diseased cells, but suffer shorter half-life so there is insufficient delivery of the radionuclide to the tumor site and quick biological clearance due to smaller size. This can be overcome by attaching nanocarriers to radioisotope. Surface modification by a polymer such as PEG (polyethylene glycol) increases half-life as well as avoiding interaction with opsonin. Another useful approach of nanoparticle mediated radiotherapy is radio-sensitizer, where nanoparticles enhance production of ROS (reactive oxygen species) like radiation, decreasing total dose, which reduces normal tissue exposure.

Therapies using medication for breast cancer including chemotherapy, hormonal therapy, targeted therapy and immunotherapy, are able to treat cancer throughout the body, especially for patients with metastases but there is a lack of selectively killing all the cancer cells. Systemic anticancer pharmaceuticals are usually hydro-phobic, untargeted and toxic compounds, which can cause severe side effects. Nanocarrier-based drug delivery systems for chemotherapeutic drugs act efficiently on multiple malignant sites. The justification of cancer nanomedicine relies on enhanced permeation (EP) and retention (R) effect and the capability of intracellular targeting due primarily to size after internalization (endocytosis) into the individual target cells. Nanoparticles are biomedically applicable if they are constructed by considering their size, shape, surface charge and coating, which affects their biodistribution. In breast cancer, nanomedicine is the most useful parameter because several side effects of breast cancer drugs while using alone can be reduced by application with nanoparticles. Factors such as tumor microenvironment (TME), breast cancer stem cells (BCSCs) that promote growth of the tumor can also be controlled by nanotherapy.

References

[1] Gianfaldoni S, Gianfaldoni R, Wollina U, Lotti J, Tchernev G and Lotti T 2017 An overview on radiotherapy: from its history to its current applications in dermatology *J. Med. Sci.* **5** 521–5

[2] Haume K, Rosa S and Grellet S *et al* 2016 Gold nanoparticles for cancer radiotherapy: a review *Cancer Nano.* **7** 1–20

[3] Hall E J and Giaccia A J 2012 *Radiobiology for the Radiologist* 7th edn (Philadelphia, PA: Lippincott Williams & Wilkins)

[4] Lee T, Yang J, Huang E, Lee C, Chan M and Liu A 2014 Technical advancement of radiation therapy *Biomed. Res. Int.* **2013** 549–60

[5] Khan F M 2010 Modern radiation therapy *The Physics of Radiation Therapy* 4th edn (Philadelphia, PA: Lippincott Williams & Wilkins)

[6] Garibaldi C, Alicja B, Fossa J, Marvaso G and Dicuonzo S *et al* 1995 Recent advances in radiation oncology *Ecancermedicalscience* **11** 785–92

[7] Baskar R, Lee K A, Yeo R and Yeoh K 2012 Cancer and radiation therapy: current advances and future directions *Int. J. Med. Sci.* **9** 193–9

[8] Zhang L, Chen H, Wang L, Liu T, Yeh J, Lu G, Yang L and Mao H 2010 Delivery of therapeutic radioisotopes using nanoparticle platforms: potential benefit in systemic radiation therapy *Nanotechnol. Sci. Appl.* **3** 159–70

[9] Mi Y, Shao Z, Vang J, Kaidar-Person O and Wang A 2016 Application of nanotechnology to cancer radiotherapy *Cancer Nanotechnol.* **7** 11

[10] Ting G, Chang C and Ell Wang H 2009 Cancer nanotargeted radiopharmaceuticals for tumor imaging and therapy *Anticancer Res.* **29** 4107–18

[11] Babaei M and Ganjalikhani M 2014 The potential effectiveness of nanoparticles as radio sensitizers for radiotherapy *Bioimpacts* **4** 15–20

[12] Retif P, Pinel S, Toussaint M, Frochot C, Chouikrat R, Bastogne T and Barberi-Heyob M Nanoparticles for radiation therapy enhancement: the key parameters *Theranostics* **5** 1030–44

[13] Klein S, Sommer A, Distel L, Neuhuber W and Kryschi C 2012 A superparamagnetic iron oxide nanoparticles as radiosensitizer via enhanced reactive oxygen species formation *Biochem. Biophys. Res. Commun.* **425** 393

[14] Kwatra D, Venugopal A and Anant S 2013 Nanoparticles in radiation therapy: a summary of various approaches to enhance radiosensitization in cancer *Transl. Cancer Res.* **2** 330–42

[15] Bray F, Ferlay J, Soerjomataram I, Siegel R, Torre L and Jemal A 2018 Global cancer statistics 2018: Globocan estimates of incidence and mortality worldwide for 36 cancers in 185 countries *CA Cancer J. Clin.* **68** 394–424

[16] Cancer staging guide-NCCN Report www.nccn.org

[17] Li H, Jin H, wan W, Wu C and Wei L 2018 Cancer nanomedicine: mechanisms, obstacles and strategies *Nanomedicine* 1639–56

[18] Tran S, Degiovanni P, Pie B and Rai P 2017 Cancer nanomedicine: a review of recent success in drug delivery *Clin. Transl. Med.* **6** 4

[19] Kutty R, Wei L and Feng S 2014 Nanomedicine for the treatment of triple-negative breast cancer *Nanomedicine (Lond.)* **9** 561–4

[20] Youn Y and Bae Y 2018 Perspectives on the past, present, and future of cancer nanomedicine *Adv. Drug Deliv. Rev.* at press

[21] Bregni C and Carlucci A 2013 Nanomedicines in cancer therapy *J. Mol. Pharm. Org. Process Res.* **1** 1

[22] Karve S, Werner M, Sukumar R, Cummings N, Copp J and Wang E *et al* 2012 Revival of the abandoned therapeutic wortmannin by nanoparticle drug delivery *Proc. Natl. Acad. Sci. USA* **109** 8230–5

[23] Wicki A, Witzigmannb D, Balasubramanian V and Huwyler J 2015 Nanomedicine in cancer therapy: challenges, opportunities, and clinical applications *J. Controlled Rel.* **200** 138–57

[24] Du J, Du X, Mao C and Wang J 2011 Tailor-made dual pH-sensitive polymerdoxorubicin. nanoparticles for efficient anticancer drug delivery *J. Am. Chem. Soc.* **133** 17560–3

[25] Bi Y, Hao F, Yan G, Teng L, Lee R and Xie J 2016 Actively targeted nanoparticles for drug delivery to tumor *Curr. Drug Metabol.* **17** 763–82

[26] Ankamwar B 2012 Size and shape effect on biomedical applications of nanomaterials *Biomedical Engineering-Technical Applications in Medicine* (Rijeka: Intech)

[27] Yoo J, Chambers E and Mitragotri S 2010 Factors that control the circulation time of nanoparticles in blood: challenges, solutions and future prospects *Curr. Pharm. Des.* **16** 2298–307

[28] Muro S, Garnacho C and Champion J *et al* 2008 Control of endothelial targeting and intracellular delivery of therapeutic enzymes by modulating the size and shape of Icam-1-targeted carriers *Mol. Ther.* **16** 1450–8

[29] Chithrani B, Ghazani A and Chan W 2006 Determining the size and shape dependence of gold nanoparticle uptake into mammalian cells *Nano. Lett.* **6** 662–8

[30] Chithrani B and Chan W 2007 Elucidating the mechanism of cellular uptake and removal of protein-coated gold nanoparticles of different sizes and shapes *Nano. Lett.* **7** 1542–50

[31] Steinmetz N 2010 Viral nanoparticles as platforms for next-generation therapeutics and imaging devices *Nanomedicine* **6** 634

[32] Villa N, Bartee E, Mohamed M, Rahman a M, Barrett J and McFadden G 2010 Myxoma and vaccinia viruses exploit different mechanisms to enter and infect human cancer cells

[33] Duncan R 2006 Polymer conjugates as anticancer nanomedicines *Virology* **401** 266–79

[34] Chari R, Martell B, Gross J and Cook S *et al* 1992 Immunoconjugates containing novel maytansinoids: promising anticancer drugs *Cancer Res.* **52** 127–31

[35] Diamantis N and Banerji U 2016 Antibody-drug conjugates an emerging class of cancer treatment *Br. J. Cancer* **114** 362–7

[36] Duncan R 2006 Polymer conjugates as anticancer nanomedicines *Nat. Rev. Cancer* **6** 688–701

[37] Villard L, Romer A, Marincek N, Brunner P, Koller M T and Schindler C *et al* 2012 Cohort study of somatostatin-based radiopeptide therapy with [(90)Y-DOTA]-TOC versus

[(90)Y-DOTA]-TOC plus [(177)Lu-DOTA]-TOC in neuroendocrine cancers *J. Clin. Oncol.* **30** 1100–6

[38] Fang J and Al-Suwayeh S 2012 Nanoparticles as delivery carriers for anticancer prodrugs *Expert Opin. Drug Deliv.* **9** 657–69

[39] Barenhol Y 2012 Doxil®—the first FDA-approved nano-drug: lessons learned *J. Control. Rel.* **160** 117–34

[40] Harrison M, Tomlinson D and Stewart S 1995 Liposomal-entrapped doxorubicin: an active agent in AIDS-related Kaposi's sarcoma *J. Clin. Oncol.* **13** 914–20

[41] Chan S, Davidson N, Juozaityte E, Erdkamp F, Pluzanska A and Azarnia N *et al* 2004 Phase III trial of liposomal doxorubicin and cyclophosphamide compared with epirubicin and cyclophosphamide as first-line therapy formetastatic breast cancer *Ann. Oncol.* **15** 1527–34

[42] Gill P, Wernz J, Scadden D, Cohen P, Mukwaya G and Roenn J *et al* 1996 Randomized phase III trial of liposomal daunorubicin versus doxorubicin, bleomycin, and vincristine in AIDS-related Kaposi's sarcoma *J. Clin. Oncol.* **14** 2353–64

[43] Gökbuget N, Hartog C-M, Bassan R, Derigs H-G, Dombret H and Greil R *et al* 2011 Liposomal cytarabine is effective and tolerable in the treatment of central nervous system-relapse of acute lymphoblastic leukemia and very aggressive lymphoma *Haematologica* **96** 238–44

[44] O'Brien S, Schiller G, Lister J, Damon L, Goldberg S and Aulitzky W *et al* 2013 High-dose vincristine sulfate liposome injection for advanced, relapsed, and refractory adult Philadelphia chromosome-negative acute lymphoblastic leukemia *J. Clin. Oncol.* **31** 676–83

[45] Frampton J 2010 Mifamurtide: a review of its use in the treatment of osteosarcoma

[46] Meyers P, Schwartz C, Krailo M, Healey J, Bernstein M and Betcher D *et al* 2008 Osteosarcoma: the addition of muramyl tripeptide to chemotherapy improves overall survival—a report from the Children's Oncology Group *J. Clin. Oncol.*

[47] Mamot C, Drummond D C, Noble C O, Kallab V, Guo Z and Hong K *et al* 2005 Epidermal growth factor receptor-targeted immunoliposomes significantly enhance the efficacy of multiple anticancer drugs *in vivo Paediatr. Drugs* **12** 141–53

[48] Park J, Hong K, Kirpotin D, Colbern G, Shalaby R and Baselga J *et al* 2002 Anti-HER2 immunoliposomes: enhanced efficacy attributable to targeted delivery *Clin. Cancer Res.* **8** 1172–81

[49] Wicki A, Rochlitz C, Orleth A, Ritschard R, Albrecht I and Herrmann R *et al* 2011 Targeting tumor-associated endothelial cells: anti-VEGFR2 immunoliposomes mediate tumor vessel disruption and inhibit tumor growth *Clin. Cancer Res.* **18** 454–64

[50] DeFrates K, Markiewicz T, Gallo P, Rack A, Weyhmiller A, Jarmusik B and Hu X 2018 Protein polymer-based nanoparticles: fabrication and medical applications *Int. J. Mol. Sci.* **19** 1717

[51] Gradishar W, Tjulandin S, Davidson N, Shaw H, Desai N and Bhar P *et al* 2005 Phase III trial of nanoparticle albumin-bound paclitaxel compared with polyethylated castor oil-based paclitaxel in women with breast cancer *J. Clin. Oncol.* **23** 7794–803

[52] Hoff D, Ervin T, Arena F, Chiorean E, Infante J and Moore M *et al* 2013 Results of a randomized phase III trial (MPACT) of weekly nab-paclitaxel plus gemcitabine versus gemcitabine alone for patients withmetastatic adenocarcinoma of the pancreas with PET and CA19-9 correlates *J. Clin. Oncol.* **31** 4005

[53] Penadés S, Davis B and Seeberger P 2015–2017 Glycans in nanotechnology *Essentials of Glycobiology* 3rd edn (New York: Cold Spring Harbor Laboratory Press) ch 58

[54] Weiss G, Chao J, Neidhart J, Ramanathan R, Bassett D and Neidhart J *et al* 2013 First-in-human phase 1/2a trial of CRLX101, a cyclodextrin-containing polymercamptothecin nanopharmaceutical in patients with advanced solid tumor malignancies *Invest. New Drugs* **31** 986–1000

[55] Davis M 2009 The first targeted delivery of siRNA in humans via a self-assembling, cyclodextrin polymer-based nanoparticle: from concept to clinic *Mol. Pharmaceut.* **6** 659–68

[56] Derakhshandeh K and Fathi S 2012 Role of chitosan nanoparticles in the oral absorption of Gemcitabine *Int. J. Pharm.* **437** 172–7

[57] Kamaly N, Xiao Z, Valencia P, Radovic-Moreno A and Farokhzad O 2012 Targeted polymeric therapeutic nanoparticles: design, development and clinical translation *Chem. Soc. Rev.* **41** 2971–3010

[58] Battig M and Wang Y 2014 *Natural and Synthetic Biomedical Polymers* 1st edn (Amsterdam: Elsevier)

[59] Plummer R, Wilson R, Calvert H, Boddy A, Griffin M and Sludden J *et al* 2011 A Phase I clinical study of cisplatin-incorporated polymeric micelles (NC-6004) in patients with solid tumors *Br. J. Cancer* **104** 593–8

[60] Ross R, Zietman A, Xie W, Coen J, Dahl D and Shipley W *et al* 2009 Lymphotropic nanoparticle-enhancedmagnetic resonance imaging (LNMRI) identifies occult lymph node metastases in prostate cancer patients prior to salvage radiation therapy *Clin. Imag.* **4** 301–5

[61] Rivera Gil P, Hühn D, del Mercato L, Sasse D and Parak W J 2010 Nanopharmacy: inorganic nanoscale devices as vectors and active compounds *Pharmacol. Res.* **62** 115–25

[62] Maggiorella L, Barouch G, Devaux C, Pottier A, Deutsch E and Bourhis J *et al* 2012 Nanoscale radiotherapy with hafnium oxide nanoparticles *Future Oncol.* **8** 1167–81

[63] Libutti S, Paciotti G, Byrnes A, Alexander H, Gannon W and Walker M *et al* 2010 Phase I and pharmacokinetic studies of CYT-6091, a novel PEGylated colloidal gold-rhTNF nanomedicine *Clin. Cancer Res.* **16** 6139–49

[64] Kovalainen M, Mönkäre J, Kaasalainen M, Riikonen J, Lehto V and Salonen J *et al* 2012 Development of porous silicon nanocarriers for parenteral peptide delivery *Mol. Pharm.* **10** 353–9

[65] DeSantis C, Ma J, Bryan L and Jemal A 2013 Breast cancer statistics *CA Cancer J. Clin.* **64** 52–62

[66] Torre L, Islami F, Siegel R, Ward E and Jemal A 2017 Global cancer in women: burden and trends *Cancer Epidemiol. Biomarkers Prev.* **26** 444–57

[67] https://seer.cancer.gov

[68] www.medicinenet.com

[69] Dillon D, Guidi A and Schnitt S 2014 Pathology of invasive breast cancer *Diseases of the Breast* 5th edn (Philadelphia, PA: Lippincott Williams & Wilkins)

[70] Dhankhar R, Vyas S, Jain A, Arora S, Rath G and Goyal A 2010 Advances in novel drug delivery strategies for breast cancer therapy *Artif. Cells Blood Substit. Immobil. Biotechnol.* **38** 230–49

[71] www.cancer.net

[72] NCCN guidelines for patient. Breast cancer invasive 2018

[73] Stewart A, Khan A and Devlin P 2010 Partial breast irradiation: a review of techniques and indications

[74] Masoud V and Pagès G 2017 Targeted therapies in breast cancer: New challenges to fight against resistance

[75] www.breastcancer.org

[76] www.cancer.gov

[77] Wu D, Si M, Xue H and Wong H 2017 Nanomedicine applications in the treatment of breast cancer: current state of the art *Br. J. Radiol.* **83** 369–78

[78] Zhang R, Wong H, Xue H, Eoh J and Wu X 2016 Nanomedicine of synergistic drug combinations for cancer therapy–Strategies and perspectives *J. Control. Rel.* **240** 489–503

[79] Lee J, Yazan L and Abdullah C 2017 A review on current nanomaterials and their drug conjugate for targeted breast cancer treatment *Int. J. Nanomed.* **12** 2373–84

[80] Adair J, Parette M, Altinoglu E and Kester M 2010 Nanoparticulate alternatives for drug delivery *ACS Nano* **4** 4967–70

[81] www.cancer.org

[82] Nahar M, Dutta T, Murugesan S, Asthana A and Mishra D *et al* 2006 Functional polymeric nanoparticles: an efficient and promising tool for active delivery of bioactives *Crit. Rev. Ther. Drug Carrier Syst.* **23** 259–318

[83] Torchilin V 2007 Nanocarriers *Adv. Drug Deliv. Rev.* **58** 1532–55

[84] Shapiro C and Recht A 2008 Side effects of adjuvant treatment of breast cancer *J. Med.* **344** 1997–2008

[85] Tang X, Loc W, Dong C, Matters G, Butler P, Kester M, Meyers C, Jiang Y and Adair J 2017 The use of nanoparticulates to treat breast cancer *Nanomedicine (Lond)* **12** 2367–88

[86] Wong H, Bendayan R, Rauth A, Xue H, Babakhanian K and Wu X 2006 A mechanistic study of enhanced doxorubicin uptake and retention in multidrug resistant breast cancer cells using a polymer-lipid hybrid nanoparticle system *J. Pharmacol. Exp. Ther.* **317** 1372–81

[87] Singh S, Singh S, Wlillard J and Singh R 2017 Drug delivery approaches for breast cancer *Int. J. Nanomed.* **12** 6205–18

[88] Franco M, Roque M and Oliveir M 2017 Active targeting of breast cancer cells using nanocarriers *Mod. Appl. Pharm. Pharmacol.* doi: 10.31031/MAPP.2017.01.000507

[89] Zheng C, Qiu L, Yao X and Zhu K 2009 Novel micelles from graft polyphosphazenes as potential anticancer drug delivery systems: drug encapsulation and *in vitro* evaluation *Int. J. Pharm.* **373** 133–40

[90] Yang Y, Wang J, Zhang X, Lu W and Zhang Q 2009 A novel mixed micelle gel with thermo-sensitive property for the local delivery of docetaxel *J. Control. Release.* **135** 175–82

[91] Ony ü ksel H, Jeon E and Rubinstein I 2009 Nanomicellar paclitaxel increases cytotoxicity of multidrug resistant breast cancer cells *Cancer Lett.* **274** 327–30

[92] Peng C, Shieh M, Tsai M, Chang C and Lai P 2008 Self-assembled star-shaped chlorin-core poly (epsilon-caprolactone)-poly (ethylene glycol) diblock copolymer micelles for dual chemo-photodynamic therapies *Biomaterials* **29** 3599

[93] Lee K, Chung H and Im S *et al* 2008 Multicenter phase II trial of Genexol-PM, a Cremophor-free, polymeric micelle formulation of paclitaxel, in patients with metastatic breast cancer *Breast Cancer Res. Treatment* **108** 241–50

[94] Park J 2002 Liposome-based drug delivery in breast cancer treatment *Breast Cancer Res.* **4** 95–9

[95] Kullberg M, Owens J and Mann K 2010 Listeriolysin O enhances cytoplasmic delivery by Her-2 targeting liposomes *J. Drug Target.* **18** 313–20

[96] Lee H J and Nan A 2012 Combination drug delivery approaches in metastatic breast cancer *J. Drug Deliv.* **2012** 915375

[97] Sengupta S, Eavarone D and Capila I *et al* 2005 Temporal targeting of tumor cells and neovasculature with a nanoscale delivery system *Nature* **436** 568–72

[98] Dhar S, Kolishetti N, Lippard S and Farokhzad O 2011 Targeted delivery of a cisplatin prodrug for safer and more effective prostate cancer therapy *in vivo Proc. Natl. Acad. Sci. USA* **108** 1850–5

[99] Balakrishnan S, Bhat F and Singh P *et al* 2016 Gold nanoparticle-conjugated quercetin inhibits epithelial-mesenchymal transition, angiogenesis and invasiveness via EGFR/ VEGFR-2-mediated pathway in breast cancer *Cell Prolif.* **49** 678

[100] Ahmed M and Douek M 2013 The role of magnetic nanoparticles in the localization and treatment of breast cancer *Biomed. Res. Int.* **2013** 281230

[101] Thorek D, Chen A, Czupryna J and Tsourkas A 2006 Superparamagnetic iron oxide nanoparticle probes for molecular imaging

[102] Bae P and Chung B 2014 Multiplexed detection of various breast cancer cells by perfluorocarbon/quantum dot nanoemulsions conjugated with antibodies *Ann. Biomed. Eng.* **34** 23–38

[103] Hrib J, Sirc J, Hobzova R, Hampejsova Z, Bosakova Z, Munzarova M and Michalek J 2015 Nanofibers for drug delivery—incorporation and release of model molecules, influence of molecular weight and polymer structure *J. Nanotechnol.* **6** 1939–45

[104] Marty M, Cognetti F and Maraninchi D *et al* 2005 Randomized phase II trial of the efficacy and safety of trastuzumab combined with docetaxel in patients with human epidermal growth factor receptor 2-positive metastatic breast cancer administered as first-line treatment: the M77001 study group *J. Clin. Oncol.* **23** 4265–74

[105] Hoare T and Kohane D 2008 Hydrogels in drug delivery: Progress and challenges *Polymer* **49** 1993–2007

[106] Han H, Mora E and Roh J *et al* 2011 Chitosan hydrogel for localized gene silencing *Cancer Biol. Ther.* **11** 839–45

[107] Dooley W, Ljung B and Veronesi U *et al* 2001 Ductal lavage for detection of cellular atypia in women at high risk for breast cancer *J. Natl. Cancer Inst.* **93** 1624–32

[108] Dave K, Averineni R, Sahdev P and Perumal O 2014 Transpapillary drug delivery to the breast *PLoS One* **9** e115712

[109] Iqbal N and Iqbal N 2014 Human epidermal growth factor receptor 2 (HER2) in cancers: overexpression and therapeutic implications *Mol. Biol. Int.* **2014** 852748

[110] www.mayoclinic.org

[111] Rochefort H, Glondu M, Sahla M, Platet N and Garcia M 2003 How to target estrogen receptor-negative breast cancer? *Endocr. Relat.* **10** 261–6

[112] Wu H, Yupeng C, Liang J, Shi B and Wu G *et al* 2005 Hypomethylationlinked activation of PAX2 mediates tamoxifen-stimulate dendometrial carcinogenesis *Nature* **438** 981–7

[113] Shao W and Brown M 2004 Advances in estrogen receptor biology: prospects for improvements in targeted breast cancer therapy *Breast Cancer Res.* **6** 39–52

[114] Michaud L 2008 Treatment-experienced breast cancer *Am. J. Health Syst. Pharm.* **65** S4–9

[115] Sheikh M, Garcia M, Pujol P, Fontana J and Rochefort H 1994 Why are estrogen-receptornegative breast cancers more aggressive than the estrogen receptor-positive breast cancers? *Invas. Metastasis* **14** 329–36

[116] Moulik S and Chatterjee A 2007 Vascular endothelial growth factor (VEGF) and tumor angiogenesis *Indian J. Med. Res.* **125** 747–51

[117] Widakowich C and Azambuja E *et al* 2007 Molecular targeted therapies in breast cancer: Where are we now? *Int. J. Biochem. Cell Biol.* **39** 1375–87

[118] Morin M 2000 From oncogene to drug: development of small molecule tyrosine kinase inhibitors as anti-tumor and anti-angiogenic agents *Oncogene* **19** 6574–83

[119] Kulik G, Klippel A and Weber M 1997 Antiapoptotic signalling by the insulin-like growth factor I receptor, phosphatidylinositol 3-kinase, and Akt *Mol. Cell Biol.* **17** 1595–606

[120] Youngren J, Gable K and Penaranda C *et al* 2005 Nordihydroguaiaretic acid (NDGA) inhibits the IGF-1 and c-erbB2/HER2/neu receptors and suppresses growth in breast cancer cells *Breast Cancer Res. Treat.* **94** 37–46

[121] Arteaga C, Kitten L and Coronado E *et al* 1989 Blockade of the type I somatomedin receptor inhibits growth of human breast cancer cells in athymic mice *J. Clin. Invest.* **84** 1418–23

[122] Surmacz E 2000 Function of the IGF-I receptor in breast cancer *J. Mammary Gland Biol. Neoplasia* **5** 95–105

[123] Khandwala H, McCutcheon I, Flyvbjerg A and Friend K 2000 The effects of insulin-like growth factors on tumorigenesis and neoplastic growth *Endocr. Rev.* **21** 215–44

[124] Mattheolabakis G, Milane L, Singh A and Amiji M 2015 Hyaluronic acid targeting of CD44 for cancer therapy: from receptor biology to nanomedicine *J. Drug Target.* **23** 605–18

[125] Swaminathan S, Roger E, Toti U, Niu L, Ohlfest J and Panyam J 2013 CD133-targeted paclitaxel delivery inhibits local tumor recurrence in a mouse model of breast cancer *J. Control Rel.* **171** 280–7

[126] Carcereri de P, Butturini E and Rigo A *et al* 2017 Metastatic breast cancer cells enter into dormant state and express cancer stem cells phenotype under chronic hypoxia *J. Cell Biochem.* **118** 3237–48

[127] Qiao S, Zhao Y and Geng S *et al* 2016 A novel double-targeted nondrug delivery system for targeting cancer stem cells *Int. J. Nanomed.* **11** 6667–78

[128] Uthaman S, Huh K and Park I 2018 Tumor microenvironment-responsive nanoparticles for cancer theragnostic applications *Biomater. Res.* **22** 22

[129] Liu P, Wang Z and Brown S *et al* 2014 Liposome encapsulated Disulfiram inhibits NFκB pathway and targets breast cancer stem cells *in vitro* and *in vivo* *Oncotarget* **5** 7471–85

[130] Pham E, Yin M and Peters C *et al* 2016 Preclinical efficacy of bevacizumab with CRLX101, an investigational nanoparticle–drug conjugate, in treatment of metastatic triple-negative breast cancer *Cancer Res.* **76** 4493–503

[131] Conley S, Baker T and Burnett J *et al* 2015 CRLX101, an investigational camptothecin-containing nanoparticle-drug conjugate, targets cancer stem cells and impedes resistance to antiangiogenic therapy in mouse models of breast cancer *Breast Cancer Res. Treat.* **150** 559–67

[132] He H *et al* 2017 Selective cancer treatment via photodynamic sensitization of hypoxia-responsive drug delivery *Nanoscale* **10** 2856–65

[133] Wiseman B and Werb B 2002 Stromal effects on mammary gland development and breast cancer *Science* **296** 1046–9

[134] Mikhaylov G, Klimpel D and Schaschke N *et al* 2014 Selective targeting of tumor and stromal cells by a nanocarrier system displaying lipidated cathepsin b inhibitor *Angew. Chem. Int. Ed. Eng.* **53** 10077–81

[135] Ernsting M, Murakami M, Undzys E, Aman A, Press B and Li S 2012 A docetaxel-carboxymethylcellulose nanoparticle outperforms the approved taxane nanoformulation, Abraxane, in mouse tumor models with significant control of metastases *J. Control Rel.* **162** 575–81

[136] Pillai G and Coronel M 2013 Scienece and technology of the emerging nanomedicines in cancer therapy: A primer for physicians and pharmacists *SAGE Open Med.* **1** 2050312113513759

[137] Gewirtz D 1999 A critical evaluation of the mechanisms of action proposed for the antitumor effects of the anthracycline antibiotics adriamycinand daunorubicin *Biochem. Pharmacol.* **57** 727–41

[138] Park J 2002 Liposome-based drug delivery in breast cancer treatment *Breast Cancer Res.* **4** 95–9

[139] Gordon A, Granai C and Rose P *et al* 2000 Phase II study of liposomal doxorubicin in platinum- and paclitaxel-refractory epithelial ovarian cancer *J. Clin. Oncol.* **18** 3093–100

[140] Pillai G 2014 Nanomedicines for cancer therapy : an update of FDA approved and those under various stages of development *SOJ Pharm. Pharm. Sci.* **1** 13

[141] Vail D, Amantea M, Colbern G, Martin F and Hilger R 2004 Working PK. Pegylated liposomal doxorubicin: proof of principle using preclinical animal models and pharmacokinetic studies *Semin. Oncol.* **31** 16–35

[142] Tahover E, Patil Y and Gabizon A 2015 Emerging delivery systems to reduce doxorubicin cardiotoxicity and improve therapeutic index: focus on liposomes *Anticancer Drugs* **26** 241–58

[143] Rau K, Lin Y and Chen Y *et al* 2015 Pegylated liposomal doxorubicin (Lipo-Dox®) combined with cyclophosphamide and 5-fluorouracil is effective and safe as salvage chemotherapy in taxane-treated metastatic breast cancer: an open-label, multi-center, non-comparative phase II study *BMC Cancer* **15** 423

[144] Torrisi R, Montagna E and Scarano E *et al* 2011 Neoadjuvant pegylated liposomal doxorubicin in combination with cisplatin and infusional fluoruracil (CCF) with and without endocrine therapy in locally advanced primary or recurrent breast cancer *Breast* **20** 34–8

[145] Gil-Gil M, Bellet M and Morales S *et al* 2015 Pegylated liposomal doxorubicin plus cyclophosphamide followed by paclitaxel as primary chemotherapy in elderly or cardiotoxicity-prone patients with high-risk breast cancer: results of the phase II CAPRICE study *Breast Cancer Res. Treat.* **151** 579–606

[146] Torrisi R, Cardillo A and Cancello G *et al* 2010 Phase II trial of combination of pegylated liposomal doxorubicin, cisplatin, and infusional 5-fluorouracil (CCF) plus trastuzumab as preoperative treatment for locally advanced and inflammatory breast cancer *Clin. Breast Cancer* **10** 483–8

[147] Ansari L, Shiehzadeh F and Taherzadeh Z *et al* 2017 The most prevalent side effects of pegylated liposomal doxorubicin monotherapy in women with metastatic breast cancer: a systematic review of clinical trials *Cancer Gene Ther.* **24** 189–93

[148] Leonard R, Williams S, Tulpule A, Levine A and Oliveros S 2009 Improving the therapeutic index of anthracycline chemotherapy: Focus on liposomal doxorubicin (Myocet™) *Breast* **18** 218–24

[149] Desai N, Trieu V and Yao Z *et al* 2006 Increased antitumor activity, intratumor paclitaxel concentrations, and endothelial cell transport of cremophor-free, albumin-bound paclitaxel, ABI-007, compared with cremophor-based paclitaxel *Clin. Cancer Res.* **12** 1317–24

[150] Chen N, Li Y, Ye Y, Palmisano M, Chopra R and Zhou S 2014 Pharmacokinetics and pharmacodynamics of nab-paclitaxel in patients with solid tumors: Disposition kinetics and pharmacology distinct from solvent-based paclitaxel *J. Clin. Pharmacol.* **54** 1097–107

[151] Gardner E, Dahut W and Scripture C *et al* 2008 Randomized crossover pharmacokinetic study of solvent-based paclitaxel and nab-paclitaxel *Clin. Cancer Res.* **14** 4200–5

[152] Untch M, Jackisch C and Schneeweiss A *et al* 2016 Nab-paclitaxel versus solvent-based paclitaxel in neoadjuvant chemotherapy for early breast cancer (GeparSepto-GBG 69): a randomized, phase 3 trial *Lancet Oncol.* **17** 345–56

[153] Tezuka K, Takashima T and Kashiwagi S *et al* 2017 Phase I study of nanoparticle albumin-bound paclitaxel, carboplatin and trastuzumab in women with human epidermal growth factor receptor 2-overexpressing breast cancer *Mol. Clin. Oncol.* **6** 534–38

[154] Nahleh Z, Barlow W and Hayes D *et al* 2016 SWOG S0800 (NCI CDR0000636131): addition of bevacizumab to neoadjuvant nab-paclitaxel with dose-dense doxorubicin and cyclophosphamide improves pathologic complete response (pCR) rates in inflammatory or locally advanced breast cancer *Breast Cancer Res. Treat.* **158** 485–95

[155] Boulikas T, Stathopoulos G, Volakakis N and Vougiouka. M 2005 Systemic Lipoplatin infusion results in preferential tumor uptake in human studies *Anticancer Res.* **25** 3031–9

[156] Boulikias T 2009 Clinical overview of Lipoplatin, a successful liposomal formulation of cisplatin *Expert Opin. Invest. Drugs* **18** 1197–218

[157] Boulikas T, Stathopoulos G, Volakakis N and Vougiouka M 2005 Systemic Lipoplatin infusion results in preferential tumor uptake in human studies *Anticancer Res.* **25** 3031–9

[158] Jehn C, Boulikas T, Kourvetaris A, Kofla G, Possinger G and Lüftner D 2008 First safety and response results of a randomized phase III study with liposomal platin in the treatment of advanced squamous cell carcinoma of the head and neck (SCCHN) *Anticancer Res.* **28** 3961–4

[159] Ravaioli A, Papi M and Pasquini E *et al* 2009 Lipoplatin™ monotherapy: a phase II trial of second-line treatment of metastatic non-small-cell lung cancer *J. Chemother.* **21** 86–90

[160] Mylonakis N, Athanasiou A and Ziras N *et al* 2010 Phase II study of liposomal cisplatin (Lipoplatin) plus gemcitabine versus cisplatin plus gemcitabine as first line treatment in inoperable (stage IIIB/IV) non-small cell lung cancer *Lung Cancer* **68** 240–7

[161] Zhang H 2016 Onivyde for the therapy of multiple solid tumors *Oncol. Targets Ther.* **9** 3001–7

[162] Hoff Von D, Li C and Wang-Gillam A *et al* 2014 NAPOLI 1: Randomized phase 3 study of MM-398 (nal-IRI), with or without 5-fluorouracil and leucovorin, versus 5-fluorouracil and leucovorin, in metastatic pancreatic cancer progressed on or following gemcitabine-based therapy *Eur. J. Cancer* **108** 78–87

[163] Chang T, Shiah H and Yang C *et al* 2015 Phase I study of nanoliposomal irinotecan (PEP02) in advanced solid tumor patients *Cancer Chemother. Pharmacol.* **75** 579–86

[164] Samyang Biopharm *Genexol P M* Available from: https://samyangbiopharm.com/eng/ProductIntroduce/injection01 (Accessed May 20, 2017)

[165] Samyang Biopharmaceuticals Corporation *Genexol-PM (paclitaxel)* Available from: https://samyangbiopharm.com/eng/ProductIntroduce/injection01 (Accessed November 29, 2016)

IOP Publishing

External Field and Radiation Stimulated Breast Cancer Nanotheranostics

Nanasaheb D Thorat and Joanna Bauer

Chapter 8

Ionizing radiation stimulated breast cancer nanomedicine

Gayatri Sharma, Jaidip M Jagtap, Abdul K Parchur and Christopher P Hansen

Ionizing radiation already plays a key role in breast cancer treatment in various modalities (x-rays, γ-rays, proton or neutron beams, etc). Radiation therapy is commonly used in combination with chemotherapy and other treatment regimens. However, these conventional techniques have several drawbacks mainly due to off target effects of the therapy and radio-resistance in the tumor itself. The severe side effects of systemic chemotherapies are well known, targeted delivery of the drugs increases the dose delivered to the tumor and minimizes side effects. Ionizing radiation damages healthy tissues at least as much as cancerous tumors; any radiation not absorbed by the tumor is not only minimizing therapeutic effect but is also causing damage to the surrounding tissues in the beam path. Various nanomedicine systems have been created to augment the efficacy of radiation therapy, either by enhancing the delivery of ionizing radiation itself or by delivering a combination therapy and exploiting the specificity of the beam to trigger a target release of therapeutic agents.

8.1 Introduction

Breast cancer is the most commonly occurring cancer in women and the second most common cancer overall. All over world, approximately, there were over 2 million new cases reported in 2018 [1]. The incidence rates vary greatly worldwide from 19.3 per 100 000 women in Eastern Africa to 89.7 per 100 000 women in Western Europe [1]. Due to the advancement in the medical field, various treatment options are available, which include surgery, chemotherapy, radiotherapy, hormonal therapy and targeted therapy. These treatments are generally given in combinations and have led to increases in the survival rate among these patients.

doi:10.1088/2053-2563/ab2907ch8 8-1

Among all these therapies, radiotherapy (RT) is one of the most important and effective methods, as almost all solid cancer patients are treated with radiotherapy alone or in combination with other therapies. Typically, RT mainly uses high-energy radiation (e.g. x-rays or γ-rays) and particle radiation, including particles such as alpha (α) or beta (β) particles, electron (e), proton, or neutron beams to target cancerous tissues and kill the cancerous cells. Each radiation modality has certain advantages, as presented in table 8.1, and selection is made on the basis of the cancer type and stage of cancer [2]. The types of radiation and penetration depth are illustrated in figure 8.1. Radiation therapy for breast cancer is generally delivered in two ways: external radiator, and internal radiator. **External radiator:** delivers high-energy x-rays, electrons and proton beams to the breast and surrounding lymph nodes. The development of computers and electronics has led to improved methods for delivering radiation. The current techniques include image guided radiotherapy, intensity-modulated radiotherapy and 3D-conformal radiotherapy [3]. **Internal radiator (brachytherapy):** radioactive source is placed near tumor for short periods of time over the course of your treatment. In breast cancer treatment, external radiation is most commonly used. However, for treatment of locally advanced breast cancer which has poor prognosis, new treatment modalities including brachytherapy are under study [4]. This therapy is generally given to treat breast

Table 8.1. Comparison of different types of ionizing radiation.

Radiation modality	Relative biological effectiveness	Dose distribution
X-ray/gamma ray	1, reference	Near surface
Electron beam	1X	Near surface
Proton beam	1.1X	At depth
Heavy-ion beam (C, He)	>3X	At depth

Types of radiation and penetration depth

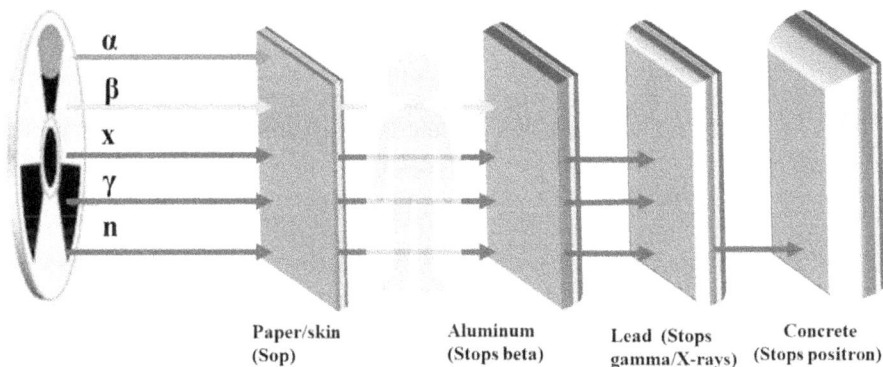

Figure 8.1. Types of radiation and penetration depth.

cancer at all stages. All breast cancer patients are provided radiation doses in fractions extended over a 4–6 week period. The standard whole breast dose is 42.5 Gy in 16 daily fractions, but this varies according to type of breast cancer and size of tumor. Despite its various advantages, the major disadvantage of radiotherapy is only high radiation beams are effective in killing tumor cells and of these high radiation beams only a fraction of it is absorbed by tumors and the remaining beam damages adjacent normal healthy tissues and causes various side effects such as radiation dermatitis, lymphedema, lung toxicity, long-term cardiac toxicity and thyroid toxicity [5]. Hence, present research is targeted to deliver targeted maximum doses to the tumor and minimize the side effects on normal tissues.

With the rapid development in the field of nanomedicine, there is great interest in studying nanomaterials to augment radiation responses and overcome radio-resistance [6]. During radiotherapy, these nanomaterials can get concentrated in the tumor and enhance the radiation effect by physical means or deliver other therapeutic units such chemotherapy and immunotherapy [7, 8]. All the nanoparticles serve as radiosensitizers by different mechanisms, such as dose enhancement, generation of radical oxygen species (ROS) and alteration of cell cycle. The nanoparticles with radiosensitization capabilities generally have the following capabilities.

Dose enhancement effect: Extensive studies have highlighted enhanced effect of radiation by high-Z materials [9]. The first excellent study was presented by Hainfeld *et al* of AuNP-enabled radiosensitization [10]. In general, high-Z atoms have the ability to concentrate high ionizing radiation and this absorbed energy leads to physical processes like Rayleigh scattering, photoelectric effect, and Compton scattering. Among these processes, photoelectric effects are positively related to $(Z/E)^3$, where E is the energy of incident x-ray and Z is the atomic number of the matter and causes generation of short-range low-energy electrons called Auger electrons [11]. Thus, various nanoparticles composed of noble metals (gold), rare-earth elements (gadolinium oxide and upconversion nanoparticles) and other nanostructures (bismuth elements, tungsten, tantalum and hafnium) take advantage of any of the above physical processes and act as radiosensitizers.

Generation of ROS: Nanoparticles augment the efficacy of RT by enhancing the production of ROS which increases oxidative stress and binds to DNA and causes damage [12]. ROS have unpaired electrons and causes injury by chemical reactions, such as hydrogen extraction, addition, disproportionation, and electron capture. The increased yield of ROS leads to single or double stranded breaks of DNA and cross-linking of DNA–DNA or DNA–protein. This bio molecular damage results in a significant inhibition of tumor growth [13–15].

Cell cycle disruption: The disruption of cell cycle is an important factor that influences radiosensitivity. The cell cycle phase also determines a cell's relative radiosensitivity. The DNA damage induced by ionizing radiation initiates signals that lead to checkpoint activation which involves critical proteins (RAD, BRCA, NBS1, ATM, CHK, p53, p21, CDK). p53 is a key protein as it coordinates DNA repair with cell cycle progression and apoptosis. p53 mediates the two major DNA damage-dependent cellular checkpoints, one at the G1–S transition and the other at the G2–M transition. The cells are radioresistant in the late S phase and are radiosensitive in the late G2 and mitotic phases [16]. Ionizing radiation with nanoparticles disrupts cell cycles and directs cells more towards radiosensitive G2/M phases.

8.2 X-rays and γ-rays radiation therapy

8.2.1 Metal based nanoparticles

Gold based nanoparticles:

Gold nanoparticles (AuNPs) have been widely used in preclinical studies for diagnosis and cancer therapy [17–20], especially in radiotherapy, because gold has a high atomic number, high inertness, good biocompatibility, easy chemical modifications and relatively strong photoelectric absorption coefficient [19]. Various studies have demonstrated the synergistic effect of AuNPs in radiation therapy using x-rays (high and low energy), γ-rays, electron beams and high-energy charged protons/carbon ions [20–23]. The size, shape, functionalization, concentration, and intracellular distribution of AuNPs can influence their effect on radiation [24]. Small size nanoparticles are perfect for enhanced permeation and retention (EPR) effect and consequently greater accumulation in tumors, but these nanoparticles are excreted out of the system quickly [25–27]. Although many cappings with different features such as bovine serum albumin (BSA), –SH PEG have exhibited many benefits such as high stability, uniform size, biocompatibility and internalization is also more [28–30]. Another approach to increase selective uptake of GNPs is conjugation of GNPs with antibodies or aptamers [31, 32]. Chattopadhyay *et al* studied radioenhancement effect by humanized anti-HER2 antibody Trastuzumab (Herceptin), PEGylated and conjugated 30 nm AuNPs [33]. *In vitro* studies on breast cancer cell line MDA-MB-361 cell line demonstrated an effective dose enhancement in the presence of gold nanoparticles using 100 kVp x-rays. The intratumoral injection of ~0.8 mg total Au was used (4.8 mg g^{-1} tumor) to MDA-MB-361 xenografts. Image guided x-ray irradiation was performed 24 h after injection at 11 Gy of 100 kVp. This dose led to a 46% reduction in tumor size relative to irradiation alone, with no damage to normal tissue [33]. Further, it has been reported that radiation enhances the invasive potential of some cancer cells due to upregulation of α5β1-integrin and fibronectin (FN) signaling. Wu *et al* conjugated (RGD)4 peptides on polyethylene glycolylated (PEGylated) AuNPs (P-AuNPs) to enable binding to integrins and internalization into cancer cells via endocytosis [34]. This study demonstrated that combination of RGD/P-AuNPs and radiation reduced cancer cell viability, invasion and increased DNA damage compared to radiation alone in MDA-MB-231 cells. Similarly, Lian *et al* synthesized fluorescent gold nanoclusters (AuNCs), with core size of <2 nm, that typically comprised a few to about 100 gold atoms and conjugated them to RGD peptide [35]. *In vitro* and *in vivo* treatment with RGD-conjugated gold nanoclusters indicated better uptake by 4T1 cells and tumor xenograft. Tumor size and tumor weight showed significant decline in RGD-conjugated gold nanoclusters and radiation treated group (figure 8.2).

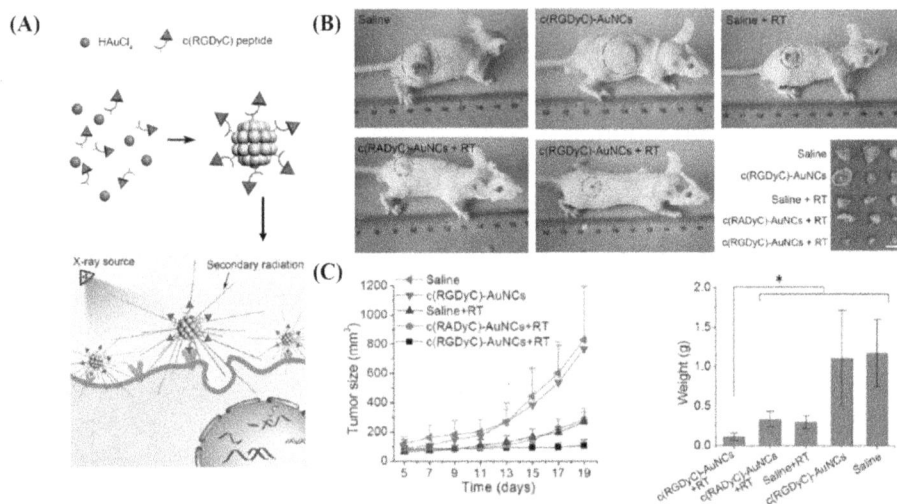

Figure 8.2. (A) Schematic representation of gold nanoclusters (AuNC) conjugated with cRGD peptide. RGD-conjugated AuNCs accumulate in αβ3-integrin-positive cancer cells and interact with incident radiation intensively, generating secondary radiation, and leading to radiation enhancement effect. *In vivo* AuNCs-enhanced radiotherapy of tumor-bearing mice. (B) Representative photographs of 4T1 tumor-bearing mice received various treatments and the tumor tissues were removed at the end of the treatments. Radiotherapy was performed at 4 h after i.v. injection of saline, c(RADyC)–AuNCs or c(RGDyC)–AuNCs. The injected dose of the AuNC = 0.1 mmol Au kg^{-1}. (C) Tumor volume and tumor weight of the mice after various treatments ($n = 3$). Tumor weight was taken after 15 days of treatment. Error bars, mean ± SD; *$P < 0.05$ Student's t-test. Reprinted from [35], copyright (2017), with permission from Elsevier.

Among other widely investigated targets, which are overexpressed on the cancer cell surfaces, nucleolin (NCL) was selected by Ghahremani *et al* [37]. AS1411 is an aptamer which uniquely binds to NCL [37]. GNCs conjugated with AS1411 aptamer (Apt–GNCs) increased their internalization by cancer cells, increased radiotherapy efficacy as mean tumors' volume decreased about 39% and reduced metastasis with overall increase in the mice survival rate of mice with breast cancer xenografts. Using specialized localization microscopy RNA Hildebrandt *et al* demonstrated proof-of-principle that Her2 (RNA) directed gold core nanoparticles showed radiation enhancement of 1.2 ratio on treated cells.

Rare-earth based nanoparticles

Elements with Z numbers ranging from 57 to 71 have been investigated for enhancing the effect of radiotherapy. Gadolinium oxides and rare-earth upconversion nanoparticles which have been commonly used as magnetic resonance imaging (MRI), CT and upconversion luminenscence imaging have been studied to determine effect on radiotherapy. Consequently, Gd-based agents show great promise for multifunctional theranostic (diagnostic and therapeutic) applications in clinical practice. Gadolinium-based nanoparticles (GdNPs) have been identified as valuable theranostic sensitizers for radiation therapy [38]. More importantly, GdNPs, such as AGuIX (activation and guidance of irradiation by x-ray) [39], exhibit diminished or no toxicity in preclinical studies employing mice and monkeys and are eliminated rapidly via the kidneys [40]. Gd-doped nanomaterials are also of focus for neutron-capture therapy. Upon irradiation of thermal neutron, Gd will emit long-range gamma rays, x-rays, Auger electrons and internal conversion electrons [41]. Recently, Gd-doped calcium phosphate nanoparticles were tested for neutron-capture therapy [42]. Such nanoparticles showed accumulation in tumors and suppressed the tumor growth upon the neutron irradiation.

Upconversion nanoparticles containing other rare-earth high-Z-elements such as ytterbium ($Z = 70$) or erbium ($Z = 68$) have also been developed for local radiation dose enhancement [43, 44]. Recently, the Shi and Bu group have developed core/ satellite upconversion nanoparticles (UCNPs, e.g. NaYbF4:Er/Gd; NaYF4:Yb/Er/ Tm@NaGdF4) for enhancing radiation damage [45–47]. These multifunctional theranostic nanoparticles could induce highly localized radiation energy to substantially enhance radiation effect under x-ray irradiation and the NIR-laser triggered thermal ablation therapy. This will be discussed further in the section on combined therapy.

8.2.2 Other high-Z-elements-based nanoparticles

Other semiconducting elements have also been analyzed for radiotherapy enhancement. Bismuth element has the highest atomic number ($Z = 83$) and its photoelectric absorption coefficient is more than AuNPS. Bismuth nanoparticles offer several advantages, they can be readily oxidized and dissolved at physiological conditions and discharged from the body as soluble bismuth ions [48–50]. A soluble bismuth ion is expected to be biologically safe at a concentration up to 50 mg mL^{-1} and bismuth ion compounds have been used as drug components for centuries. The Ming Su group synthesized red blood cell membrane-modified bismuth (i.e. F-RBC bismuth) nanoparticles [51]. The red blood cell membrane coating provides long blood circulation time and folate acts as tumor targeting agent. The radiosensitizing effect of these nanoparticles in x-ray radiation therapy for breast cancer was determined. The breast cancer cell line 4T1 cells were incubated with F-RBC bismuth nanoparticles (0–100 mg ml^{-1}) and x-ray radiated. The survival fractions of 4T1 cells were determined to further assess the radiosensitization effect of nano-particles. Figure 8.3 shows that at 100 mg ml^{-1} concentration along with x-ray radiation only 21.5% survived. Figure 8.3 also demonstrates that x-ray and bismuth nanoparticles interaction generates more free radicals for cancer cells damage, and physiological condition helps dissolve bismuth nanoparticles after treatment. DNA damage effect is determined by levels of γ-H2AX fluorescence (red fluorescence) in cells nuclei. Figure 8.3(C) shows high levels of γ-H2AX foci within nuclei of cells (blue fluorescence) incubated with F-RBC bismuth nanoparticles after x-ray radiation. The *in vivo* radio-sensitizations of bismuth nanoparticles were studied using 4T1 tumor-bearing BALB/C mice. Mice bearing 4T1 tumors were randomly divided into six groups (five each group): PBS alone group, PBS with radiation group, F-RBC bismuth nanoparticles no radiation group, bismuth nanoparticles with radiation group, RBC bismuth nanoparticles with radiation group, and F-RBC

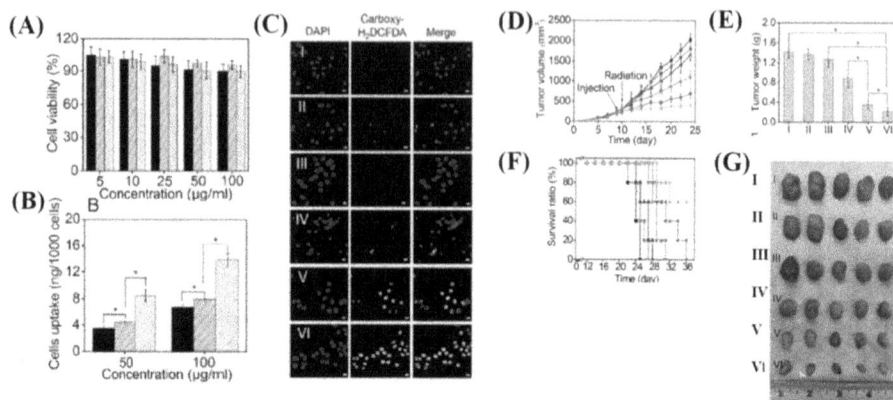

Figure 8.3. (A) Viabilities of 4T1 cells after treatment with bismuth nanoparticles (black), RBC bismuth nanoparticles (dense), and F-RBC bismuth nanoparticles (sparse). (B) Cellular uptake of bismuth nanoparticles (black), RBC bismuth nanoparticles (dense) and F-RBC bismuth nanoparticles (sparse) by 4T1 cells after 24 h of treatment. (C) Representative fluorescent images of cells after following treatment with PBS alone (I), PBS with radiation (II), F-RBC bismuth no radiation (III), bismuth with radiation (IV), RBC bismuth with radiation (V), and F-RBC bismuth with radiation (VI). Cells were stained with DAPI (blue) and carboxy-H2DCFDA (green). Scale bar = 20 mm. (D) Changes of tumor volume and (E) tumor weight in 4T1 tumor-bearing mice with different treatments: PBS alone (black curve), PBS with radiation (red curve), F-RBC bismuth nanoparticles no radiation (blue curve), bismuth nanoparticles with radiation (pink curve), RBC bismuth nanoparticles with radiation (green curve), and F-RBC bismuth nanoparticles with radiation (orange curve). (F) Surviving proportion of six different groups after various treatments ($n = 5$). (G) Photograph of tumors dissected after different treatments: PBS alone (I), PBS with radiation (II), F-RBC bismuth nanoparticles no radiation (III), bismuth nanoparticles with radiation (IV), RBC bismuth nanoparticles with radiation (V), and F-RBC bismuth nanoparticles with radiation (VI). Scale bar = 50 mm. Reprinted from [51], copyright (2018), with permission from Elsevier.

bismuth nanoparticles with radiation group. These mice were intravenously injected with 100 ml of 4 mg ml^{-1} nanoparticles and irradiated with x-ray radiation at a dose of 9 Gy after 24 h. The changes in tumor volume were monitored and are shown in figure 8.3(D). The results clearly indicated that tumor volume significantly decreased in animals treated with RBC and F-RBC bismuth nanoparticles along with x-ray radiation. Furthermore, body weight and the surviving ratio after various treatments (figure 8.3(C)) also indicated increased effect of bismuth nanoparticles along with radiation treatment. The average tumor weight of mice treated with F-RBC bismuth nanoparticles and exposed to radiation was much reduced as compared to mice treated with PBS alone (figure 8.3). Tumor slices of mice treated with F-RBC bismuth nanoparticles and radiation showed the obvious decrease of tumor cells density compared to those of other mice, further indicating the enhanced x-ray radiation therapy, which was consistent with the observed inhibition on tumor growth [51].

Tantalum ($Z = 73$) is a nontoxic and bio-inert element and as a high-Z element, x-ray attenuation coefficient is larger than gold [52]. Song *et al* synthesized polyethylene glycol (PEG) stabilized perfluorocarbon (PFC) nano-droplets decorated with TaOx nanoparticles (TaOx@PFC–PEG) as a multifunctional RT sensitizer [6]. The obtained TaOx@PFC–PEG nanoparticles on one hand can absorb x-ray by TaOx to concentrate radiation energy within tumor cells, on the other hand after saturating PFC with oxygen will act as an oxygen reservoir to gradually release oxygen and improve tumor oxygenation. The radiosensitizing effects of TaOx@PFC–PEG@O_2 were determined on breast cancer 4T1 cells. The DNA damage assay after treatment of above mentioned nanoparticles and 6 Gy x-ray irradiation exhibited a higher level of DNA damages. *In vivo* photoacoustic (PA) imaging indicated oxygenated hemoglobin level of tumors after injection of above mentioned nanoparticles is more. Finally, *in vivo* study with 4T1 mice model investigated the effect of nanoparticles on radiotherapy. Mice with 4T1 tumors were divided into five groups and treated with PBS; TaOx@PFC–PEG@O_2; x-ray irradiation alone; TaOx–PEG + RT; TaOx@PFC–PEG@O_2 + RT, respectively. The tumor growth curves and average tumor weights indicate stronger inhibition of tumor growth in TaOx@PFC–PEG@O_2 + RT than TaOx–PEG + RT. Further, TUNEL and H&E staining revealed that treatment by TaOx@PFC–PEG@O_2 + RT and TaOx–PEG + RT (Group 4) induced tumor cell apoptosis and necrosis. Therefore, TaOx@PFC–PEG@O_2 caused long-lasting radiotherapeutic effects due to high-Z element and also high oxygen releasing capability.

Hafnium-based nanoparticles
Hafnium-based nanoparticles have significant radiosensitization as well as good biocompatibility [53]. Most of the inorganic nanoparticles cause toxicity, if retained for long term. Liu group recently synthesized nanoscale metal–organic frameworks (NMOFs) composed with hafnium (Hf4+) and tetrakis (4-carboxyphenyl) porphyrin (TCPP) for enhanced radiotherapy. Hf4+ with strong x-ray attenuation ability could enhance the radiotherapeutic effect and TCPP is a photosensitizer to allow photodynamic therapy (PDT) (figure 8.4) [54]. Those NMOFs with PEG coating show efficient tumor homing upon intravenous injection, and show efficient clearance from the mouse body, minimizing concerns regarding their possible long-term toxicity. The remarkable combined radiotherapy and PDT was demonstrated in a 4T1 breast cancer model. The difference in hypoxic region in 4T1 tumor treated only with RT by x-ray irradiation and one with NMOFs and RT is clearly visible (figure 8.4). The Balb/c mice with 4T1 tumors were randomly divided into six groups (five mice per group): (i) untreated control, (ii) i.v. injection with NMOF–PEG, (iii) x-ray irradiation, (iv) i.v. injection with NMOF–PEG + Laser, (v) i.v. injection with NMOF–PEG + x-ray, (vi) i.v. injection with NMOF–PEG + x-ray + Laser. The dosage of NMOF–PEG was same in all groups ([TCPP]: 24 mg kg^{-1}, [Hf]: 12.5 mg kg^{-1}). The x-ray radiation dose during RT was 6 Gy, while PDT was introduced by the 661 nm light irradiation (5 mW cm^{-2} for 1 h) 8 h post RT treatment to allow recovery of tumor oxygenation. The mice treated with the combined therapy (group vi, NMOF–PEG + RT + PDT) showed maximum inhibition of tumor growth (figure 8.4). Hematoxylin and eosin (H&E) staining of tumor slices (figure 8.4) further confirmed combined RT & PDT had a remarkable antitumor effect.

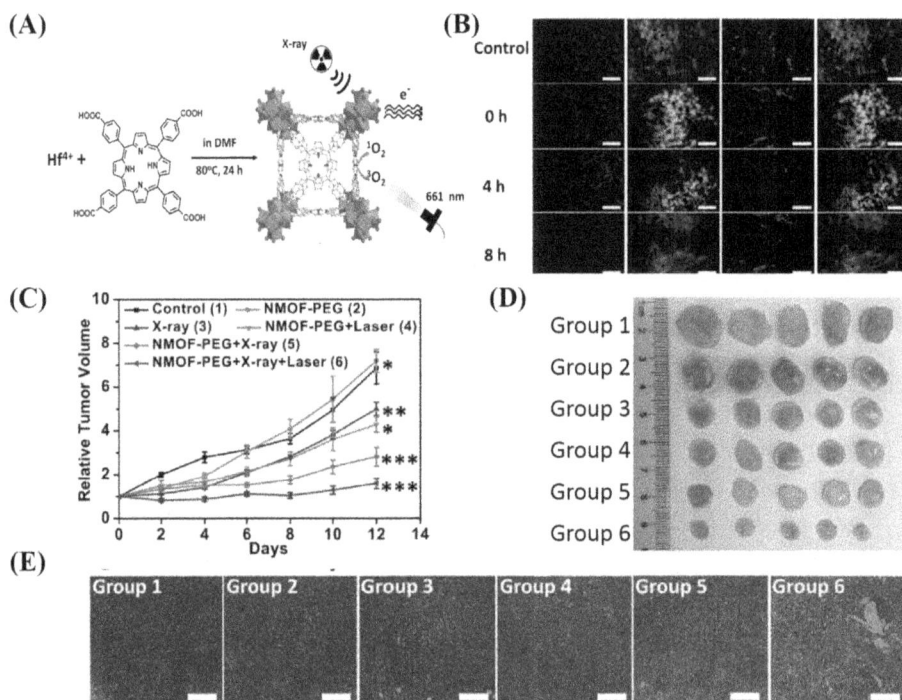

Figure 8.4. (A) Schematic representation of Hafnium (Hf4+) nanoparticles and processes of light-triggered SO generation by tetrakis (4-carboxyphenyl) porphyrin (TCPP) for PDT, as well as x-ray absorbance by Hf for enhanced RT. (B) Representative immunofluorescence images of tumor slices at different times post RT. The nuclei, blood vessels and hypoxia areas were stained with DAPI (blue), anti-CD31 antibody (red), and anti-pimonidazole antibody (green), respectively. (C)Tumor growth curves of different groups after various treatments ($n = 5$). P values were calculated by Tukey's post-test (***$p < 0.001$, **$p < 0.01$, or *$p < 0.05$). (D) Photograph of tumors from mice collected after receiving various treatments for 14 days. (E) H&E stained images of tumors collected 2 days after receiving various treatments. Scale = 200 mm. Reprinted from [54], copyright (2016), with permission from Elsevier.

8.2.3 Non-high-Z-elements

Silver-based nanoparticles

Silver nanoparticles ($Z = 47$) serve as radiosensitizers, not only due to their x-ray absorbance induced photoelectric or Auger effect [55, 56] but also as a result of the release of Ag+ ions to act as an oxidative agent to capture electrons and increase ROS production inside the cells [56–58], inhibitory action on the efflux activity of drug-resistant cells [59], and reactivity with glutathione (GSH) molecules [60]. Many research studies on different cancers demonstrated that AgNPs could serve as radiosensitizers and enhancers for radiotherapy [58, 61, 62]. Lu and coworkers reported marked x-ray irradiation enhancement on human breast adenocarcinoma MDA-MB-231 cells [63]. Swanner *et al* demonstrated significant cytotoxic and radiosensitizing effects of AgNPs on triple-negative breast cancer (TNBC) cells.

AgNPs induced more DNA and oxidative damage in TNBC cells treated with x-ray radiation [64].

Iron-based nanoparticles

Iron oxide nanoparticles (IONs) including inorganic paramagnetic iron oxide (or magnetite) nanoparticles, or superparamagnetic iron oxide nanoparticles (SPIONs) [65] are highly biocompatible and have been extensively explored for applications in magnetic resonance (MR) imaging, drug delivery, and magnetic hyperthermia therapy [66]. Interestingly, although the atomic number of iron (Fe, $Z = 26$) is relatively low, IONs when used in combination with low-linear energy transfer (LET) kV and MV x-rays, exhibit x-ray induced radiosensitization effect of cancer cells [67]. It was proposed that the released Fe3+ ions and the active surfaces of iron oxide nanoparticles could possess a strong catalysis effect under x-ray irradiation to generate ROS within cancer cells, acting as radiosensitizer to enhance the efficacy of RT [68, 69]. Klein *et al* compared the impact of x-rays on uncoated, citrate-coated, and malate-coated SPION when added to MCF-7 cells. They observed that citrate-coated SPIONs on exposure to 1 Gy x-ray radiation demonstrated an increase of up to 300% in the fluorescence intensity of the ROS reporter dichlorofluorescein diacetate (DCF-DA) [70].

Silicon-based nanoparticles

A recent *in vitro* study revealed that surface-oxidized silicon nanoparticles may increase the impact of x-radiation on the formation of reactive oxygen species (ROS) for clinically relevant doses [71, 72]. The SiO_x shell is considered to enhance x-ray induced generation of oxygen radicals (OII, HO_2, O_2) in aqueous solutions. Similarly, as for SPIONS, it was observed that surface-oxidized silicon nanoparticles may also increase the impact of x-radiation by the formation of reactive oxygen species (ROS). Klein *et al* in their study with silica nanoparticles (SiO_2NPs) demonstrated that functionalization of silica NPs with an amine function can significantly increase ROS production due to their positive surface charge, which facilitates NPs accumulation in the membranes of the endoplasmic reticulum, vesicles and especially mitochondria [73]. The presence of SiO_2NPs in the membranes also induced a membrane lipid peroxidation, improving the radiosensitizing effect [73]. The SiO_x shell is considered to enhance x-ray induced generation of oxygen radicals (OH, HO_2, O_2) in aqueous solutions. Similarly, as with SPIONS, it was observed that surface-oxidized silicon nanoparticles may also increase the impact of x-radiation by the formation of reactive oxygen species (ROS). Klein *et al* in their study with silica nanoparticles (SiO_2NPs) demonstrated that functionalization of silica NPs with an amine function can significantly increase ROS production due to their positive surface charge, which facilitates NP accumulation in the membranes of the endoplasmic reticulum, vesicles and especially mitochondria. The presence of SiO_2NPs in the membranes also induced a membrane lipid peroxidation, improving the radiosensitizing effect.

8.3 Nanomaterials delivering radioisotope for internal radioisotope therapy

The major obstacles in administration of radioisotopes for the treatment of cancer are their rapid elimination and nonspecific widespread distributions into normal tissues, which result in reduced efficacy and increased risks in side effects [9]. Hence, radioisotope-labeled nanoparticles have been in development to avoid unwanted circulation in the whole body and increase accumulation in the tumor [74]. Radioisotopes that can be used for radiotherapy are mainly divided into three categories including α, β and Auger particle emitters. The penetration and biological effect of these different particle emitters is mentioned in table 8.1. Vanpouille-Box *et al* generated lipid nanocapsules loaded with ^{188}Re (β-emitting radioisotope) and implanted in the brain to demonstrate improved survival rates to 83% in the rat orthotopic glioma model [75]. Another example of combining β-emitters with organic nanoparticles is ^{131}I-labeled multifunctional dendrimers for targeted SPECT imaging and radiotherapy of folic acid receptors overexpressing C6 xenografted tumor model [76]. Zhou *et al* used β-emitter ^{64}Cu-labeled copper sulfide nanoparticles and demonstrated that even after 24 h of intratumoral injection more than 90% of the nanoparticles were restricted in 4T1 breast tumor-bearing mice [77]. The combined radiotherapy and photothermal therapy (PTT) resulted in significant inhibition in tumor growth of the subcutaneous BT474 breast cancer model. As well as this, there was reduced growth of tumor initiating cells and hence, this led to reduction in lung metastasis.

Guryev *et al* [78] reported upconversion nanoparticles (UCNP) coupled to two therapeutic agents: β-emitting radionuclide yttrium-90 (90Y) fractionally substituting yttrium in UCNP, and a fragment of the exotoxin A derived *pseudomonas aeruginosa* genetically fused with a targeting designed ankyrin repeat protein (DARPin) specific to HER2 receptors. They tested the efficacy of these nanoparticles using the SKBR3 breast cancer cell line on mice, bearing HER2-positive xenograft tumors. Figure 8.5 demonstrates the schematic of the hybrid bio-functional nanocomplex and its potential *in vivo* as well *in vitro* therapeutic effect on breast cancer.

In another study, Yook *et al* [79] applied 30 nm-diameter gold nanoseeds modified with PEG chains linked to DOTA for complexing the β-emitter ^{177}Lu and panitumumab for EGFR targeting. This nanocomplex was injected into the tumors of CD-1 athymic mice having a subcutaneous EGFR-positive MDA-MB-468 human breast cancer xenograft model. Figure 8.6 shows the schematic of nanocomplex and SPECT/CT images after intratumoral injection of ^{177}Lu-T/NT-AuNP. It has been shown that Au nanoseed treatment was effective for inhibiting the tumor growth in mice and no organ toxicity was observed. Since, EGFR targeting in this study did not demonstrate any special benefit, this therapy can be widely used for locally advanced breast cancer without any specific phenotype.

Essler *et al* generated conjugates of alpha particle emitting radionucleotide (^{225}Ac and ^{213}Bi) with nucleolin targeted peptide [80]. Therapeutic efficacy and

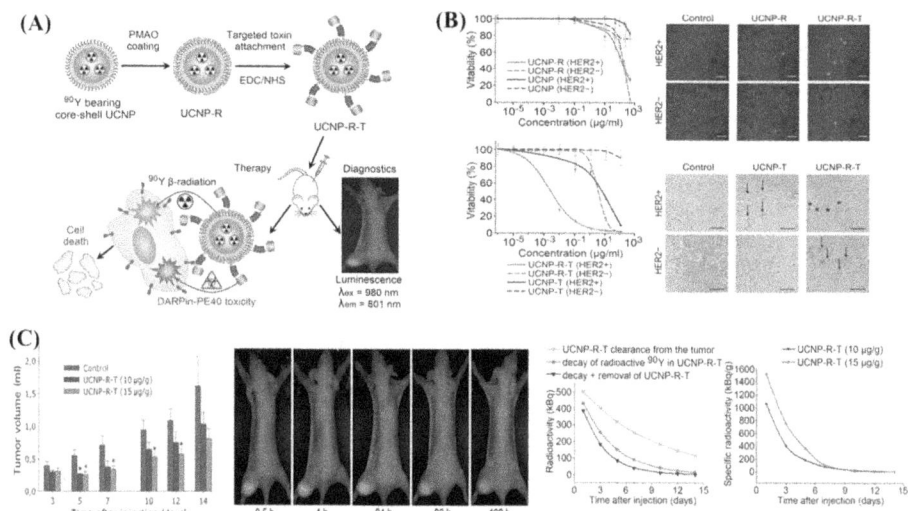

Figure 8.5. (A) Schematic of the hybrid bio-functional nanocomplex UCNP-R-T. R, radioactive; T, toxic; targeting module, targeted toxin DARPin-PE40. (B) In *in vitro* interaction of UCNP-R-T with eukaryotic cells demonstrates: the viability of HER2+and HER2− with UCNP-R and UCNP, LSCM images showing binding of UCNP-R-T to cells and, effect of UCNP-R-T and UCNP-T on cell morphology. (C) In *in vivo* therapeutic effect of UCNP-R-T: At doses of 10 and 15 μg g^{-1} on the tumor volume in mice where the tumor volume was normalized with respect to day 0 and data presented as mean $\pm$ SEM (*statistical value $P < 0.05$ by t-test), Retention of nanocomplex in human breast adenocarcinoma SK-BR-3 xenografts on athymic mice fluorescence imaged with *ex/em* = 980 nm/485–831 nm at dose 10 μg g^{-1}, Dynamics of the radioactivity/ specific confirms decrease with time in the tumor. Reprinted from [78], copyright 2018 National Academy of Sciences.

toxicity of ^{225}Ac-DOTA-F3 in comparison with that of ^{213}Bi-DTPA-F3 was determined through clonogenic assays of MDA-MB-435. The *in vivo* effect was tested in mice bearing intraperitoneal MDA-MB-435 xenograft tumors. This study concluded that therapy with both ^{225}Ac-DOTA-F3 (half-life 10 days) and ^{213}Bi-DTPA-F3 (half-life 46 min) helped to increase the survival of mice bearing peritoneal carcinoma with mild renal toxicity.

Radiotherapy with an ionizing x-ray remains a prevalent way for all solid cancer treatment. Although this is a powerful method to treat breast cancer, it is rarely sufficient alone to cause systemic tumor rejection. This treatment modality is generally combined with chemotherapy and now with immunotherapy. Hence, there have been studies in designing nanoparticles which can provide combined benefits [81].

8.4 Combined therapy

Chemo-radiotherapy

Platinum-based chemotherapeutic agents such as cisplatin, carboplatin, and oxali-platin, etc, result in the cell cycle arresting, DNA replication inhibition, cellular apoptosis as well as Auger electron generation due to the high-Z number of

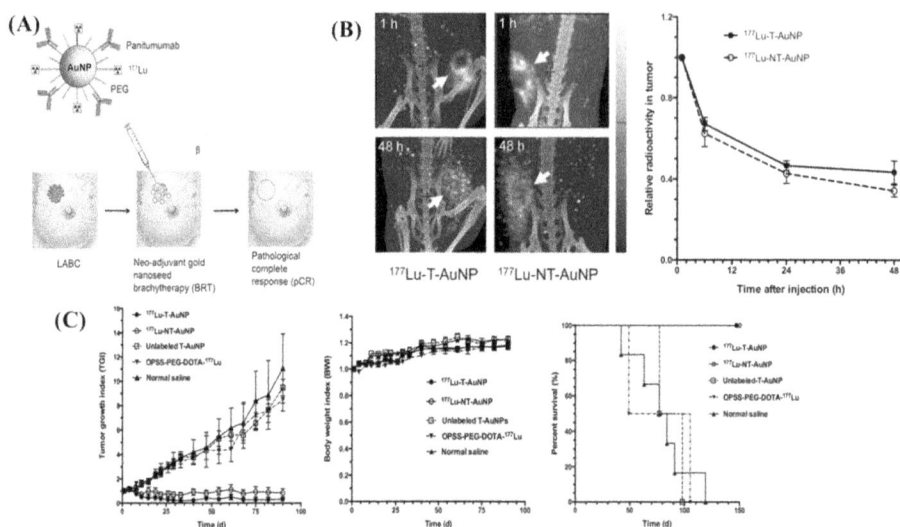

Figure 8.6. (A) Schematic of targeted 30 nm-diameter gold nanoseeds with PEG chains linked to panitumumab and to complex ^{177}Lu and steps in radiation treatment of tumors. (B) SPECT/CT images of pelvis of CD-1 athymic mice with subcutaneous MDA-MB-468 human breast cancer xenografts (white arrows) after intratumorally 177Lu-T/NT-AuNP injection and relative tumor radioactivity. (C) Tumor growth, body weight growth and percentage curve of survival. This research was originally published in [79], copyright SNMMI.

platinum ($Z = 78$). The liposomal formulation of cisplatin was found to have improved drug delivery and a greater radiosensitizing effect for *in vivo* treatment of carcinoma, and exhibited significantly less toxicity compared with free cisplatin in lung cancer [82]. There are relatively few studies which have investigated the synergistic effect of platinum-based chemotherapeutic agents. Further studies have explored ther effect of the combination of platinum-based chemotherapeutic agents with other metallic radio sensitizers. Cui *et al* examined radiation enhancement effects of AuNPs and cisplatin, individually and in combination and compared both *in vitro* and *in vivo* effects using the triple-negative breast cancer model. They reported that AuNPs and cisplatin together significantly enhanced the effects of fractionated irradiation without any toxicity [83]. Further, with the aim to decrease side effects of the related single modality treatments, to increase cytotoxicity and more localization on breast tumor cells, Islamian *et al* reported new superparamagnetic mesoporous hydroxyapatite nanocomposites conjugated with 1 mM doxorubicin and 0.5 mM 2-deoxy-D-glucose. The treatment T47D and SKBR3 cells with these nanoparticles and 1 and 2 Gy gamma rays showed reduced viability as compared to single treatment with radiation or anthracyclin antibiotics [84]. These are just a few examples to emphasize that radiosensitizers can be combined with chemotherapeutic agents to obtain a synergistic effect on treatment.

Photothermal-radiotherapy

NIR-laser-induced photothermal therapy (PTT) converts NIR laser into heat and thermally ablates cancer cells [85, 86]. It is widely known that mild hyperthermia is capable of increasing vascular permeability and improves the oxygen level in the tumor and thus alleviates the hypoxia caused after radiotherapy. Thus, it is believed that the combination of PTT and radiotherapy can be beneficial in the treatment of breast cancer. The most common is the combination of various Au-derived nanostructures (Au nanospikes, Au@Pt nanodendrites, PB@Au coresatellite nanoparticles, and Au@FeS core–shell nanoparticles). Other than these Au nanostructures, various other combinations have been studied such as silica-coated upconversion nanoparticles as the core and ultra-small CuS nanoparticles as the satellites [87]. All these nanostructures have been explored but are either complex or not cost-effective. Recently, Bi has been identified as a good alternative to gold as discussed in the section of other high-Z-elements. Yu *et al* reported Bi nanoparticles (~40 nm) capped with thiol ligands (Bi–SR) and surface-modified PEG. These nanoparticles had strong NIR absorbance and high photothermal conversion efficiency. The treatment of breast cancer in the 4T1 mic model with these nanoparticles and 4 Gy x-ray radiation resulted in inhibited growth of tumors as compared to single treatment (either PTT/x-ray radiation) [50].

Chemo-photodyanmic-radiotherapy
The photodynamic therapy (PDT) destroys cancer cells by generating cytotoxic singlet oxygen under laser treatment [88]. This singlet oxygen can further intensify the DNA damage effect generated by chemotherapy and radiation therapy. Hence, the combination of chemotherapy/radiotherapy/PDT (chemo-/radio-/photodynamic therapy) can have enhanced anticancer therapeutic effects [89, 90]. To achieve this objective, Fan *et al* generated core–shell nanoparticles with Gd-UCNPs as core and mesoporous silica shell (UCMSNs) for the co-delivery of a radio-/photo-sensitizer hematoporphyrin (HP) and a radiosensitizer/chemodrug docetaxel (Dtxl) [43] (figure 8.7). The treatment of the murine 4T1 breast cancer model with intratumoral injection of UCMSNs–HP–Dtxl (16 mg ml^{-1}, 150 ml) in PBS solution and x-ray (8 Gy, 5 min) and NIR light (980 nm, 1.5 W cm^{-2}, 30 min, 1 min interval) irradiation showed the complete eradication of the tumor (figure 8.7). Hence, the synergistic chemo-/radio-/photodynamic tri-modal therapy was effective in the complete eradication of the tumor but systemic delivery of these nanoparticles was not effective [43].

Radioimmunotherapy using nanoparticles
Radiotherapy (RT) is known to have a local immune-stimulatory effect. The abscopal effect has been noticed after radiation but this is not a regular phenomenon and is totally dose dependent [91]. This suggests that RT as a single agent is not sufficient to trigger an effective antitumor immune response in all cancer patients. Hence, now, combined RT and immunotherapy is being tested among cancer patients [92, 93]. Immunotherapies including checkpoint inhibitors and adoptive transfers of T or natural killer (NK) cells are being tested with RT. Hence, with the idea to maximize the ionizing effect of radiation to tumor tissues and least effect in adjoining normal tissues which can boost the antitumor response and nanoparticles treatment has been combined with immunotherapies. Recently, Wenbin Lin's group

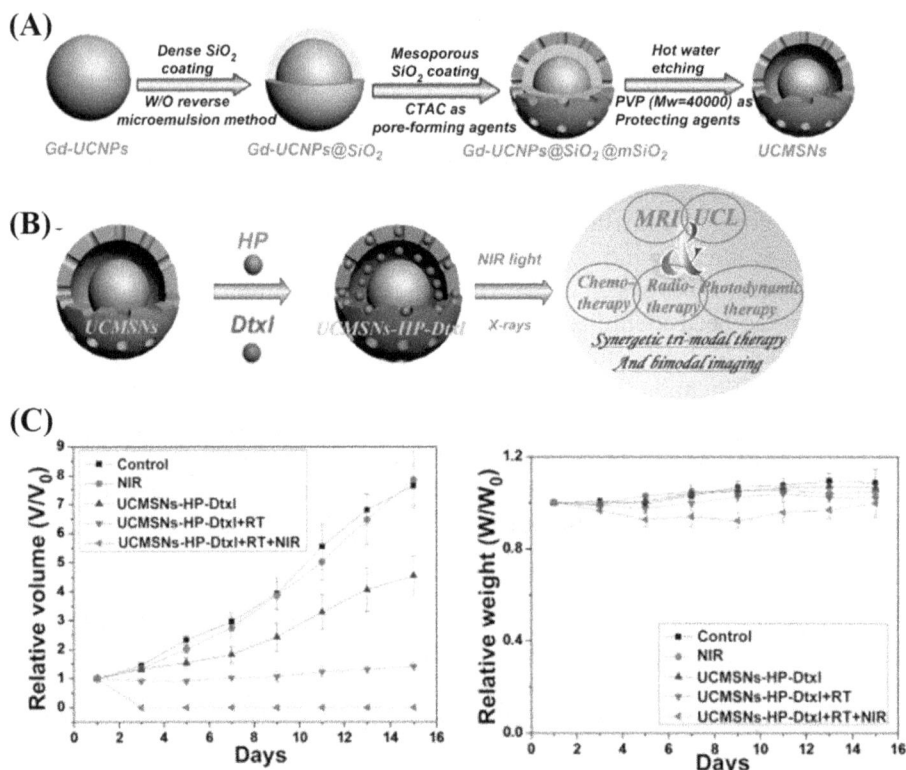

Figure 8.7. (A) Schematic illustration of the synthesis process of UCMSNs. Gd-UCNPs were prepared. (B) Schematic diagram of HP/Dtxl co-loaded UCMSNs loaded with hematoporphyrin (HP) and chemodrug docetaxel (Dtxl). By the co-delivery of HP/Dtxl via UCMSNs, synergetic chemo-/radio-/photodynamic tri-modal therapy as well as magnetic/upconversion luminescent (MR/UCL) bimodal imaging can be achieved. *In vivo* evaluation of synergetic chemo-/radio-/photodynamic therapy on 4T1 tumor-bearing mice after intra-tumoral injection of UCMSNs–HP–Dtxl. (C) Tumor growth and relative weight changes of 4T1 tumor-bearing mice over 15 days after the corresponding treatments was followed. Reprinted from [43], copyright (2014), with permission from Elsevier.

has studied the effect Hf-based nanoscale metal–organic frameworks (nMOFs) in the combination of low-dose RT and anti-programmed death-ligand 1 antibody on breast and various other cancer models. They reported that combining nMOF-mediated RT with immune checkpoint blockade elicits systemic antitumor immunity [94]. Radioimmunotherapy is currently in its infancy for clinical applications but has the potential to make significant contributions for the treatment of breast cancer.

8.5 Conclusions

The interest in nanomedicine in the treatment of cancer has grown after the success of Doxil or Abraxane. Numerous nanoparticles have been designed either as radiosensitizers or radioprotectors themselves or to mediate the delivery of a payload of another radiation modifying material. While the excitement about the impact of nanotechnology on radiation oncology is growing, research in this field is still growing with most studies confined to proof-of-principle experiments and modeling. Among all the preclinical studies, only two NPs have been translated to clinical radiosensitizers, namely NBTXR37 developed by Nanobiotix (Paris, France), which is a hafnium-based intratumorally administered NP, and AGuIX developed by NH Ther Aguix (Lyon, France), a gadolinium-based systemically administered NP. The clinical translation of nanoparticle-based strategies in radiation oncology is still evolving, but a detailed understanding of the physical and biological phenomenon of nanoparticle and radiation interactions will surely increase the potential of application in cancer treatment. With that understanding more of these nanomedicine systems need to be brought into clinical studies where the true efficacy of the system in human patients can be evaluated and begin to improve clinical outcomes in breast cancer patients.

References

[1] Siegel R L, Miller K D and Jemal A 2019 Cancer statistics, 2019 *CA Cancer J. Clin.* **69** 7–34

[2] Imai R, Kamada T, Araki N and Working Group for B and Soft Tissue Sarcomas 2016 Carbon ion radiation therapy for unresectable sacral chordoma: an analysis of 188 cases *Int. J. Radiat. Oncol. Biol. Phys.* **95** 322–7

[3] DeNardo S J and Denardo G L 2006 Targeted radionuclide therapy for solid tumors: an overview *Int. J. Radiat. Oncol. Biol. Phys.* **66** S89–95

[4] Roddiger S J *et al* 2006 Neoadjuvant interstitial high-dose-rate (HDR) brachytherapy combined with systemic chemotherapy in patients with breast cancer *Strahlenther. Onkol.* **182** 22–9

[5] Darby S C *et al* 2013 Risk of ischemic heart disease in women after radiotherapy for breast cancer *N. Engl. J. Med.* **368** 987–8

[6] Song G, Cheng L, Chao Y, Yang K and Liu Z 2017 Emerging nanotechnology and advanced materials for cancer radiation therapy *Adv. Mater.* **29**

[7] Peer D, Karp J M, Hong S, Farokhzad O C, Margalit R and Langer R 2007 Nanocarriers as an emerging platform for cancer therapy *Nat. Nanotechnol.* **2** 751–60

[8] Chatterjee D K, Wolfe T, Lee J, Brown A P, Singh P K, Bhattarai S R, Diagaradjane P and Krishnan S 2013 Convergence of nanotechnology with radiation therapy-insights and implications for clinical translation *Transl. Cancer Res.* **2** 256–68

[9] Porcel E, Liehn S, Remita H, Usami N, Kobayashi K, Furusawa Y, Le Sech C and Lacombe S 2010 Platinum nanoparticles: a promising material for future cancer therapy? *Nanotechnology* **21** 85103

[10] Hainfeld J F, Dilmanian F A, Slatkin D N and Smilowitz H M 2008 Radiotherapy enhancement with gold nanoparticles *J. Pharm. Pharmacol.* **60** 977–85

[11] Jeremic B, Aguerri A R and Filipovic N 2013 Radiosensitization by gold nanoparticles *Clin. Transl. Oncol.* **15** 593–601

[12] Cheng N N, Starkewolf Z, Davidson R A, Sharmah A, Lee C, Lien J and Guo T 2012 Chemical enhancement by nanomaterials under X-ray irradiation *J. Am. Chem. Soc.* **134** 1950–3

[13] Misawa M and Takahashi J 2011 Generation of reactive oxygen species induced by gold nanoparticles under x-ray and UV irradiations *Nanomedicine* **7** 604–14

[14] Jaramillo T F, Baeck S H, Cuenya B R and McFarland E W 2003 Catalytic activity of supported Au nanoparticles deposited from block copolymer micelles *J. Am. Chem. Soc.* **125** 7148–9

[15] Zheng Y, Hunting D J, Ayotte P and Sanche L 2008 Radiosensitization of DNA by gold nanoparticles irradiated with high-energy electrons *Radiat. Res.* **169** 19–27

[16] Pawlik T M and Keyomarsi K 2004 Role of cell cycle in mediating sensitivity to radiotherapy *Int. J. Radiat. Oncol. Biol. Phys.* **59** 928–42

[17] Li J, Gupta S and Li C 2013 Research perspectives: gold nanoparticles in cancer theranostics *Quant. Imaging Med. Surg.* **3** 284–91

[18] Lim Z Z, Li J E, Ng C T, Yung L Y and Bay B H 2011 Gold nanoparticles in cancer therapy *Acta Pharmacol. Sin.* **32** 983–90

[19] Jain S, Hirst D G and O'Sullivan J M 2012 Gold nanoparticles as novel agents for cancer therapy *Br. J. Radiol.* **85** 101–13

[20] Chithrani D B, Jelveh S, Jalali F, van Prooijen M, Allen C, Bristow R G, Hill R P and Jaffray D A 2010 Gold nanoparticles as radiation sensitizers in cancer therapy *Radiat. Res.* **173** 719–28

[21] Liu Y, Zhang P, Li F, Jin X, Li J, Chen W and Li Q 2018 Metal-based nanoenhancers for future radiotherapy: radiosensitizing and synergistic effects on tumor cells *Theranostics* **8** 1824–49

[22] Haume K, Rosa S, Grellet S, Smialek M A, Butterworth K T, Solov'yov A V, Prise K M, Golding J and Mason N J 2016 Gold nanoparticles for cancer radiotherapy: a review *Cancer Nanotechnol.* **7** 8

[23] Hainfeld J F, Slatkin D N and Smilowitz H M 2004 The use of gold nanoparticles to enhance radiotherapy in mice *Phys. Med. Biol.* **49** N309–15

[24] Chithrani B D, Ghazani A A and Chan W C 2006 Determining the size and shape dependence of gold nanoparticle uptake into mammalian cells *Nano Lett.* **6** 662–8

[25] Zhang X D *et al* 2014 Enhanced tumor accumulation of sub-2 nm gold nanoclusters for cancer radiation therapy *Adv. Healthc. Mater.* **3** 133–41

[26] Zhang X D, Luo Z, Chen J, Shen X, Song S, Sun Y, Fan S, Fan F, Leong D T and Xie J 2014 Ultrasmall Au(10-12)(SG)(10-12) nanomolecules for high tumor specificity and cancer radiotherapy *Adv. Mater.* **26** 4565–8

[27] Zhang X D *et al* 2015 Ultrasmall glutathione-protected gold nanoclusters as next generation radiotherapy sensitizers with high tumor uptake and high renal clearance *Sci. Rep.* **5** 8669

[28] Zhou W, Cao Y, Sui D, Guan W, Lu C and Xie J 2016 Ultrastable BSA-capped gold nanoclusters with a polymer-like shielding layer against reactive oxygen species in living cells *Nanoscale* **8** 9614–20

[29] Choi C H, Alabi C A, Webster P and Davis M E 2010 Mechanism of active targeting in solid tumors with transferrin-containing gold nanoparticles *Proc. Natl. Acad. Sci. U. S. A.* **107** 1235–40

[30] Shah N B, Vercellotti G M, White J G, Fegan A, Wagner C R and Bischof J C 2012 Blood-nanoparticle interactions and *in vivo* biodistribution: impact of surface PEG and ligand properties *Mol. Pharm.* **9** 2146–55

[31] Dam D H, Lee J H, Sisco P N, Co D T, Zhang M, Wasielewski M R and Odom T W 2012 Direct observation of nanoparticle-cancer cell nucleus interactions *ACS Nano* **6** 3318–26

[32] Wu X, Chen J, Wu M and Zhao J X 2015 Aptamers: active targeting ligands for cancer diagnosis and therapy *Theranostics* **5** 322–44

[33] Chattopadhyay N, Cai Z, Kwon Y L, Lechtman E, Pignol J P and Reilly R M 2013 Molecularly targeted gold nanoparticles enhance the radiation response of breast cancer cells and tumor xenografts to X-radiation *Breast Cancer Res. Treat.* **137** 81–91

[34] Wu P H, Onodera Y, Ichikawa Y, Rankin E B, Giaccia A J, Watanabe Y, Qian W, Hashimoto T, Shirato H and Nam J M 2017 Targeting integrins with RGD-conjugated gold nanoparticles in radiotherapy decreases the invasive activity of breast cancer cells *Int. J. Nanomed.* **12** 5069–85

[35] Liang G, Jin X, Zhang S and Xing D 2017 RGD peptide-modified fluorescent gold nanoclusters as highly efficient tumor-targeted radiotherapy sensitizers *Biomaterials* **144** 95–104

[36] Hildenbrand G *et al* 2018 Dose enhancement effects of gold nanoparticles specifically targeting RNA in breast cancer cells *PLoS One* **13** e0190183

[37] Ghahremani F, Kefayat A, Shahbazi-Gahrouei D, Motaghi H, Mehrgardi M A and Haghjooy-Javanmard S 2018 AS1411 aptamer-targeted gold nanoclusters effect on the enhancement of radiation therapy efficacy in breast tumor-bearing mice *Nanomedicine* **13** 2563–78

[38] Du F *et al* 2017 Engineered gadolinium-doped carbon dots for magnetic resonance imaging-guided radiotherapy of tumors *Biomaterials* **121** 109–20

[39] Detappe A, Lux F and Tillement O 2016 Pushing radiation therapy limitations with theranostic nanoparticles *Nanomedicine* **11** 997–9

[40] Sancey L *et al* 2015 Long-term *in vivo* clearance of gadolinium-based AGuIX nanoparticles and their biocompatibility after systemic injection *ACS Nano* **9** 2477–88

[41] Franken N A, Bergs J W, Kok T T, Kuperus R R, Stecher-Rasmussen F, Haveman J, Van Bree C and Stalpers L J 2006 Gadolinium enhances the sensitivity of SW-1573 cells for thermal neutron irradiation *Oncol. Rep.* **15** 715–20

[42] Mi P *et al* 2015 Hybrid calcium phosphate-polymeric micelles incorporating gadolinium chelates for imaging-guided gadolinium neutron capture tumor therapy *ACS Nano* **9** 5913–21

[43] Fan W *et al* 2014 A smart upconversion-based mesoporous silica nanotheranostic system for synergetic chemo-/radio-/photodynamic therapy and simultaneous MR/UCL imaging *Biomaterials* **35** 8992–9002

[44] Park Y I *et al* 2012 Theranostic probe based on lanthanide-doped nanoparticles for simultaneous *in vivo* dual-modal imaging and photodynamic therapy *Adv. Mater.* **24** 5755–61

[45] Liu J, Liu Y, Bu W, Bu J, Sun Y, Du J and Shi J 2014 Ultrasensitive nanosensors based on upconversion nanoparticles for selective hypoxia imaging *in vivo* upon near-infrared excitation *J. Am. Chem. Soc.* **136** 9701–9

[46] Liu Y, Liu Y, Bu W, Xiao Q, Sun Y, Zhao K, Fan W, Liu J and Shi J 2015 Radiation-/hypoxia-induced solid tumor metastasis and regrowth inhibited by hypoxia-specific upconversion nanoradiosensitizer *Biomaterials* **49** 1–8

[47] Xiao Q *et al* 2013 A core/satellite multifunctional nanotheranostic for *in vivo* imaging and tumor eradication by radiation/photothermal synergistic therapy *J. Am. Chem. Soc.* **135** 13041–8

[48] Hossain M and Su M 2012 Nanoparticle location and material dependent dose enhancement in X-ray radiation therapy *J. Phys. Chem. C Nanomater. Interfaces* **116** 23047–52

[49] Wei B, Zhang X, Zhang C, Jiang Y, Fu Y Y, Yu C, Sun S K and Yan X P 2016 Facile synthesis of uniform-sized bismuth nanoparticles for ct visualization of gastrointestinal tract in vivo ACS Appl. Mater. Interfaces **8** 12720–6

[50] Yu N, Wang Z, Zhang J, Liu Z, Zhu B, Yu J, Zhu M, Peng C and Chen Z 2018 Thiol-capped Bi nanoparticles as stable and all-in-one type theranostic nanoagents for tumor imaging and thermoradiotherapy Biomaterials **161** 279–91

[51] Deng J, Xu S, Hu W, Xun X, Zheng L and Su M 2018 Tumor targeted, stealthy and degradable bismuth nanoparticles for enhanced X-ray radiation therapy of breast cancer Biomaterials **154** 24–33

[52] Oh M H, Lee N, Kim H, Park S P, Piao Y, Lee J, Jun S W, Moon W K, Choi S H and Hyeon T 2011 Large-scale synthesis of bioinert tantalum oxide nanoparticles for X-ray computed tomography imaging and bimodal image-guided sentinel lymph node mapping J. Am. Chem. Soc. **133** 5508–15

[53] Maggiorella L, Barouch G, Devaux C, Pottier A, Deutsch E, Bourhis J, Borghi E and Levy L 2012 Nanoscale radiotherapy with hafnium oxide nanoparticles Future Oncol. **8** 1167–81

[54] Liu J et al 2016 Nanoscale metal-organic frameworks for combined photodynamic & radiation therapy in cancer treatment Biomaterials **97** 1–9

[55] Wu H, Lin J, Liu P, Huang Z, Zhao P, Jin H, Wang C, Wen L and Gu N 2015 Is the autophagy a friend or foe in the silver nanoparticles associated radiotherapy for glioma? Biomaterials **62** 47–57

[56] Kleinauskas A, Rocha S, Sahu S, Sun Y P and Juzenas P 2013 Carbon-core silver-shell nanodots as sensitizers for phototherapy and radiotherapy Nanotechnology **24** 325103

[57] Tamborini M, Locatelli E, Rasile M, Monaco I, Rodighiero S, Corradini I, Franchini M C, Passoni L and Matteoli M 2016 A combined approach employing chlorotoxin-nanovectors and low dose radiation to reach infiltrating tumor niches in glioblastoma ACS Nano **10** 2509–20

[58] Zheng Q, Yang H, Wei J, Tong J L and Shu Y Q 2013 The role and mechanisms of nanoparticles to enhance radiosensitivity in hepatocellular cell Biomed. Pharmacother. **67** 569–75

[59] Kovacs D et al 2016 Silver nanoparticles modulate ABC transporter activity and enhance chemotherapy in multidrug resistant cancer Nanomedicine **12** 601–10

[60] Shen D F, Wu S S, Wang R R, Zhang Q, Ren Z J, Liu H, Guo H D and Gao G G 2016 A silver(I)-estrogen nanocluster: GSH sensitivity and targeting suppression on HepG2 cell Small **12** 6153–9

[61] Liu P, Jin H, Guo Z, Ma J, Zhao J, Li D, Wu H and Gu N 2016 Silver nanoparticles outperform gold nanoparticles in radiosensitizing U251 cells in vitro and in an intracranial mouse model of glioma Int. J. Nanomed. **11** 5003–14

[62] Huang P, Yang D P, Zhang C, Lin J, He M, Bao L and Cui D 2011 Protein-directed one-pot synthesis of Ag microspheres with good biocompatibility and enhancement of radiation effects on gastric cancer cells Nanoscale **3** 3623–6

[63] Lu R, Yang D, Cui D, Wang Z and Guo L 2012 Egg white-mediated green synthesis of silver nanoparticles with excellent biocompatibility and enhanced radiation effects on cancer cells Int. J. Nanomed. **7** 2101–7

[64] Swanner J, Mims J, Carroll D L, Akman S A, Furdui C M, Torti S V and Singh R N 2015 Differential cytotoxic and radiosensitizing effects of silver nanoparticles on triple-negative breast cancer and non-triple-negative breast cells Int. J. Nanomed. **10** 3937–53

[65] Huang G, Chen H, Dong Y, Luo X, Yu H, Moore Z, Bey E A, Boothman D A and Gao J 2013 Superparamagnetic iron oxide nanoparticles: amplifying ROS stress to improve anticancer drug efficacy *Theranostics* **3** 116–26

[66] Sun C, Lee J S and Zhang M 2008 Magnetic nanoparticles in MR imaging and drug delivery *Adv. Drug Deliv. Rev.* **60** 1252–65

[67] Johannsen M, Thiesen B, Gneveckow U, Taymoorian K, Waldofner N, Scholz R, Deger S, Jung K, Loening S A and Jordan A 2006 Thermotherapy using magnetic nanoparticles combined with external radiation in an orthotopic rat model of prostate cancer *Prostate* **66** 97–104

[68] Hauser A K, Mitov M I, Daley E F, McGarry R C, Anderson K W and Hilt J Z 2016 Targeted iron oxide nanoparticles for the enhancement of radiation therapy *Biomaterials* **105** 127–35

[69] Klein S, Sommer A, Distel L V, Neuhuber W and Kryschi C 2012 Superparamagnetic iron oxide nanoparticles as radiosensitizer via enhanced reactive oxygen species formation *Biochem. Biophys. Res. Commun.* **425** 393–7

[70] Klein S, Sommer A, Distel L V, Hazemann J L, Kroner W, Neuhuber W, Muller P, Proux O and Kryschi C 2014 Superparamagnetic iron oxide nanoparticles as novel x-ray enhancer for low-dose radiation therapy *J. Phys. Chem.* B **118** 6159–66

[71] Ahmad J, Ahamed M, Akhtar M J, Alrokayan S A, Siddiqui M A, Musarrat J and Al-Khedhairy A A 2012 Apoptosis induction by silica nanoparticles mediated through reactive oxygen species in human liver cell line HepG2 *Toxicol. Appl. Pharmacol.* **259** 160–8

[72] Petrache Voicu S N, Dinu D, Sima C, Hermenean A, Ardelean A, Codrici E, Stan M S, Zarnescu O and Dinischiotu A 2015 Silica nanoparticles induce oxidative stress and autophagy but not apoptosis in the MRC-5 cell line *Int. J. Mol. Sci.* **16** 29398–416

[73] Klein S, Dell'Arciprete M L, Wegmann M, Distel L V, Neuhuber W, Gonzalez M C and Kryschi C 2013 Oxidized silicon nanoparticles for radiosensitization of cancer and tissue cells *Biochem. Biophys. Res. Commun.* **434** 217–22

[74] Avitabile E, Bedognetti D, Ciofani G, Bianco A and Delogu L G 2018 How can nanotechnology help the fight against breast cancer? *Nanoscale* **10** 11719–31

[75] Vanpouille-Box C *et al* 2011 Tumor eradication in rat glioma and bypass of immunosuppressive barriers using internal radiation with (188)Re-lipid nanocapsules *Biomaterials* **32** 6781–90

[76] Zhu J, Zhao L, Cheng Y, Xiong Z, Tang Y, Shen M, Zhao J and Shi X 2015 Radionuclide (131)I-labeled multifunctional dendrimers for targeted SPECT imaging and radiotherapy of tumors *Nanoscale* **7** 18169–78

[77] Zhou M, Zhao J, Tian M, Song S, Zhang R, Gupta S, Tan D, Shen H, Ferrari M and Li C 2015 Radio-photothermal therapy mediated by a single compartment nanoplatform depletes tumor initiating cells and reduces lung metastasis in the orthotopic 4T1 breast tumor model *Nanoscale* **7** 19438–47

[78] Guryev E L *et al* 2018 Radioactive ((90)Y) upconversion nanoparticles conjugated with recombinant targeted toxin for synergistic nanotheranostics of cancer *Proc. Natl. Acad. Sci. U. S. A.* **115** 9690–5

[79] Yook S, Cai Z, Lu Y, Winnik M A, Pignol J P and Reilly R M 2016 Intratumorally Injected 177Lu-labeled gold nanoparticles: gold nanoseed brachytherapy with application for neo-adjuvant treatment of locally advanced breast cancer *J. Nucl. Med.* **57** 936–42

[80] Essler M, Gartner F C, Neff F, Blechert B, Senekowitsch-Schmidtke R, Bruchertseifer F, Morgenstern A and Seidl C 2012 Therapeutic efficacy and toxicity of 225Ac-labelled vs. 213Bi-labelled tomor-homing peptides in a preclinical mouse model of peritoneal carcinomatosis *Eur. J. Nucl. Med. Mol. Imaging* **39** 602–12

[81] Tsoutsou P G, Zaman K, Lluesma S M, Cagnon L, Kandalaft L and Vozenin M-C 2018 Emerging opportunities of radiotherapy combined with immunotherapy in the era of breast cancer heterogeneity *Front. Oncol.* **8** 609

[82] Zhang X, Yang H, Gu K, Chen J, Rui M and Jiang G L 2011 In vitro and *in vivo* study of a nanoliposomal cisplatin as a radiosensitizer *Int. J. Nanomed.* **6** 437–44

[83] Cui L, Her S, Dunne M, Borst G R, De Souza R, Bristow R G, Jaffray D A and Allen C 2017 Significant radiation enhancement effects by gold nanoparticles in combination with cisplatin in triple negative breast cancer cells and tumor xenografts *Radiat. Res.* **187** 147–60

[84] Pirayesh Islamian J, Hatamian M, Aval N A, Rashidi M R, Mesbahi A, Mohammadzadeh M and Asghari Jafarabadi M 2017 Targeted superparamagnetic nanoparticles coated with 2-deoxy-d-gloucose and doxorubicin more sensitize breast cancer cells to ionizing radiation *Breast* **33** 97–103

[85] Cheng L, Wang C, Feng L, Yang K and Liu Z 2014 Functional nanomaterials for phototherapies of cancer *Chem. Rev.* **114** 10869–939

[86] Meng Z, Wei F, Wang R, Xia M, Chen Z, Wang H and Zhu M 2016 NIR-laser-switched *in vivo* smart nanocapsules for synergic photothermal and chemotherapy of tumors *Adv. Mater.* **28** 245–53

[87] Tian Q, Tang M, Sun Y, Zou R, Chen Z, Zhu M, Yang S, Wang J, Wang J and Hu J 2011 Hydrophilic flower-like CuS superstructures as an efficient 980 nm laser-driven photothermal agent for ablation of cancer cells *Adv. Mater.* **23** 3542–7

[88] Spyratou E, Makropoulou M, Mourelatou E A and Demetzos C 2012 Biophotonic techniques for manipulation and characterization of drug delivery nanosystems in cancer therapy *Cancer Lett.* **327** 111–22

[89] Ma M, Chen H, Chen Y, Wang X, Chen F, Cui X and Shi J 2012 Au capped magnetic core/mesoporous silica shell nanoparticles for combined photothermo-/chemo-therapy and multimodal imaging *Biomaterials* **33** 989–98

[90] Wang T, Zhang L, Su Z, Wang C, Liao Y and Fu Q 2011 Multifunctional hollow mesoporous silica nanocages for cancer cell detection and the combined chemotherapy and photodynamic therapy *ACS Appl. Mater. Interfaces* **3** 2479–86

[91] Reynders K, Illidge T, Siva S, Chang J Y and De Ruysscher D 2015 The abscopal effect of local radiotherapy: using immunotherapy to make a rare event clinically relevant *Cancer Treat. Rev.* **41** 503–10

[92] Barker C A and Postow M A 2014 Combinations of radiation therapy and immunotherapy for melanoma: a review of clinical outcomes *Int. J. Radiat. Oncol. Biol. Phys.* **88** 986–97

[93] Tang C *et al* 2017 Ipilimumab with stereotactic ablative radiation therapy: phase I results and immunologic correlates from peripheral T cells *Clin. Cancer Res.* **23** 1388–96

[94] Ni K, Lan G, Chan C, Quigley B, Lu K, Aung T, Guo N, La Riviere P, Weichselbaum R R and Lin W 2018 Nanoscale metal-organic frameworks enhance radiotherapy to potentiate checkpoint blockade immunotherapy *Nat. Commun.* **9** 2351

Chapter 9

Strengths and limitations of physical stimulus in breast cancer nanomedicine

Rakesh M Patil, Joanna Bauer and Nanasaheb D Thorat

9.1 Introduction

9.1.1 Cancer

Cancer is one of the leading causes of deaths globally; the World Health Organisation (WHO) attributed approximately 8.2 million mortalities due to cancer in the year 2012 which constituted 13% of all deaths [1]. In spite of many new and innovative strategies developed for effective detection and treatment of various cancers, global cancer statistics show that the incidence of cancer has increased worldwide [2]. Cancer develops due to deregulated division of an abnormal cell or from the changes that cause normal cells to acquire abnormal functions and has the potential to invade other parts of the body (Cancer Metastasis). Major factors for cellular abnormality include inherited mutations or mutation induced by environmental factors such as viruses, x-rays, ultraviolet (UV) light, chemicals, tobacco, etc, and lifestyle factors such as obesity, stress, lack of physical exercise, etc. Almost 100 different types of cancer affect humans and it was found that most of them are not due to a single factor. All evidence suggests that cancer results from a series of molecular events that fundamentally alter the normal function and properties of cells [3]. Lung cancer, prostate cancer, colorectal cancer and stomach cancer are the most common types of cancer in men while breast cancer, colorectal cancer, lung cancer and cervical cancer are the most common types in women.

Cancer can be treated by a variety of methods, including surgery, chemotherapy and radiotherapy. Chemotherapy is the most widely used therapy method where chemotherapeutic agents (anticancer drugs) are injected or orally delivered to kill rapidly growing cells. Most of these agents interfere with normal DNA replication, primarily blocking the cells to complete the S phase of the cell cycle. In addition

doi:10.1088/2053-2563/ab2907ch9

there are some chemotherapeutic agents which cause extensive DNA damage as well as inhibiting the formation of spindle fibers. Although chemotherapeutic agents are relatively specific, they also kill normal cells which are dividing very rapidly, such as gastrointestinal tract cells, bone marrow cells and hair follicles. This causes some of the side effects of chemotherapy; including gastrointestinal distress, low white blood cell count and hair loss [3]. These side effects are due to non-specific tissue biodistribution of chemotherapeutic drugs. Moreover, because of the systemic distribution, only a small portion of drugs reach tumor tissues, leading to relatively low drug efficacy [4]. In some patients drug resistance has emerged as a major obstacle limiting the therapeutic efficacy of chemotherapeutic agents, allowing tumors to evade chemotherapy. Delivering therapeutically active molecules to a target site is a challenging task. Poor aqueous solubility and non-specific targeting of chemotherapeutic agents are limitations of current treatment methods. Nanostructures, especially multifunctional nanostructures, have the potential to overcome these limitations by acting as carriers for therapeutic agents such as drugs, genes, etc, targeted to specific cancer cells and as imaging agents [3]. Chemotherapy, despite its severe side effects, is still one of the most effective treatments for tumor therapy owing to its ability to eradicate disseminated cancer cells and retard recurrence [5].

High-energy radiation in radiation therapy kills cancer cells by either directly damaging DNA or by generating reactive oxygen species (ROS) preventing cellular division. It is of two forms: brachytherapy where the radioactive source (in pellets) is placed close to the tumor, e.g. uterine cancer, and teletherapy, where the patient is irradiated from a source placed some distance away from the body, e.g. skeletal cancer. However, radiation therapy also kills the normal cells through which the radiation has to pass to reach the tumor. For a better result a combination of surgery and radiation therapy is advisable. A major drawback of radiation therapy is that certain cancers develop radiation resistance due to the expression of specific ROS scavengers. In addition, not all cancers are eligible for radiation therapy due to the adjacency of radiosensitive organs [3].

Surgery, is the oldest treatment method for cancer and is the most effective approach to remove solid tumors (benign tumors as well as in the early stages of cancer). Surgical resection is often not advised in cases of large tumors, locations close to other vital structures and the presence of distant metastases (high grade tumors) requiring assistance treatment. Surgery has no great effect if the tumor has already spread to other organs [3]. Several types of surgeries are available now and it is also used in combination with other treatment methods to eradicate cancer.

Despite the advantages of all these methods individually or in combination these treatment methods are not the absolute solution to kill all cancer cells. The reasons behind treatment failure are, in most cases, that cancer is commonly detected at a later stage; conventional chemotherapy has been disappointing in efficacy due to multidrug resistance (MDR) and severe side effects and cancer therapeutic biological agents, as well as the new class of anticancer drugs, are commonly unstable in *in vivo* circulation, with rapid degradation and inactivation before reaching the target site. For these reasons, early detection, effective diagnosis and effective treatment of cancer need to be optimized to increase the survival rate and decrease cancer associated deaths [6].

9.1.2 Breast cancer

Among all the types of cancer, breast cancer is the most ubiquitous type of malignant neoplasm among women with more than a million new cases yearly. After skin cancer, breast cancer is the most common cancer diagnosed in women in the United States (US). Breast cancer can occur in both men and women, but it is far more common in women. In the US the incidence rate of breast cancer in women is 12.5% and the risk of an individual dying from breast cancer is 1 in 35. It is the primary cause of mortality among women aged 45–55 years and is the second leading cause of cancer induced death [7].

Breast cancer is a type of tissue cancer that develops from breast tissue and mainly involves the inner layer of milk glands or lobules (lobular carcinomas) and ducts (ductal carcinomas). The primary symptoms of breast cancer may include a lump in the breast (thickening that feels different from the surrounding tissue), a change in breast shape, size and appearance, dimpling of the skin, fluid coming from the nipple, a newly inverted nipple and a red or scaly patch of skin. The risk factors for developing breast cancer include being female, age, high hormone level, obesity, lack of physical exercise, ionizing radiations, family history, beginning of periods at a younger age, first child at an older age, never been pregnant, economic status, iodine deficiency in diet, etc [7]. About 5%–10% of cases are due to genes inherited from a person's parents. A number of inherited mutated genes that can increase the likelihood of breast cancer have been identified. The most well-known are BRCA1 (breast cancer gene 1) and BRCA2 (breast cancer gene 2), both of which significantly increase the risk of both breast and ovarian cancer.

In breast cancer treatments, most often women require complete tissue removal. Today's treatment methods which are widely used for breast cancer are chemotherapy, radiotherapy, and hormone therapy. Breast cancer is confirmed by taking a biopsy of the concerning lump in the breast. After the diagnosis of cancer, further tests are done to determine if the cancer has spread beyond the breast or not and which treatment is better for the patient. Due to lacunas in the breast cancer treatments and the tremendous potential in nanotechnology a lot of interest has been generated in recent years to develop nanomedicine to eradicate the breast cancer with new modalities.

9.1.3 Nanomedicine

After the commencement of nanotechnology in 1959, the field of nanomedicine has developed very rapidly and we are now successfully approaching solutions to various challenges. The field of nanoscience and nanotechnology has a very high potential to overcome today's challenges most probably in the field of theranostics (the combination of therapy and diagnostics). A few of the current challenges of theranostics include preparation of drug nanocarriers which are inert (biologically), easily internalize into the cells and remain intact until they reach their desired target with a high specificity [8]. The unique properties of nanoparticles (optical, electrical, magnetic, etc) make them promising candidates for a range of applications in nanomedicine, ranging from *in vitro* diagnostic assays to *in vivo* localized imaging, drug delivery and therapy [1–5]. Gold nanoparticles (GNPs) and quantum dots are used for high resolution imaging due to their optical properties [9], while magnetic nanoparticles (MNPs) are used in magnetic resonance imaging (MRI) techniques for diagnosing cancer [10] and magnetic fluid hyperthermia therapy [11] to kill cancer cells. The use of nanoparticles in nanomedicine requires a molecular level under-standing of the interactions between nanoparticles and cells in a physiological environment.

9.1.4 Nanomedicine for cancer treatment

Recent development of nanomedicine for cancer treatment conquers various challenges in this field. The advancement of nanomedicine passed through various achievements starting from gold nanoparticles, polymeric nanoparticles, quantum dots, fullerenes, MNPs, etc, to the clinically approved nanomedicines for chemo-therapy. In addition, engineered biocompatible endogenous protein based nano-formulations are currently constructed to deliver cancer therapeutic and diagnostic agents simultaneously. Protein nanoformulations are commonly incorporated with dyes, contrast agents, drug payloads or inorganic nanoclusters, serving as imaging guided combinatorial cancer therapeutics [6]. The success of a Food and Drug Administration (FDA) approved protein based nanoformulation called Abraxane (albumin based paclitaxel nanoparticles) generates high interest in this field. The rationale for developing nanomaterial for cancer treatment includes; (i) multi-functionality, (ii) increased potency and multivalency, (iii) increased selectivity for targets, (iv) theranostic potential, (v) altered pharmacokinetics, (vi) controlled synthesis, (vii) controlled release of cancer therapeutic agent, (viii) novel properties and interactions, (ix) lack of immunogenicity and (x) enhanced physical stability [3].

9.2 Nanomedicine for tumor targeting

The ability of nanoparticles and macromolecules to passively accumulate in solid tumors and enhance therapeutic effects in comparison with conventional anticancer agents has resulted in the development of various multifunctional nanomedicines including liposomes, polymeric micelles and MNPs. Further modifications of these nanoparticles has improved their characteristics in terms of tumor selectivity, circulation time in blood, enhanced uptake by cancer cells, and sensitivity to tumor microenvironment. These 'smart' systems have enabled highly effective delivery of drugs, genes, shRNA, radioisotopes and other therapeutic molecules at the diseased/targeted site [12]. Nanoparticle mediated delivery of cancer therapeutic agents can be done either by passive targeting or by active targeting. Active targeting utilizes specific biomarkers such as cell surface receptors for specifically targeting to tumor tissue. In the case of passive targeting, the drug loaded nanoparticles entered into the tumor tissue owing to the enhanced permeability and retention effect (EPR) provided by the anatomical and pathophysiological abnormality of tumor vasculature. Active and passive targeting is described in detail in subsequent sections. Both passive targeting and actively enhanced cellular internalization play significant roles in tumor-targeted therapy [4].

9.2.1 Active targeting

Active drug targeting is based on the use of targeting ligands, like antibodies and peptides, which specifically bind to receptor structures expressed on the target cells [13]. Active drug targeting is generally implemented to improve target cell recognition and target cell uptake and not to improve overall tumor accumulation. Folate [14], transferrin [15] and galactosamine [16] are examples of targeting ligands routinely used in active targeting. Figure 9.1 shows the conceptual and realistic models for passive and active drug targeting. To date, only antibody based nanomedicines, such as Zevalin, Mylotarg, Ontak and Bexxar have been approved for clinical use in spite of significant progress made at the preclinical level [13].

9.2.2 Passive targeting

Passive targeting is a consequence of the EPR effect; EPR is thought to exploit unique characteristics such as leaky tumor vasculature and poor lymphatic drainage of many rapidly growing solid tumors in order to promote the accumulation of nanocarriers in tumor tissue. The particle size, surface charge, and surface modification of nanocarriers are particularly important factors in passive targeting [4]. However, the resulting therapeutically relevant local concentrations of

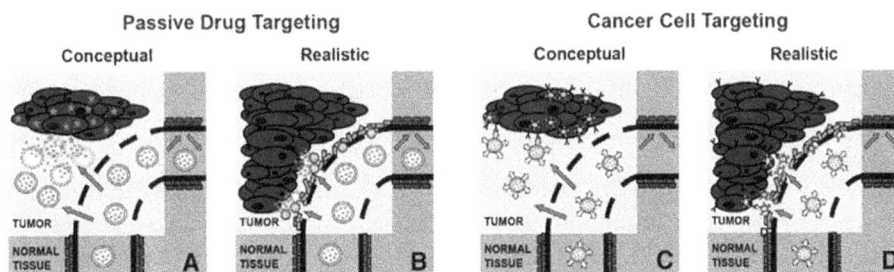

Figure 9.1. Conceptual and realistic models for passive and active drug targeting. (A, B) In passive drug targeting, it is often mistakenly assumed that all tumor blood vessels are leaky, and that all tumors possess leaky blood vessels. This might be the case in rapidly growing tumor models in rodents, but is definitely not the case in humans, where substantially enhanced vascular leakiness is observed only in certain specific tumors (e.g. in Kaposi sarcoma), and where only certain parts of tumors are hyperpermeable. (C, D) Active drug targeting to cancer cells is often mistakenly assumed to be able to increase overall tumor accumulation. This, however, cannot be the case, since nanomedicines enter the tumor interstitium via passive extravasation. After this, especially in physiologically relevant (i.e. slowly growing; comparable to the clinical situation) tumors, they need to cross several pericyte based, smooth muscle based and fibroblast based cell layers before they are able to bind to cancer cells. Furthermore, even if they are able to reach cancer cell-containing compartments, their penetration deep (-er) into the tumor is limited by the binding site barrier. Consequently, active cancer cell targeting is considered to be useful only for improving cellular uptake, as well as for targeting certain specific cell types within a solid tumor. Reproduced with permission from [13]. Copyright © 2012 Elsevier B.V.

anticancer agents are often insufficient to cause tumor regression and complete elimination. The reasons are: poor drug release inside tumor cells as well as the presence of drug efflux pumps, poor perfusion of inner regions of solid tumors as well as vascular barrier, high interstitial fluid pressure and dense intercellular matrix are the main intratumoral barriers that impair drug delivery and impede uniform distribution of nanomedicines throughout a tumor [12].

9.3 Stimuli responsive/triggered nanomedicine for cancer theranostics

Tumor cells in an environment with sub-lethal drug concentration could result in acquired resistance. To circumvent this problem, rapid drug release is crucial to provide optimum drug concentrations and destroy tumor cells before they acquire the capability to exclude chemotherapeutics. Thus, stimuli responsive nanoparticles have been developed through the design of materials recognizing endogenous or exogenous stimuli to rapidly trigger drug release [5, 17].

The principle of stimuli responsive systems is based on active and passive targeting. After injecting intravenously (or by another administration mode) nano-carriers can extravasate through the leaky architecture of tumors and accumulate at the tumor lesion via passive targeting or active targeting and then the delivery systems can be activated by single or several specific triggers using endogenous or exogenous stimuli to release the bioactive cargoes at the intended sites [18]. In other words, stimuli responsive nanomedicines are desirable for achieving site specific accumulation and triggered drug release in response to slight changes in physico-chemical properties in pathological conditions or to exogenous stimuli. A smart response of the nanomedicine external (physical) or internal stimuli allows better localization of the system in the desired biological compartment, controlled release of payload at the location of the pathological event and rapidly addressing/imaging the pathological event [19].

This chapter deals with the discussion of such novel nanomedicine formulations that are responsive to various physical stimuli and which have been exploited for cancer therapy and simultaneously can be used for diagnosis. This chapter does not elaborate equally important internal stimuli used in cancer therapy. Thus, the physical stimuli included in this chapter are light (UV–visible and NRI), magnetic field (MF) and ultrasound in detail and to some extent other physical stimuli such as radiofrequency and ionizing radiation.

9.3.1 Endogenous/internal stimuli responsive nanomedicine

Nanomedicines can be developed for target specific delivery on the basis of biological signals to improve specificity and minimize side effects. Such biological signals in cancer are like the pH difference in tumor microenvironment, over-expression of some receptors or enzymes, etc. Depending on these changes nano-medicines are developed to respond to internal stimuli. The nanomedicines responsive to internal stimuli can be classified as pH responsive nanomedicines, redox-responsive nanomedicines and enzyme responsive nanomedicines.

9.3.1.1 pH responsive nanomedicine

It is well-known that a pH gradient exists between intracellular and extracellular compartments in tumors. The pH gradient is due to the rapid proliferation and growth of tumor cells exceeding blood transportation, causing an inadequate supply of nutrients and oxygen, and thus generating lactic acid due to glycolysis [20]. Given the acidic tumor microenvironment, pH responsive nanomedicine is considered as the most skillful strategy to kill cancer cells, especially the MDR cancer cells. Figure 9.2 shows a schematic illustration of the dynamic response progress of nanoparticles. These nanomedicines remain stable at physiological pH, but collapse in an acidic microenvironment due to hydrolysis of acid labile bonds or protonation of chemical groups. This phenomenon results in release of their cargo into the cytosol, which promotes cytoplasmic drug concentrations to increase several folds. The increased concentrations of drug exceed the capacity of the efflux transporters [5, 21]. pH responsive iron oxide nanoparticles bicoated with sodium alginate and hydroxyapatite were shown to exhibit a controlled drug release profile for the

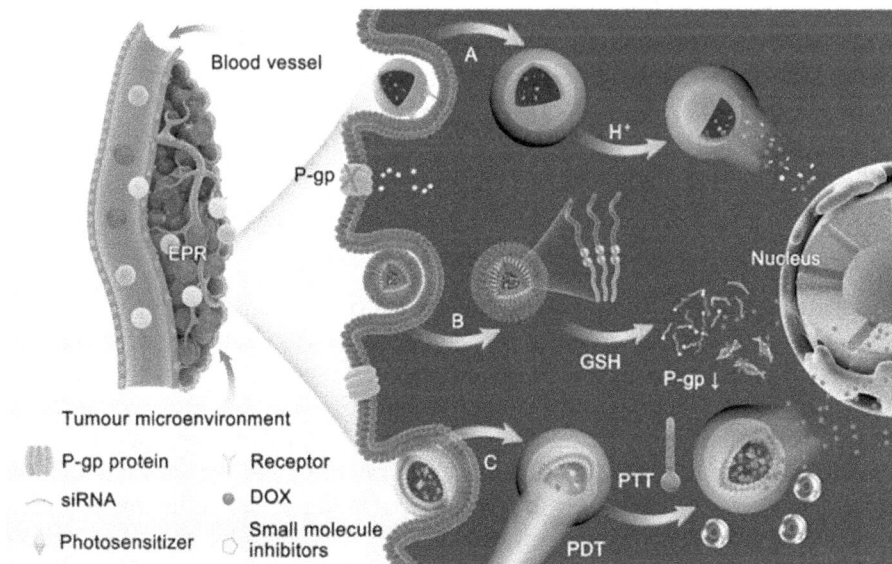

Figure 9.2. Schematic diagram of the dynamic response progress of intelligent nanoparticles. (A) Example of pH responsive nanoparticles loaded with small molecule inhibitors disintegrating in the acidic endosomal environment with triggered release of the cargo to inhibit P-gp function. (B) Redox-responsive nanoparticles with loaded siRNA and drug collapsing under GSH rich condition via disulfide bond cleavage. (C) Light triggered nanoparticles with photosensitizer could increase the temperature in tumor sites via photothermal therapy (PTT) or generate ROS, leading to endosomal disruption and abundant drug release. Reproduced with permission from [5] under the terms of the Creative Commons Attribution Non-Commercial License. Copyright © 2019 Ivyspring International Publisher.

hydrophobic drugs curcumin and 6-gingerol and may offer a potential platform for cancer therapy [22].

9.3.1.2 Redox-responsive nanomedicine

The abnormal metabolism of cancer cells plays a crucial role in affording anabolic energy demands. The glutathione (GSH) concentration in cancer cells is much higher (100–1000 times) than in the extracellular fluids, especially in MDR cancer cells [5, 23]. Due to this quality of cancer cells redox-responsive delivery systems have gained great attention for intracellular drug release via thiolysis in the presence of GSH. The reducible linkers commonly utilized are disulfide bonds, thioether bonds and diselenide bonds. The most common and simplest method is disulfide bond linkage and it can be inserted into the material of nanocarriers, acting as a linker between two blocks of polymers [24, 25], as well as a linkage to bind agents or ligands to nanocarriers.

9.3.1.3 Enzyme responsive nanomedicine

The degradation of tumor extracellular matrix is responsible for metastasis and proliferation of cancer cells. Matrix metalloproteinase (MMP) and other proteolytic enzymes secreted by cancer cells degrade the membrane basement, resulting in thin, interrupted and even defective states. In addition, these enzymes enhance metastasis and make cancer difficult to eradicate, thereby causing tumor resistance. Since some enzymes are over expressed in cancer cells especially in MDR cancer cells, such as protease [26], phospholipase [27] and glycosidase [28], enzyme responsive nanomedicine is an emerging strategy to deliver chemotherapeutics to tumor cells selectively and effectively. MMP is one of the most promising approaches utilized for controlled release of enzyme responsive systems [5].

9.3.2 Exogenous/external stimuli responsive nanomedicine

As discussed earlier, currently used clinical cancer treatment strategies (surgery, radiotherapy and chemotherapy) have many limitations. Surgery, which is highly invasive and can rarely eliminate all tumor cells especially when metastasis has already happened. Chemotherapy and radiotherapy have severe toxic side effects, limited therapeutic efficacy, being able to induce drug resistance and cause long term damage of immune systems [29]. Thus, development of new generations of cancer treatment methods has been an urgent need. Among many different new cancer therapeutic approaches being explored in the past decade, therapies that are responsive to external physical stimuli usually by utilizing nanoscale agents that are capable of responding to light, MF, ultrasound and others have received substantial interest from many research groups [29–34]. Excitingly, physical stimuli responsive therapies have demonstrated their abilities to enhance therapeutic efficacies and reduce unwanted side effects in many preclinical animal studies. In addition, these therapies have shown a number of unique advantages compared with conventional chemotherapy. Many physical stimuli responsive therapies can spatially control the therapeutic effect only in the tumor region by focusing external stimuli specifically on the tumor without causing much damage to normal tissues; many physical stimuli responsive therapies can not only be used to directly kill cancer cells, but are also useful to trigger or enhance other different cancer therapies to achieve the desired synergistic therapeutic effects via different mechanisms and certain types of physical stimuli may be able to control the drug internalization or drug release in the tumor, and greatly improve the efficacy of chemotherapy drugs and reduce their systemic toxicity [29].

9.3.2.1 Light triggered nanomedicine

Electromagnetic radiation in visible, near-infrared and longer wavelengths is an attractive source for transferring energy to living tissue, due to its non-ionizing nature. The most common therapeutic method modulated with light is hyperthermia or PTT. From the ancient times heat has been used to treat cancer; one of the first treatment methods used was the use of the glowing tip of an iron rod to treat breast cancer [35]. Radiofrequency, microwaves, ultrasound and laser irradiations are now used to induce heat at the tumor site.

The first photosensitizer was approved in 1994 by the US FDA, after which the advent of photo responsive therapy made an inspiring impact on the field of cancer therapy owing to its non-invasiveness and spatiotemporally controllable ability [18]. Light activated nanomedicines are created using optically active substances such as functional dyes, metals and light sensitive polymers that are capable of degrading or releasing their cargo. Light responsive polymers have been designed to respond to UV, visible light and near-infrared (NIR) light [36]. Once exposed to irradiation in a specific wavelength range directly, photosensitizer molecules absorb the energy from light and turn into a highly unstable state. The absorbed energy then transferred to surrounding oxygen molecules, generating reactive oxygen species (ROS) to damage nearby biomolecules or convert absorbed energy into heat, raising local temperature [37]. In addition, the energy would be released by emitting photons that possess lower power and eventually, the activated molecules return to the ground state [18]. Figure 9.3 shows a schematic illustration of light responsive nanoparticles. The goal of this technology is to precisely control the time and location of particle degradation within a body and thereby minimize side effects.

9.3.2.2 Ultrasound triggered nanomedicine

Ultrasound imaging devices are ubiquitous in clinics. The wide penetration, low cost and low safety concerns of ultrasound technology presents a promising path for modulating nanomedicine mediated treatments. However, this field has made limited progress due to lack of sufficient advancements in nanoparticles engineered to efficiently absorb sonic energy. Ultrasound modulation for cancer treatment is mostly used to enhance drug uptake into the tumor tissue [35]. Ultrasound exposure at frequencies of 0.5–5 MHz generates cavitational, thermal and acoustic radiation forces which can be used to increase accumulation and extravasation of nano-medicines in localized tumors, depending on exposure time and conditions [39]. Figure 9.4 shows a schematic illustration of various physical stimuli employed for responsive nanosystems.

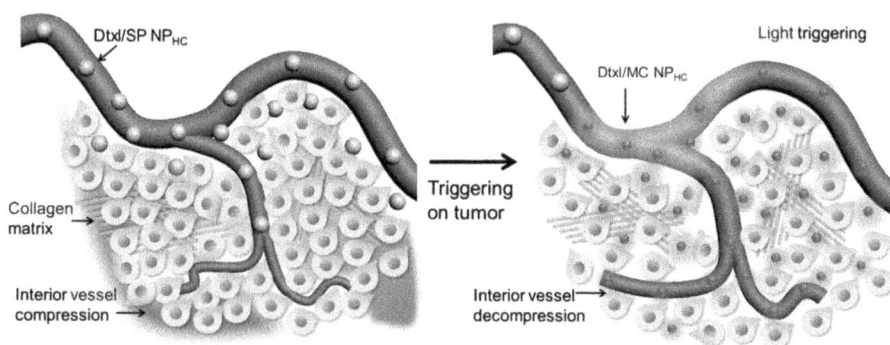

Figure 9.3. Schematic illustration of light responsive nanoparticles. Emerging delivery systems can be designed to degrade and release their therapeutic cargo upon light activation using UV, visible, or NIR light. Reproduced with permission from [38] under the terms of the Creative Commons Attribution Non-Commercial License. Copyright freely available online through the PNAS open access option.

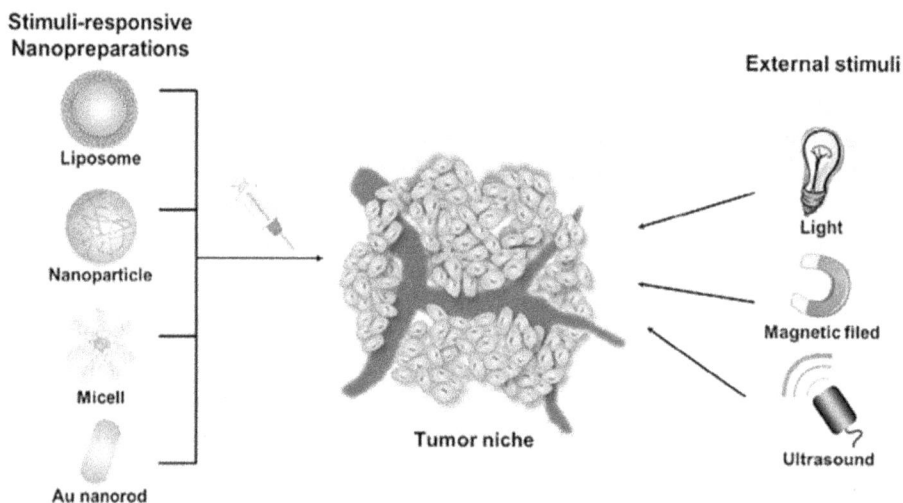

Figure 9.4. Schematic illustrations summarized various external stimuli employed for responsive nanosystems following systemic administration of different nanopreparations. Reproduced with permission from [18] under the terms of the Creative Commons Attribution Non-Commercial License. Copyright 2016 Shenyang Pharmaceutical University. Production and hosting by Elsevier B.V.

9.3.2.3 Magnetic field triggered nanomedicine

MF is another type of widely used physical force to trigger therapeutic functions of MNPs, which could be heated up under an alternating (AMF) [40], or be attracted to the tumor lesion under a locally applied external MF. Heated under a high-frequency MF, MNPs trigger drug release or produce hyperthermia/ablation of tissues [41]. In particular, oxide based spinel ferrites [42] are considered to be very promising for MRI and hyperthermia treatment. Iron oxides are good candidates for magnetic fluid hyperthermia (MFH) due to their well-known biocompatibility. The successful implication of MNPs for cancer hyperthermia therapy depends on

both the intrinsic magnetization properties of MNPs and their biophysical properties such as biocompatibility, colloidal stability and specific absorption rate (SAR) at
physiological pH. These attributes of MNPs can be achieved through their surface
functionalization using chemical modifications [43–46]. In MF triggered nanomedicine, MF driven nanoparticles collide with vessel walls in tumors more
intensively, extravasate and move towards the magnet despite the elevated interstitial fluid pressure, showing magnetically enhanced retention [12].

In addition to light, ultrasound and magnetic field; radiofrequency with wavelengths much longer than the optical wavelength, as well as x-rays with wavelengths
several orders of magnitude below that of light, have also been utilized to trigger
cancer therapy with the help of specially designed nanoagents [29].

9.4 Strengths and limitations of physical stimulus in breast cancer nanomedicine

9.4.1 Light triggered nanomedicine in breast cancer therapy

PTT and photodynamic therapy (PDT) are two different types of light triggered
therapies. PTT usually involves NIR absorbing agents to kill cancer cells under light
irradiation by effectively converting light energy into heat, raising the local
temperature of the tumor [47, 48]. Recently, many reports have discovered that
mild photothermal heating could enhance the cellular uptake of drugs, or trigger the
release of drug molecules from the nanocarriers inside cells, to enhance anticancer
efficacy of chemotherapeutics in a synergistic manner [49]. On the other hand, PDT
depends on the photosensitizers that can transfer the absorbed optical energy to
surrounding oxygen molecules, generating cytotoxic singlet oxygen ($^1O_2$) or ROS to
damage nearby biomolecules with critical functions and then kill cancer cells [50].
Many nanoscale theranostic agents have been developed for both types of light
triggered therapies [29].

9.4.1.1 Photothermal therapy

Solid tumors with several physiological barriers, such as high interstitial pressure
and a dense extracellular matrix, to a great extent affect the uptake of nanoparticles
and tumor penetration. PTT has attracted much attention to overcome these
barriers for hyperthermia damage to cancer cells, enhanced tumor penetration of
nanoparticles as well as triggered release of cargoes [5, 51].

The disordered and imperfect vasculature of tumor tissue leads to lower pH and
oxygen level within the tumor environment, which makes tumors more sensitive to
hyperthermia than normal tissues. In a temperature range of 40 °C–45 °C tumor
cells would undergo mitochondrial swelling, protein denaturation and membrane
rupture, etc, and at the same temperature range normal cells do not sustain
significant injury [52]. For effective and safe PTT, biocompatible photothermal
agents must be nontoxic, exhibit high absorbance in NIR window and great tumor
homing abilities are required. There are several nanomaterials that could respond to
NIR and transform light energy into thermal energy.

Inorganic photothermal agents which are explored extensively include different nanostructures of noble metals [53–55], carbon nanomaterials [56], transition metal sulfides and oxides [57, 58]. Among these responsive materials gold nanoparticles have been widely investigated due to their excellent light energy to heat transfer efficiency and inertness. GNPs can strongly absorb light from the NIR region through localized surface plasmon resonance and release the amount of heat for the ablation of surrounding cells [18].

Organic NIR absorbing conjugated polymers such as polyaniline and polypyrrole have also been widely used as PTT agents in recent years [29, 59, 60]. Currently, nanocarriers with NIR dyes incorporated, such as micelles, liposomes, porphysomes and protein based photothermal agents have been explored to achieve safe and effective PTT. It is worth noting that although PTT has demonstrated its great potential to effectively ablate tumors in preclinical animal experiments, it has not yet been formally approved for clinical use, although there are a few ongoing clinical trials with gold nanoagents [29].

Chen *et al* reported a new theranostic 'Abraxane' nanoformulation that is composed of human serum albumin (HSA), Paclitaxel (PTX) and FDA approved NIR dye indocyanine green (ICG). Interestingly, the *in vitro* application of the nanoformulation on 4T1 murine breast cancer cells showed strong NIR absorbance, and appeared to be an effective photothermal agent to induce rapid heating under NIR laser irradiation (808 nm). The HSA–ICG–PTX incubation plus laser irradiation was able to induce the most effective cancer cell death, showing an obvious synergistic effect [49]. Hirsch *et al* used GNPs for magnetic resonance guided NIR-PTT. They performed *in vitro* tests with human breast carcinoma cells, which were incubated with the GNPs and then irradiated with an 820 nm cw NIR laser ($35\,W\,cm^{-2}$) for 7 min. The experiment showed irreversible cell damage of the cells, whereas control cells (without GNPs) irradiated with the same power, showed no cell damage [61].

Riley *et al* evaluated a dual PTT/PDT strategy for treatment of triple negative breast cancer (TNBC) cells mediated by a powerful combination of silica–gold core–shell nanoshells (NSs) and palladium 10,10-dimethyl-5,15-bis(pentafluorophenyl) biladiene based (Pd[DMBil1]-PEG750) photosensitizers, which enable PTT and PDT, respectively. Figure 9.5 shows a schematic of nanoshells used to induce PTT; (b) a scanning electron micrograph showing the size and monodispersity of NS; (c) the chemical structure of the photosensitizers used to mediate PDT; (d) a scheme showing the proposed therapeutic strategy of dual PTT/PDT to treat TNBC. The group found that the dual therapy works synergistically to induce more cell death than therapy alone and further, determined that low doses of light can be applied to primarily induce apoptotic cell death, which is vastly preferred over necrotic cell death [62].

Liu *et al* fabricated functional chlorin gold nanorods (Ce6–AuNR@SiO$_2$–d-CPP), aiming to treat breast cancer by PTT/PDT. The nanostructure was developed by synthesizing Au nanorods as the photothermal conversion material and by coating the PEGylated mesoporous SiO$_2$ as the shell for entrapping photosensitizer Ce6 and for linking the D-type cell penetrating peptide (d-CPP). The function of

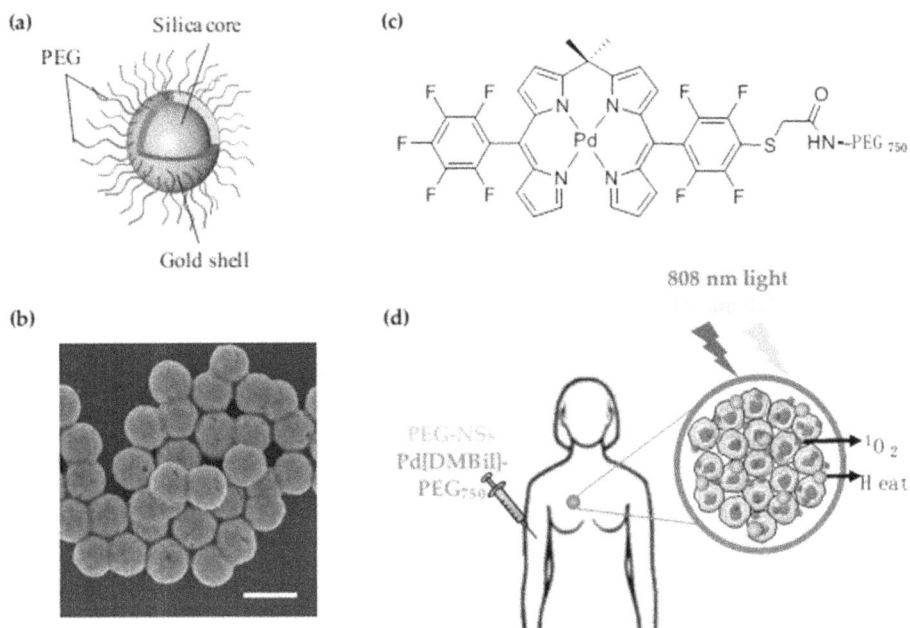

Figure 9.5. (a) Schematic of poly(ethylene glycol) nanoshells (PEG-NSs) used to induce PTT; (b) scanning electron micrograph showing the size and monodispersity of NS. Scale = 200 nm; (c) chemical structure of Pd [DMBil1]–PEG750 photosensitizers used to mediate PDT; (d) scheme showing the proposed therapeutic strategy of dual PTT/PDT to treat TNBC. Reproduced with permission from [62] under the terms of the Creative Commons Attribution Non-Commercial License. Copyright 2018 by the authors. Licensee MDPI, Basel, Switzerland.

Ce6–AuNR@SiO$_2$–d-CPP was verified on human breast cancer MCF-7 cells and MCF-7 cells xenografts in nude mice. Under combinational treatment of PTT and PDT they found that, the Ce6–AuNR@SiO$_2$–d-CPP has strong cytotoxicity and apoptosis inducing effects in breast cancer cells *in vitro* and a robust treatment efficacy in breast cancer bearing nude mice. They further demonstrated that the uptake mechanism involved is the energy consuming caveolin mediated endocytosis and Ce6–AuNR@SiO$_2$–d-CPP in PTT/PDT mode could induce apoptosis by multiple pathways in breast cancer cells [63].

9.4.1.2 Photodynamic therapy

PDT is another light triggered strategy to treat cancer. There are three prerequisites to be satisfied for PDT light, photosensitizers and oxygen. Hence, sufficient oxygen content is an indispensable element for an optimal efficacy. However, PTT turned photon energy into substantial heat diffused to the tissues nearby, so that PTT is independent from oxygen concentration within tumors, which offered an alternative way to excise the deeper core of solid tumors [64].

The photosensitizer, such as pyrolipid, chlorin e6 (Ce6), NaYF$_4$:Yb/Tm–TiO$_2$ inorganic photosensitizers [5, 65] oligo(p-phenylene vinylene) derivative (OPV) [66], is a key component in the light sensitive nanocarrier. They can be activated by light,

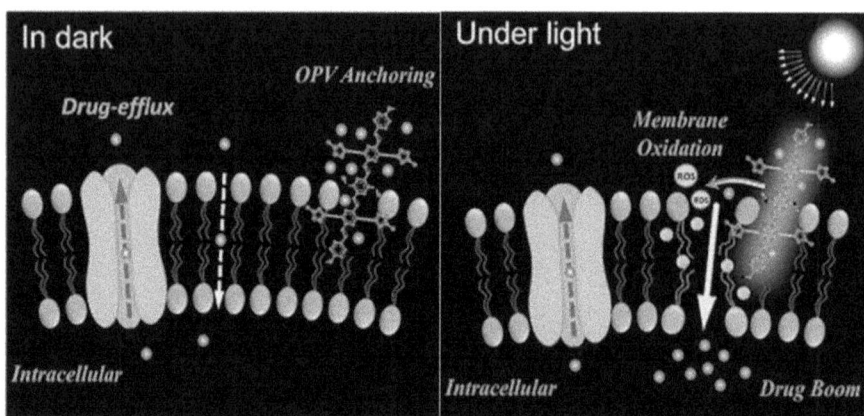

Figure 9.6. The mechanism of light triggered drug uptake approach to reverse drug resistance of resistant cells and ROS generation by photosensitizer. Reproduced with permission from [66]. Copyright 2014 WILEY-VCH Verlag GmbH & Co. KGaA, Weinheim.

producing reactive oxygen species (ROS), which can not only kill cancer cells directly but also induce oxidation of membranes and affect the permeability of cell membranes, which is beneficial for anticancer drug penetration [5]. Figure 9.6 shows the mechanism of light triggered drug uptake approach to reverse drug resistance of resistant cells and ROS generation by photosensitizer.

Activation of photosensitizers with specific wavelengths of light leads to energy transfer cascades that ultimately generate cytotoxic ROS which can induce cell death by local tissue apoptosis and necrosis [67]. Cell death subroutines are strongly connected with successful therapy outcome. At the cellular level, PDT has been shown to induce multiple cell death subroutines (as shown in figure 9.7) that can be accidental (necrosis) or natural (apoptosis). Necrotic cell death is an uncontrollable form of cell death corresponding to the physical disassembly of the plasma membrane caused by extreme physical, chemical or mechanical actions. On the other hand, apoptic cell death results from the activation of one or more signal transduction modules and hence can be pharmacologically or genetically modulated, at least to some extent. PDT includes apoptosis and different mechanisms of regulated necrosis subroutines of cell death [68]. Therefore, harnessing this functional mechanism in tumor therapy provides a safe and controllable way to selectively eradicate tumors with reduced systemic toxicity and side effects on healthy tissues.

Lovell *et al* prepared a liposome-like nanovesicle by porphyrin bilayers named porphysome, which possesses excellent biocompatibility, structure dependent fluorescence quenching and high absorption of NIR light [69]. In one of his recent publications, porphysomes were conjugated with folate as a targeting-triggered activatable nano sized beacon for PDT. Before folate mediated endocytosis by tumor cells, the intact porphysomes could transfer photon energy to thermal energy, but the nanoparticles would switch back to efficient photodynamic activity once they were internalized and disrupt the nanostructure intracellularly, thus this

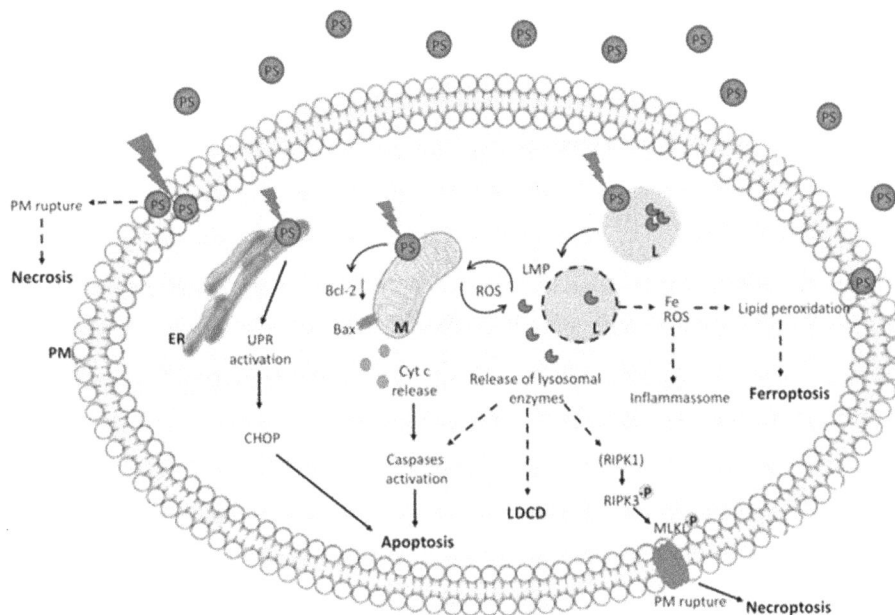

Figure 9.7. Overview of cell death subroutines that can be elicited by PDT. The most described locations of different photosensitizers (PS) are the plasma membrane (PM), endoplasmic reticulum (ER), mitochondria (M) or the lysosome (L). Depending on its localization, after activation by light (red light in bolt) it can directly damage the PM causing unregulated necrosis or culminate in one or more regulated cell death mechanisms. UPR: unfolded protein response; LMP: lysosome membrane permeabilization; Fe: iron; ROS: reactive oxygen species; –P: phosphate group presented in the active forms of RIPK3 and MLKL on necroptosis pathway; LDCD: lysosomal dependent cell death. Reproduced with permission from [68] under the terms of the Creative Commons Attribution Non-Commercial License. Copyright 2019 The Author(s).

biocompatible responsive system showed potent potential for clinical application [70]. Zinc phthalocyanine (ZnPc), a lipophilic photosensitizer which can localize to membrane and have strong phototoxicity upon NIR irradiation, is anchored into liposomes with a given molar ratio to construct membrane fusogenic liposomes (MFLs). Compared to the non-fusogenic liposomes, cells treated with MFLs showed significantly lower viability in the MTT test, which means photosensitizers localized in the plasma membrane would induce more membrane disruption and cell death upon irradiation [18].

Zeng *et al* fabricated a bi-functional nanoprobe based on DOX loaded $NaYF_4$: Yb/Tm–TiO_2 inorganic photosensitizers (FA–NPs–DOX) for *in vivo* NIR triggered inorganic PDT and enhanced chemotherapy to overcome the MDR in breast cancers. They used the up-conversion luminescence (UCL) performance of $NaYF_4$:Yb/Tm converting NIR into UV lights. ROS were triggered from TiO_2 inorganic photosensitizers for PDT under the irradiation of a 980 nm laser. In addition, nanocarrier delivery and folic acid (FA) targeting promoted the cellular uptake and accelerated the release of DOX in drug sensitive MCF-7 and resistant MCF-7/ADR cells and found that by the combination of enhanced chemotherapy and NIR triggered inorganic PDT, the viability of MCF-7/ADR cells could decrease by 53.5%, and the inhibition rate of MCF-7/ADR tumors could increase up to 90.33%, compared with free DOX. Thus they demonstrated that, the MDR of breast cancers could be obviously overcome by enhanced chemotherapy and NIR triggered inorganic PDT of FA–NPs–DOX nanocomposites under the excitation of a 980 nm laser [65].

In a recent study, Bharathiraja et al fabricated polypyrrole nanoparticles by employing bovine serum albumin–phycocyanin complex. The nanoformulation did not cause any noticeable toxicity to MDA-MB-231 and HEK-293 cells. The obtained nanoparticles effectively killed MDA-MB-231 cells in a dual way upon laser illumination, one was through phycocyanin propagated reactive oxygen species (PDT) upon laser illumination and in another way it eradicated the treated cells by converting optical energy into heat energy (PTT) [71].

9.4.1.3 Combination of PTT/PDT with chemotherapy

Phototherapies may be combined with other therapeutic approaches such as chemotherapy, gene therapy, surgery, radiotherapy and even immune therapy, to realize synergistic antitumor effects. In recent years, many NIR absorbing nano-carriers have been engineered to load chemotherapeutic agents to realize combined PTT and chemotherapy. As a typical example, photothermal agents, such as gold nanorods, graphene, carbon nanotubes, CuS nanoparticles and many others were coated with mesoporous silica, in which chemotherapeutic molecules are encapsulated [29, 72–74]. On the other hand, many photothermal agents such as carbon nanomaterials and conjugated polymers could serve as drug loading carriers by themselves.

There have been a number of different mechanisms when combining PTT with other therapeutic approaches (e.g. chemotherapy). These are (1) it has been well documented that the NIR induced photothermal effect is able to trigger drug release from the mesoporous silica shell, or the surface of NIR absorbing nanocarriers. (2) It has been found that the photothermal effect of NIR absorbing nanocarriers may also be able to enhance the cellular uptake ability of those nanocarriers by improving the cell membrane penetrability under mild photothermal heating. Such an effect could be utilized to promote the intracellular delivery of chemotherapeutic agents and photosensitizer molecules for photothermally enhanced chemotherapy, PDT respectively. (3) While photothermal heating at temperatures over 50 °C would be able to effectively ablate tumor cells, it has been revealed that mild hyperthermia (e.g. below 45 °C) may be able to change the tumor microenvironment such as increasing the blood flow, oxygen levels in the tumor, the perfusion and permeability of the tumor vasculature, then improving the accumulation of materials at the tumor site [29].

The combined delivery of photothermal and chemotherapeutic agents is an emerging strategy to overcome drug resistance in treating cancer and controlled light responsive drug release is a proven tactic to produce a continuous therapeutic effect for a prolonged duration. Khatun et al prepared a nanogel with a combination of light responsive graphene, chemotherapeutic doxorubicin (DOX) and pH sensitive disulfide bond linked hyaluronic acid (called a graphene–DOX conjugate in a hyaluronic acid nanogel) that exerts an activity with multiple effects: thermo and chemotherapeutic, real-time non-invasive imaging and light glutathione responsive controlled drug release. They injected the nanogels and observed their subsequent distribution in numerous organs, including kidney, lung, spleen, liver and the tumor and found significant increases in fluorescence intensity in the tumors after laser irradiation. The nanogels circulated systemically in the vasculature and released the doxorubicin payload at the sites of the tumor. To further observe the therapeutic efficacy of the nanogel, tumor-bearing mice were treated with saline, doxorubicin, doxorubicin plus laser irradiation, nanogel alone and the nanogel plus laser irradiation. The mice treated with saline and doxorubicin did not show tumor regression. In contrast, mice treated with the nanogel showed tumor regression in 18 days, indicating that the combination arrested tumor growth [75].

The combination of both PTT and chemotherapy has attracted increasing attention. Photothermal agent CuS and anticancer drug DOX integrated into a thermosensitive composite labeled as 'G' to fabricate an on-demand drug release and photothermal therapy system G–Cus–DOX. Because the low critical solution temperature (42 °C) of the composite 'G', once G–CuS–DOX is exposed to NIR laser at 915 nm, the boosted thermal energy would ablate the surrounding tumor cells and the high temperature can melt the composite 'G' for the release of DOX [76]. Li *et al* constructed docetaxel (DTX) loaded micellar nanomedicines coloaded with NIR dye IR820 for PTT/PDT/chemotherapy of breast cancer. Lyp-1, a tumor homing peptide, was introduced into the nanosystems to construct the active targeting nanomedicine. The tumor cells targeting and anticancer performances of this nanosystem have been studied *in vitro* (4T1 breast cancer cell lines) and *in vivo* (4T1 tumor model). The results demonstrated that Lyp-1 modification has enhanced the tumor targeting delivery of DTX and IR820. By combining PTT and PDT, DTX nanomedicine efficiently inhibited the growth and metastasis of breast cancer in mice [77].

9.4.1.4 Limitations of light triggered nanomedicine

Phototherapy triggered by light is a powerful tool and has demonstrated its high efficacy in treating tumors in many preclinical animal experiments. The limited light penetration depth in tissues has already been a concern when using light to trigger cancer treatment. As a result, the future applications of phototherapies may be restricted only to the treatment of superficial tumors or tumors accessible by endoscopy techniques [29]. In addition, light responsive systems need illumination to activate the light sensitive nanomaterials; however, not all regions of the spectrum are appropriate for clinical use, which is also a major drawback of light treatments [18].

In PTT, the *in vivo* penetration depth of NIR light is dependent on a variety of factors including the degree of light scattering and absorption within tissue. The heating of the tissue depends on the intensity of NIR light at the given point, the absorptive cross-section of the nanoparticles, the distribution and concentration of nanoparticles within the tissue and the degree of NIR absorption by chromatophores in the surrounding tissue [3]. On the other hand, UV light has the highest power which can damage surrounding tissue and skin. Particles that are responsive to visible light instead of UV offer a less harmful alternative [36].

Although GNPs have an outstanding performance in PTT, there are still several flaws that restrict their further clinical application, such as the NIR absorbance peak of GNPs which is related to the particle size and morphology and after a long period of laser irradiation the GNPs NIR absorbance peak would diminish due to the 'melting effect', which weakened their conversion efficiency [18, 78]. In addition, the main problem of using small spherical GNPs for PTT lies in the position of the resonance peak which, depending on the size of the nanoparticle, is located between 520 and 580 nm. But in this wavelength range healthy tissue, specifically hemoglobin and oxygenated hemoglobin, strongly absorbs light and converts the light energy into heat. Thus, healthy tissue can also be damaged during the PTT, making spherical gold NPs less ideal for PTT [35].

In the case of PDT most of the photosensitizers used currently are activated by visible light from 600–700 nm, which has limited tissue penetration. The efficacy of PDT to treat internal tumors or tumors with large sizes has thus been a challenge [29]. In addition, the lack of stability, solubility and biological utility of photosensitizers limited the application of PDT. The short half life of ROS generated based on photodynamic mechanism and the limited scope to it to diffuse, limits the use of PDT and also the efficacy of PDT highly depends on the subcellular location of the photosensitizers.

9.4.2 Ultrasound triggered nanomedicine in breast cancer therapy

9.4.2.1 Ultrasound triggered drug release

The initial uses of ultrasound waves in the medical field were limited to the thermal ablation of solid tumors and as a diagnostic tool. Recent advances at the preclinical stage have allowed the use of ultrasound as a powerful tool to improve drug delivery when the agent is administered encapsulated inside a nanoparticle [79]. Ultrasound possesses several superiorities such as non-invasiveness, low cost, good penetration depth by tuning the frequency, duty cycle and time of exposure, as well as the absence of ionizing irradiation.

In the process of drug delivery to tumor cells, whether by passive or active targeting, when the drug carrier reaches the target site, ultrasound may be used in order to trigger the release of the drug. Ultrasound waves move through a medium, transferring energy from one element to the next in a manner referred to as propagation. Two important parameters of ultrasound waves are the frequency (number of sinusoidal cycles occurring per second) and the power density (power per cross-sectional area) [80].

Ultrasound triggered drug delivery of nanomedicine may be helpful for the time and location specific release of anticancer drugs, thus decreasing the side effects. In drug delivery the effects of ultrasound may be either thermal or mechanical. Thermal effects occur due to the absorption of thermal energy by the tissues or cells and the mechanical effects occur due to oscillating bubbles, acoustic cavitation, wave pressure and acoustic streaming. Acoustic cavitation occurs due to the oscillation, growth and collapse of gas bubbles in the medium due to the varying pressure caused by the ultrasound wave. Cavitation may be divided into two categories, namely inertial and stable. Stable cavitation involves constant oscillation of the gas bubble and occurs when the pressure amplitude is low while inertial cavitation involves unstable oscillation of the bubble after which it grows rapidly in size and implodes violently [80–83].

Shock waves resulting from the implosion of the gas bubble due to inertial cavitation, lead to shear stresses that can damage surrounding cells. Moreover, when inertial cavitation occurs, a sonic jet of liquid is produced which pierces the surface of the nearby vesicle (e.g. a micelle, liposome or another drug carrier). This phenomenon results in increased permeability of cells and in turn enhanced passive or active drug targeting. In addition, oscillating bubbles and cavitation, open up the drug carrier and lead to the release of the encapsulated drug at the targeted site. For example, when pulsated ultrasound is applied on polymeric micelles as nanocarriers, there is an ON period and an OFF period; as we know polymeric micelles are capable of self-assembly, any amount of drug that has not entered the tumor during the ON period, can be re-encapsulated in the carrier during the OFF period and circulate in the bloodstream once again [80]. Diaz de la Rosa et al in their study showed that DOX encapsulated P105 micelles, when exposed to ultrasound released DOX in the environment of the tumor and then the remaining DOX was re-encapsulated in the absence of ultrasound. This helps in reducing the interaction of the drug with non-tumor tissues, successfully enhancing chemotherapeutic drug delivery while reducing unwanted side effects [84]. Husseini et al studied the effect of ultrasound intensity on drug release from micelles and found that higher intensities resulted in higher release of drug at all frequencies of ultrasound studied [85].

Li et al studied the ultrasound triggered release of a new sonosensitizer (sinoporphyrin sodium, DVDMS) loaded into liposome–microbubble complexes (DLMBs) as a possible candidate to enhance sonodynamic therapy (SDT) against breast cancer, in vitro and in vivo. In this study, composite liposome–MBs were synthesized as DVDMS carriers; DVDMS-loaded liposomes were linked onto the surface of MBs via the biotin–avidin linkage. The impact of ultrasound on the release of DVDMS in vitro on MDA-MB-231 breast cancer cells was evaluated. DL or DLMBs were exposed to ultrasound with various intensities and then dialyzed. The relative release content of the entrapped DVDMS increased significantly with the increase of ultrasonic intensity. The cytotoxicity of free DVDMS, DL, or DLMBs alone was observed after incubation with cells for 24 h in the dark; no obvious cytotoxic effects were detected. The rate of cell death caused by ultrasound alone was less than 10%. After sonication they found that approximately 30.43% and 24.93% of cells were killed incubated with free DVDMS and DL respectively and the cell viability decreased to 47.82% in the DLMBs group. Additionally, to investigate the sonodynamic antitumor efficacy of composite DLMBs with ultra-sound, a 4T1 mouse mammary tumor model was utilized. Both free DVDMS and DL with ultrasound inhibited tumor growth (volume and weight) to a certain extent. Importantly, in the mice treated with DLMBs plus ultrasound, tumor growth was greatly retarded with increased dosage indicating a greater SDT effect against breast cancer [86].

Baghbani *et al* in a similar way studied the ultrasonic nanotherapy of breast cancer using novel alginate stabilized perfluorohexane nanodroplets. They synthesized DOX loaded multifunctional nanodroplets (DOX–NDs) via a nanoemulsion process and evaluated *in vitro* and *in vivo* cytotoxicity and antitumor activity using human breast carcinoma tumor xenografts inoculated in BALB/c mice. Tumor therapy using DOX–NDs combined with sonication (DOX–ND–Ultrasound) resulted in strong *in vivo* antitumor activity characterized by tumor regression which could be because of on-demand efficient ultrasound aided drug release from nanodroplets in tumor tissue under the action of ultrasound. Interestingly they found that the DOX concentration in the tumor area for the DOX–ND–US treated group reached $10.9 \, \mu g \, g^{-1}$ after sonication for 5 min, which was 5.2 fold higher compared to the nonsonicated DOX–NDs group [87].

9.4.2.2 Ultrasound guided imaging

The diagnostic imaging techniques available currently for diagnosis of tumors include optical imaging, ultrasound imaging computed tomography (CT), MRI, positron emission tomography (PET) imaging, etc. Among these techniques ultrasound imaging has its unique advantages such as real-time imaging, low cost, high safety and portability [88]. Diagnostic ultrasound imaging is nowadays routinely performed with ultrasound contrast agents (UCAs). Typically, UCAs are gas filled microcapsules with a diameter between 1 and 10 mm. The microcapsule shell material may be of a broad variety of molecules, including lipids, proteins, polysaccharides or synthetic polymers. UCAs scatter ultrasound efficiently and they also respond to low energy ultrasound by emitting harmonic frequencies, resulting in contrast enhancement with respect to the surrounding tissue. Alternatively, air filled microcapsules with fluorinated polymer shells have also been used for this purpose [89]. In addition, integration of multifunctional nanoparticles with UCAs, nanotheranostics could be constructed for ultrasound imaging guided therapy. When the UCAs are exposed to ultrasound, they start to cavitate. Microstreams and shock waves would be generated when the ultrasound exposure is high enough, leading to a local release of drugs loaded on the UCAs. At the same time, temporary perforate cell membranes (sonoporation) may occur to result in the intracellular delivery of the released drugs. Furthermore, the microjets and shock waves could also permeabilize blood vessels, allowing the release of high molecular weight drugs or nanomedicines. On the other hand, high intensity focused ultrasound (HIFU) can trigger thermal effect to produce hyperthermia for direct thermal ablation therapy and enhanced drug delivery [29].

In a study, Rapoport *et al* constructed DOX loaded perfluoropentane (PFP) nanoemulsion having low boiling point (29 °C) using a biodegradable block copolymer poly(ethylene glycol)–poly(L-lactide) (PEG–PLLA). Upon intravenous injection into mice, DOX–PFP nanoemulsions could selectively extravasate into the tumor sites, producing a strong durable ultrasound contrast via PFP evaporation to form nano/microbubbles, as PFP has low boiling point it would be vaporized at physiological temperature. In the meantime, under ultrasound irradiation DOX was released from those nano/microbubbles to offer significant chemotherapeutic efficacy. Similar nanotheranostic platforms under the stimulation of ultrasound were applied with PTX loaded nanoemulsions for ovarian, breast and pancreatic tumors [90]. Thus, ultrasound triggered therapy also showed excellent therapeutic efficacy, indicating the potential of the multifunctional nanoemulsions for cancer theranostics.

In comparison with other physical stimuli, ultrasound shows deep tissue penetration, and it is safe for healthy tissues. Ultrasound is a convenient diagnostic imaging modality and it may also be utilized to trigger the local release of therapeutic drugs in the tumor or to directly ablate tumors by hyperthermia if high intensity ultrasound irradiation is applied, making ultrasound a versatile tool for physically stimulated/triggered nanotheranostics.

9.4.2.3 Limitations of ultrasound triggered nanomedicine
Ultrasound therapy is a powerful medical tool, as such it does not have any direct effects on healthy or normal cells but ultrasound heating which can lead to irreversible tissue changes follows an inverse time temperature relationship. Depending on the temperature gradients, the effects from ultrasound exposure can include mild heating, coagulative necrosis, tissue vaporization, or all three [91]. In ultrasound triggered drug release UCAs or microbubbles are frequently used, their relatively large size (hundreds of nm or even at the micrometer scale), limits the extravasculature while on the other hand, if the bubble sizes are too small then their acoustic properties will largely be affected. Due to the large size, microbubbles have a rather short half life. After injection, they will circulate a few times and then get stuck in the lungs where gas exchange occurs. Consequently, microbubble ultrasound triggered drug delivery will be mainly restricted to cardiovascular targets and to tumor endothelia.

There are three major disadvantages of ultrasound as a treatment method for cancer. The first is the incapability of ultrasound to penetrate air filled organs. This will limit the ability of ultrasound with tumors located in areas such as the lungs, intestines or bladder. The second disadvantage is the tumor location, particularly if there is no acoustic window for the ultrasound to reach the tissue of interest. For example, if there is a structure obscuring the tumor with a high absorption coefficient, such as bone, the acoustic wave may be unable to reach its intended target. The third major disadvantage to ultrasound is the long treatment times.

9.4.3 Magnetic field triggered nanomedicine in breast cancer therapy

Magnetic field (MF) is another type of widely used external physical stimulus to trigger or enhance therapies against cancer. MNPs, such as superparamagnetic iron oxide nanoparticles (SPIONPs) with varied sizes and morphologies, after surface modification, could not only serve as contrast probes for MRI, but also act as the therapeutic agent by themselves or as a delivery platform when they are loaded with other therapeutic agents [29].

Among various nanoparticles, IONPs with unique properties are being evaluated to improve the treatment options available for breast cancer. Besides already being approved by the FDA as contrast for MRI, they are being used in various ways ranging from nanocarriers for therapeutics, overcoming therapy resistance to killing cancer by hyperthermia [2].

9.4.3.1 Magnetic field induced magnetic fluid hyperthermia

MFH is a well-known therapy method triggered by AMF, which is able to induce heating of ferromagnetic or superparamagnetic nanoparticles located inside this field to ablate cancer cells. As a technology which has been investigated for several decades by numerous research groups, MFH is now being tested in several different clinical trials and has been found to be effective in treating various cancers including prostatic cancer, breast cancer, malignant glioma, cholangiocarcinoma, oral cancer, esophageal cancer and so on [92–95].

MFH is an artificially induced heat treatment of a disease especially designed for cancer, using temperatures ranging between 42 °C–47 °C. Generally, a temperature below 45 °C induces apoptotic cell death, because cancer cells are susceptible to heat at about 43 °C, while most of the normal cells remain undamaged [96]. MFH for cancer treatment involves injecting a fluid containing MNPs directly into tumors. When placed in an AMF with frequencies similar to those of FM radio signals, the nanoparticles generate heat and destroy the tumors. This is a minimally invasive procedure (unlike laser, microwave, and ultrasound hyperthermia), which prevents unnecessary heating to surrounding healthy tissues as the localized MNPs only absorb the MF [97]. Figure 9.8 shows popular multipurpose magnetic nanosystems in MFH and magnetically guided drug delivery. The MFH effect can not only be used to directly kill cancer cells, but also may be able to trigger the release of chemotherapeutic agents loaded on theranostic MNPs.

In our recent study, we synthesized superparamagnetic nanoparticles consisting of a perovskite $L_{0.3}Sr_{0.7}MnO_3$ core modified with a covalently linked chitosan shell. These MNPs were then used for *in vitro* MFH treatment on MCF-7 breast cancer cell line and interestingly found that the cell viability is reduced up to 40% within 120 min of exposure to AMF. During the experiment the temperature of the MNPs cell solution is maintained between 44 °C–45 °C. The exposure of the cells to only AMF (absence of MNPs) did not show any significant cytotoxic effect on cell viability, suggesting that the AMF used in the current study is not harmful to normal cells [96]. In a similar study, we synthesized magnetic core/shell nanostructures of Fe_3O_4 nanoparticles coated with oleic acid and betaine–HCl, *in vitro* MFH treatment on MCF-7 breast cancer cell line showed about 95% cell death in 90 min after magnetic hyperthermia treatment. The mechanism of cell death was found to be elevated ROS generation in cells after exposure to core/shells in external AMF [99]. Many similar publications for the treatment of breast cancer are available in the literature regarding *in vitro* and *in vivo* studies of MNPs.

Kossatz *et al* had done magnetic hyperthermia of breast cancer using innovative nanoparticles which display a high heating potential and are functionalized with a cell internalization agent and a chemotherapeutic agent to increase cell death. The

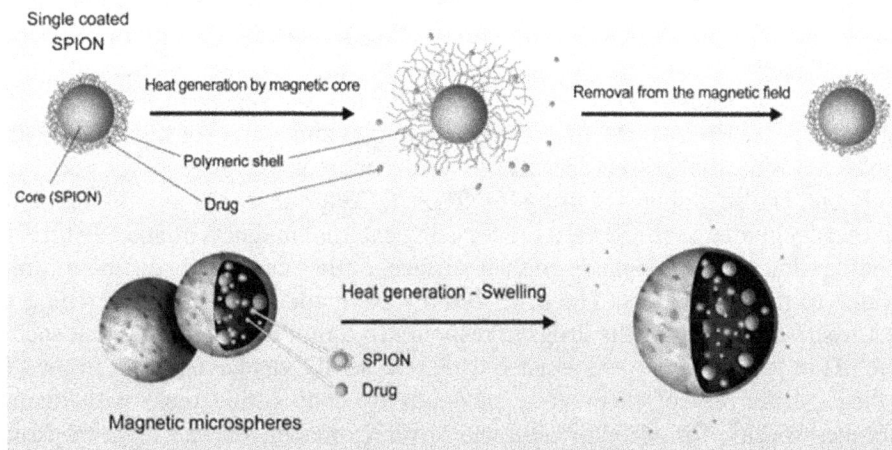

Figure 9.8. Popular multipurpose nanosystems in magnetic hyperthermia and magnetically guided drug delivery. Reproduced with permission from [98]. Copyright 2011 Elsevier B.V.

SPIONs (MF66) were electrostatically functionalized with nucant multivalent pseudopeptide (N6L; MF66–N6L), DOX (MF66–DOX) or both (MF66–N6L–DOX) and their cytotoxic potential was assessed in breast adenocarcinoma cell line, MDA-MB-231. Therapeutic efficacy was analyzed on subcutaneous MDA-MB-231 tumor-bearing female athymic nude mice. In their experimental results, all the nanoparticle variants showed an excellent heating potential around 500 W g^{-1} Fe in the alternating magnetic field (AMF, conditions: $H = 15.4$ kA m^{-1}, $f = 435$ kHz) and found the gradual inter- and intracellular release of the ligands with nanoparticle uptake in cells was increased by the N6L functionalization. Additionally they observed that MF66–DOX and MF66–N6L–DOX in combination with hyperthermia were more cytotoxic to breast cancer cells than the respective free ligands and substantial tumor growth inhibition after intratumoral injection of the nanoparticles *in vivo* [100].

9.4.3.2 Magnetic field induced tumor targeting

Magnetic tumor targeting is a unique strategy to enhance tumor accumulation of therapeutic nanoparticles in the targeted tumor region to improve therapeutic efficacy in cancer. In the 1960s, Freeman *et al* proposed that MNPs could accumulate in a particular location of the body under an additional magnetic field [101]. In the following years, a large number of reports have demonstrated that many materials based on magnetic nanoparticles could be used for magnetic tumor targeting. In 1996, Lubbe *et al* reported the first clinical trial study on magnetic drug targeting by testing chemotherapeutic bound magnetic fluids in 14 cancer patients, and demonstrated the acceptable safety of such a new cancer treatment method [102]. Here in this section for simplicity or to understand the theme only some examples associated with IONPs are given. However, a very large amouont of data is available on this context in the literature.

IONPs have been extensively used as drug delivery systems (DDS) for chemotherapeutic drugs such as DOX, tamoxifen (TMX), quercetin, rhodium (II) citrate, gallic acid, etc. The drugs are loaded in such a manner that the functionality of the drug is not compromised. DDS are more effective in killing cancer cells compared to drugs alone due to the more effective targeting and entry of the IONP–drug complexes inside the cancer cells. This results in reduced drug doses and lesser side effects. The advantages of using IONPs as DDS are that these complexes can be magnetically guided to the desired tissues using external magnets or actively targeted by conjugating targeting ligands on their surface, or they can simply diffuse to tumor sites due to the EPR effect. The drug–IONPs are made such that on reaching the target tissues, they release the drugs in response to tumor microenvironment such as acidic pH or by magnetic field assisted drug release. IONPs have also been used for the targeted delivery of therapeutic payloads by conjugating them with ligands/ antibodies specific for receptors/antigens, over expressed on the targeted cancer cells [2].

Lee *et al* successfully proposed on-demand drug delivery that uses magneto-thermally responsive DOX encapsulated supramolecular magnetic nanoparticles (DOX@SMNPs) for highly effective *in vivo* cancer treatment. DOX@SMNPs could generate heat under an AMF, which on the other hand triggered the release of DOX on-demand. Highly effective *in vivo* cancer therapy was realized in a mouse-tumor model by this strategy [103].

Nanocomplexes formed by loading chemotherapeutic drugs such as DOX, TMX and quercetin onto IONPs were found to be efficiently internalized by MCF-7 cells (human breast cancer cell line) to show dose dependent cytotoxicity [104–106]. Similarly, gallic acid loaded onto IONPs coated with PEG or PVA was found to be more toxic to breast cancer cell lines (MCF-7) than free gallic acid, with no significant toxicity on normal cells [107]. Rhodium (II) citrate–IONP complexes were found to be more specific and cytotoxic to cancer cells (MCF-7 and 4T1 carcinoma breast cells) than normal cells (MCF-10A) at doses lower than free rhodium (II) citrate. This may be due to a faster metabolism and hence a higher uptake of micronutrients such as iron in cancer cells. The carboxylic groups present in IONPs may also help in easy transport across the cell membrane through the proteic thiol groups [108].

Besides *in vitro* cytotoxicity, rhodium (II) citrate–IONP complexes caused significant reduction in tumor volume, higher tumor necrosis in Balb/c mice bearing orthotopic 4T1 breast carcinoma, compared to free rhodium (II) citrate or IONP–citrate complexes. These NP complexes were not toxic to the mice but due to effective delivery, accumulation and inhibition of glycolysis by citrate, they could effectively reduce tumor volume [109]. Intratumoral injection of L-ascorbic acid (vitamin C) coated MNPs in murine Ehrlich ascitis carcinoma (ESC) bearing mouse led to effective accumulation in tumor tissues with significant reduction in tumor size, increase in necrotic areas than intraperitoneal administration. Both intra-tumoral and intraperitoneal injection led to an enhanced expression of p53 and p16, which led to apoptosis of ESC cells. IONP complexes were also found helpful in overcoming drug resistance [110]. DOX loaded IONPs were more efficiently taken up by DOX resistant breast cancer cells (1 µM DOX resistant MCF-7) compared with free drug, thereby increasing the efficacy of the drug. Therefore, MNPs can increase the efficiency of chemotherapeutic drugs by reducing their dose due to direct accumulation in cancer cells and spare toxicity to the normal cells and tissues [2].

In general, MF induced tumor targeting would be more efficient compared with simple passive tumor accumulation of nanoparticles relying on the EPR of tumors. In comparison to active tumor targeting by conjugating tumor-specific binding ligands on nanoparticles, magnetic tumor targeting as a physical process is independent of the types of molecular receptors expressed in the tumor, and may be applicable to a wide range of different solid tumors.

9.4.3.3 Magnetic field guided MRI

Zhang *et al* tagged phosphatidylserine (PS) targeting monoclonal antibodies (MAb) to SPIONPs to help target and bind to PS exposed tumor vessels to increase accumulation of SPIONPs in breast tumors and enhance tumor contrast. This increases sensitivity and specificity of detection of breast cancer at an early stage [111]. In a similar study, Faraj *et al* tagged Endoglin/CD105 MAb to single wall carbon nanotubes (SWCNTs) loaded with IONPs which allowed enhanced targeting to 4T1 breast tumors, increasing sensitivity of MRI detection and delivery of higher doses of thernostics nanocarrier SWCNTs [112]. IONPs tagged with epidermal growth factor receptor-MAb (EGFR-MAb) have been used for dual imaging i.e. fluorescence molecular tomography and MRI to enhance breast cancer detection. IONPs–EGFR-MAb could bind efficiently to EGFR expressing human breast adenocarcinoma cell lines MDA-MB-231 and MDA-MB-231 tumors in mouse model to enhance MRI detection [113]. Zhang *et al* tagged IONPs/GNPs complex with CD105 antibody to detect tumor angiogenesis in tumor-bearing mice by MRI [114]. Jafari *et al* used bombesin tagged IONPs to enhance MRI detection of T47D breast cancer cells due to specific binding of bombesin to T47D cells. *In vivo* studies performed using breast tumor mouse model showed higher accumulation of these nanocomplexes in tumors to significantly enhance MRI detection [115]. SPIONPs conjugated with vascular endothelial growth factor (VEGF) could accumulate more efficiently in tumor tissues, thereby enhancing uptake and retention in tumor tissues compared to VEGF receptor targeted SPIONPs–liposomes and free SPIONPs. Therefore, they could be used to enhance MRI detection of breast cancer tissues [116].

9.4.3.4 Limitations of magnetic field triggered nanomedicine

Compared with PTT which uses light energy to burn cancer, MFH obviously offers greatly improved tissue penetration. However, the heating efficiency of MFH is often lower than that of PTT and thus high concentrations of MNPs in the tumor are often needed to realize effective heating. Therefore, in general it would be difficult to deliver a sufficient amount of MNPs into the tumor by intravenous injection to realize effective MFH tumor ablation [29]. Also compared with focusing light on tumors, it is not as convenient to focus the MF specifically to the tumor region, thus possibly causing more non-specific heating of surrounding healthy tissue when MFH is conducted.

Nanocarriers for effective AMF induced tumor targeting should show rapid responses under external AMF and be able to circulate in the blood for a long period of time post intravenous injection. However, MF induced tumor targeting of MNPs appears to be rather effective on subcutaneous tumor models in mice; it may be technically challenging to place a strong magnet nearby a tumor inside the human body. Therefore, the potential of such a strategy for clinical translation is yet to be further demonstrated [29].

MF is widely used in clinical application, i.e. in MRI; it is a relatively mature technology but also has shortcomings like the degradation problem of noble metal contrast agents and the relatively higher treatment cost than other diagnosis tools like CT imaging [18].

9.4.4 Radiation triggered nanomedicine in breast cancer therapy

Ionizing radiation therapy is a commonly used cancer treatment in the clinic to destroy cancer cells by high-energy radiation such as x-rays and gamma-rays. Irradiation of tumors with high-energy photons or ion beams results in irreparable damage to cancer cells and the surrounding vasculature leading to tumor death [117]. To enhance the radiation therapy, many nanoparticles containing heavy elements which are able to absorb high-energy ionizing radiation have been synthesized. Zhang *et al* used PEG coated GNPs with different sizes to enhance radiation therapy. The *in vitro* and *in vivo* results showed that, PEG coated GNPs with sizes of 12.1 nm and 27.3 nm showed high radiosensitivity and could be used to enhance the efficacy of radiotherapy induced by gamma radiation [118].

In addition, to enhance the radiation therapy with x-ray absorbing nanoparticles, x-ray triggered cancer therapy was employed in which x-ray luminescent nano-particles are used to enable x-ray triggered PDT [119]. Liu *et al* synthesized LaF$_3$: Tb^{3+} meso-tetra(4-carboxyphenyl)porphine (MTCP) nanoparticle for PDT under x-ray irradiation [120]. Under x-ray irradiation, the photosensitizer MTCP was activated by visible light emitted from x-ray luminescent nanoparticles. The generated singlet oxygen or ROS then killed cancer cells via the photodynamic effect. Importantly, such strategies could be used for deep tumor treatment, due to the deep tissue penetration ability of x-rays.

X-ray-triggered liposomal drug release is a relatively new concept in comparison with other target specific drug release systems. The mechanism responsible for inducing the destabilization of the liposomal membrane is the radiosensitization effect. Figure 9.9 shows the schematic of mechanisms associated with radiation that can be used to destabilize liposomes. Radiosensitizers, such as GNPs, enhance the local radiation dose through the increased absorption of low and medium energy x-rays and subsequent ejection of reactive secondary electrons [117].

Upon irradiation, GNPs in water produce hydroxyl radicals, although there have been no direct studies examining the mechanisms of radiosensitization in conjunction with liposomes. Theoretically, the local production of hydroxyl radicals and

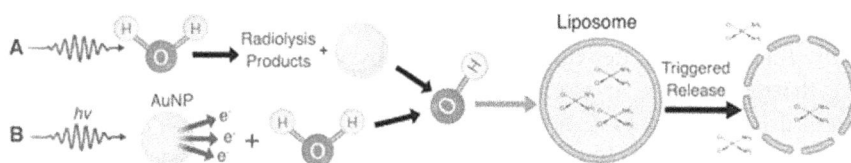

Figure 9.9. Schematic of the mechanisms associated with radiation that can be used to destabilize liposomes. The mechanisms are as follows: (A) the interaction of radiation with water to produce radiolysis products that can interact with AuNPs to amplify hydroxy radical production. These radicals can then destabilize liposome bilayers for triggered drug release. (B) The interaction of radiation with AuNPs to produce secondary electrons (such as Compton scattering and Auger electrons) which can then interact with water to produce hydroxy radicals. These radicals can then destabilize liposome bilayers for triggered drug release. Reproduced with permission from [117] under the terms of the Creative Commons Attribution Non-Commercial License. Copyright 2019 by the authors. Licensee MDPI, Basel, Switzerland.

secondary electrons mediated by embedded radiosensitizers should cause lipid peroxidation and liposomal bilayer destabilization when irradiated, thereby triggering drug release [117, 121].

9.4.4.1 Limitations of radiation triggered nanomedicine
Radiotherapy is non-specific and can damage healthy tissue along the path of the photons and normal tissue surrounding the tumor, careful consideration must be taken to limit the dose of radiation delivered during therapy. In the case of x-ray triggered drug release, the mechanism is still in its early stages of development, therefore a better understanding of GNPs embedded liposomes, their toxicities, and the effects of radiotherapy fractionation is needed before clinical translation.

9.4.5 Local hyperthermia triggered nanomedicine in breast cancer therapy

Hyperthermia or thermotherapy is a type of cancer treatment in which body tissue is exposed to high temperatures (42 °C–46 °C). As we know, high temperatures can damage and kill cancer cells, usually with minimal injury to normal tissues. Hyperthermia has been extensively used to initiate the temperature dependent changes such as the phase change of thermosensitive nanotheranostic to understand the immediate release of the encapsulated drug molecules within the tumor.

Lipids with proper phase transition temperature which should be slightly higher than the body temperature can be used to construct thermosensitive liposomes for drug release. Dipalmitoyl phosphatidylcholine (DPPC) with a transition temperature at 41 °C, is an ideal lipid to prepare thermosensitive liposomes as a hyperthermia controlled drug delivery vehicle [122]. Chen *et al* evaluates the feasibility of a thermoresponsive bubble-generating liposomal system without lysolipids for tumor-specific chemotherapy. The key component in this liposomal formulation is its encapsulated ammonium bicarbonate (ABC), which is used to actively load DOX into liposomes and trigger a drug release when heated locally. When temperature increased to 42 °C, ABC would decompose into CO_2 and NH_3, generating gas bubbles, increasing the permeability of the lipid bilayer and inducing rapid DOX release. This study demonstrates that the thermoresponsive bubble-generating liposomal system is a highly promising carrier for tumor-specific chemotherapy, especially for local drug delivery mediated at hyperthermic temperatures [123].

Lyso-thermosensitive liposomal DOX (ThermoDox) is a temperature sensitive liposomal drug delivery system approved for clinical trials that selectively accumulates in tumors. During ThermoDox/hyperthermia therapy, the tumor is heated locally with microwaves while the rest of the body remains at a normal temperature. The purpose of this study is to evaluate the effects of ThermoDox in combination with therapeutic heating of the chest wall in the treatment of recurrent regional breast cancer [124, 125].

Thermo-chemotherapy is such a combined modality treatment, generally well tolerated by patients, for malignant tumors of the mammary gland, the whole gastrointestinal tract, the lungs, the urogenital tract, the skin, bones and soft tissues as well as for advanced malignant tumors of the mouth and neck. Thermo-chemotherapy is particularly useful when conventional treatment methods do not have any effect or have ceased to have an effect. Hyperthermia alone has been used to kill tumor cells directly [75].

9.4.5.1 Limitations of local hyperthermia triggered nanomedicine

Traditional local heating methods, such as the use of heating pads or microwaves, have many drawbacks including low heat transfer efficiency (for the heating pad method), non-specific heating of the whole area. Local hyperthermia can cause pain at the site, infection, bleeding, blood clots, swelling, skin burns, blistering and damage to muscles, and nerves near the treated area. Therefore, it would be interesting to combine those thermosensitive drug delivery systems with the previously mentioned physical stimulus approaches, such as light, ultrasound and AMF to induce selective heating of the tumor with minimal heating of surrounding healthy tissues, for effective yet specific cancer treatment [29]. As mentioned earlier, hyperthermia alone has been used to kill tumor cells directly. However, hyperthermia is typically invasive and non-uniform and it requires sophisticated temperature control to achieve the desired effect while avoiding damage to normal tissue [75].

9.4.6 Radiofrequency triggered nanomedicine in breast cancer therapy

Radiofrequency (RF) ranging from 10 kHz to 900 MHz wavelengths may also be utilized to kill cancer cells by the hyperthermia effect. In comparison with PTT triggered by NIR light, RF has better tissue penetration, enabling treatment of deeply rooted tumors. In fact, FDA has already approved radiofrequency ablation for hepatic carcinoma therapy. Recently, a number of different nanomaterials which are able to absorb RF energy have been explored to enhance the RF ablation cancer therapy. GNPs have been found to be ideal for RF assisted cancer therapy [37]. Raoof *et al* used antibody conjugated GNPs for RF induced cancer therapy. GNPs when exposed to an external RF field, after being internalized by cancer cells through receptor mediated endocytosis would absorb RF and dissipate energy as heat to kill cancer cells [126]. In addition, Gannon *et al* showed that SWNTs have also been found to be an effective agent to enhance RF ablation therapy [127].

Kulkarni *et al* synthesized chitosan coated $La_{0.7}Sr_{0.3}MnO_3$ nanoparticles (C–LSMO) with chitosan facilitated doxorubicin entrapment (DC–LSMO nanoparticles) which showed drug release upon a 5 min RF exposure. They further checked the effect of these nanoparticles on MCF-7 and MDA-MB-231 breast cancer cells after 5 min RF exposure. The results showed a significant decrease in cell viability to 73% and 88% for both the cell lines, respectively, as compared to hyperthermia alone. Moreover, DC–LSMO nanoparticles successfully restricted the migration of metastatic MDA-MB-231 breast cancer cells [128]. Rejinold *et al* demonstrated that the breast cancer cells could selectively internalize hemocompatible, 170 ± 20 nm sized curcumin encapsulated chitosan-graft-poly (*N*-vinyl caprolactam) nanoparticles containing gold nanoparticles (Au–CRC–TRC). In their research on 4T1 breast cancer cells they found that the Au–CRC–TRC nanoparticles were predominantly accumulated within cytoplasm. After optimum RF exposure at 40 W for 5 min, Au–CRC–TRC nanoparticles absorbed and dissipated energy as heat in the range of 42 °C, which is the lower critical solution temperature of the chitosan-graft-poly (*N*-vinyl caprolactam), causing controlled curcumin release and inducing apoptosis to 4T1 breast cancer cells. Further, the tumor localization studies on orthotopic breast cancer models revealed that Au–CRC–TRC nanoparticles could selectively accumulate at the primary and secondary tumors as confirmed by *in vivo* live imaging followed by *ex vivo* tissue imaging and HPLC studies. These preclinical results shed light on the feasibility of nanoparticles as a better tumor targetable nanomedicine for RF assisted breast cancer treatment modalities [128, 129].

9.4.6.1 Limitations of radiofrequency triggered nanomedicine
The current RF ablation is limited by many drawbacks such as non-specific and invasive heating. RF can cause damage to normal tissue along the path and surrounding the tumor such as skin burns, muscle burns, discoloration of the skin and skin puckering, etc.

9.5 Discussion and general comment

Stimuli responsive nanomedicine delivery systems have made rapid progress in the past decade. These strategies exhibit tremendous therapeutic and detection potency for cancer at both research and clinical levels. In particular, the nanomedicines triggered by physical stimulus have very high potential compared to internal stimuli regarding delivery and release of anticancer drugs. These nanomedicines can be controlled by controlling the physical stimulus such as its power or force, time of exposure, duty cycle, etc. In addition, they offer unprecedented control over spatio temporal drug release and delivery profiles leading to superior *in vitro* and *in vivo* theranostic efficiency. Thus stimuli responsive nanomedicines hold incredible promise in overcoming the limitations and drawbacks of conventional drug delivery strategies.

This chapter has explored the strength and limitations of the physical stimuli which are now regularly used in research for cancer treatment especially in breast cancer. Some of the nanomedicine formulations triggered by physical force are now in clinical and preclinical stages. Nanomedicine products such as Doxil and Abraxane and ThermoDox have already been extensively used for breast cancer therapy with favorable clinical outcomes. Advances in systemic treatment have led to improved overall survival in patients with metastatic breast cancer. Nanomedicine is a promising alternative for breast cancer treatment. In the coming years promising nanotherapeutic strategies and devices may be developed with better understanding of the molecular biology of breast cancer. Researchers may get insight from these strategies to design and develop nanomedicine that is more tailored for breast cancer to achieve further improvements in cancer specificity, antitumorigenic effect, antimetastasis effect and drug resistance reversal effect.

Acknowledgment

The author Rakesh Patil is thankful to Dr K V Kulkarni, In-charge Director, Directorate of Forensic Science Laboratory, Kalina, Mumbai and Dr (Mrs) S V Ghumatkar, Deputy Director, RFSL, Kolhapur for their constant encouragement, motivation, valuable and kind support. The project leading to this work has received funding from the European Union's Horizon 2020 research and innovation program under the Marie Sklodowska-Curie grant agreement 751903.

References

[1] Wicki A, Witzigmann D, Balasubramanian V and Huwyler J 2015 *J. Control. Release* **200** 138–57
[2] Thoidingjam S and Tiku A B 2017 *Adv. Nat. Sci.: Nanosci. Nanotechnol.* **8** 023002
[3] Cai L-T and Sheng Z-H 2015 *Cancer Biol. Med.* **12** 141–2
[4] Wang S, Huang P and Chen X 2016 *ACS Nano* **10** 2991–4
[5] Zhou L, Wang H and Li Y 2018 *Theranostics* **8** 1059–74
[6] Gou Y, Miao D, Zhou M, Wang L, Zhou H and Su G 2018 *Front. Pharmacol.* **9** 421
[7] Ataollahi M R, Sharifi J, Paknahad M R and Paknahad A 2015 *J. Med. Life* **8** 6–11
[8] Patil R M and Shete P B 2019 *Hybrid Nanostructures for Cancer Theranostics* (Amsterdam: Elsevier) pp 63–86
[9] Almeida J P M, Chen A L, Foster A and Drezek R 2011 *Nanomedicine* **6** 815–35
[10] Kang J H and Chung J-K 2008 *J. Nucl. Med.* **49** 164s–79s
[11] Shete P B, Patil R M, Thorat N D, Prasad A, Ningthoujam R S, Ghosh S J and Pawar S H 2014 *Appl. Surf. Sci.* **288** 149–57
[12] Durymanov M O, Rosenkranz A A and Sobolev A S 2015 *Theranostics* **5** 1007–20
[13] Lammers T, Kiessling F, Hennink W E and Storm G 2012 *J. Control. Release* **161** 175–87
[14] Wang S and Low P S 1998 *J. Control. Release* **53** 39–48
[15] Qian Z M, Li H, Sun H and Ho K 2002 *Pharmacol. Rev.* **54** 561–87
[16] Seymour L W, Ferry D R, Anderson D, Hesslewood S, Julyan P J, Poyner R, Doran J, Young A M, Burtles S and Kerr D J 2002 *J. Clin. Oncol.* **20** 1668–76
[17] Liu D, Yang F, Xiong F and Gu N 2016 *Theranostics* **6** 1306–23
[18] Yao J, Feng J and Chen J 2016 *Asian J. Pharm. Sci.* **11** 585–95

[19] Cabane E, Zhang X, Langowska K, Palivan C G and Meier W 2012 *Biointerphases* **7** 1–27
[20] Karimi M *et al* 2016 *Chem. Soc. Rev.* **45** 1457–501
[21] Kanamala M, Wilson W R, Yang M, Palmer B D and Wu Z 2016 *Biomaterials* **85** 152–67
[22] Manatunga D C, de Silva R M, de Silva K M N, de Silva N, Bhandari S, Yap Y K and Costha N P 2017 *Eur. J. Pharm. Biopharm.* **117** 29–38
[23] Moreira A F, Dias D R and Correia I J 2016 *Microporous Mesoporous Mater.* **236** 141–57
[24] Xu P, Yu H, Zhang Z, Meng Q, Sun H, Chen X, Yin Q and Li Y 2014 *Biomaterials* **35** 7574–87
[25] Nguyen C T, Tran T H, Amiji M, Lu X and Kasi R M 2015 *Nanomedicine: Nanotechnol. Biol. Med.* **11** 2071–82
[26] Van Rijt S H, Bölükbas D A, Argyo C, Datz S, Lindner M, Eickelberg O, Königshoff M, Bein T and Meiners S 2015 *ACS Nano* **9** 2377–89
[27] Su C W, Chen S Y and Liu D M 2013 *Chem. Commun.* **49** 3772–4
[28] Calatrava-Pérez E, Bright S A, Achermann S, Moylan C, Senge M O, Veale E B, Williams D C, Gunnlaugsson T and Scanlan E M 2016 *Chem. Commun.* **52** 13086–9
[29] Chen Q, Ke H, Dai Z and Liu Z 2015 *Biomaterials* **73** 214–30
[30] Apetoh L, Ghiringhelli F, Tesniere A, Obeid M and Ortiz C *et al* 2007 *Nat. Med.* **13** 1050–9
[31] Goldberg S N, Kamel I R, Kruskal J B, Reynolds K, Monsky W L, Stuart K E, Ahmed M and Raptopoulos V 2002 *Am. J. Roentgenol.* **179** 93–101
[32] Mitragotri S 2005 *Nat. Rev. Drug Discov.* **4** 255–60
[33] Tran N and Webster T J 2010 *J. Mater. Chem.* **20** 8760
[34] Wang C, Tao H, Cheng L and Liu Z 2011 *Biomaterials* **32** 6145–54
[35] Urban C, Urban A S, Charron H and Joshi A 2013 *Transl. Cancer Res.* **2** 292–308
[36] Morachis J M, Mahmoud E A and Almutairi A 2012 *Pharmacol. Rev.* **64** 505–19
[37] Chen Q, Ke H, Dai Z and Liu Z 2015 *Biomaterials* **73** 214–30
[38] Tong R, Chiang H H and Kohane D S 2013 *Proc. Natl. Acad. Sci. U. S. A.* **110** 19048–53
[39] Mo S, Coussios C-C, Seymour L and Carlisle R 2012 *Expert Opin. Drug Deliv.* **9** 1525–38
[40] Patil R M, Shete P B, Thorat N D, Otari S V, Barick K C, Prasad A, Ningthoujam R S, Tiwale B M and Pawar S H 2014 *J. Magn. Magn. Mater.* **355** 22–30
[41] Patra S, Roy E, Karfa P, Kumar S, Madhuri R and Sharma P K 2015 *ACS Appl. Mater. Interfaces* **7** 9235–46
[42] Gupta A K and Gupta M 2005 *Biomaterials* **26** 3995–4021
[43] Monopoli M P, Åberg C, Salvati A and Dawson K A 2012 *Nat. Nanotechnol.* **7** 779–86
[44] Kim D-H, Rozhkova E A, Ulasov I V, Bader S D, Rajh T, Lesniak M S and Novosad V 2010 *Nat. Mater.* **9** 165–71
[45] Di Corato R, Espinosa A, Lartigue L, Tharaud M, Chat S, Pellegrino T, Ménager C, Gazeau F and Wilhelm C 2014 *Biomaterials* **35** 6400–11
[46] Majeed J, Pradhan L, Ningthoujam R S, Vatsa R K, Bahadur D and Tyagi A K 2014 *Colloids Surf. B Biointerfaces* **122** 396–403
[47] Yang K, Zhang S, Zhang G, Sun X, Lee S-T and Liu Z 2010 *Nano Lett.* **10** 3318–23
[48] Huang X, El-Sayed I H, Qian W and El-Sayed M A 2006 *J. Am. Chem. Soc.* **128** 2115–20
[49] Chen Q, Liang C, Wang C and Liu Z 2015 *Adv. Mater.* **27** 903–10
[50] Hong E J, Choi D G and Shim M S 2016 *Acta Pharm. Sin. B* **6** 297–307
[51] Li Z, Wang H, Chen Y, Wang Y, Li H, Han H, Chen T, Jin Q and Ji J 2016 *Small* **12** 2731–40
[52] Reinhold H S and Endrich B 1986 *Int. J. Hyperthermia* **2** 111–37

[53] Tian Q, Tang M, Sun Y, Zou R, Chen Z, Zhu M, Yang S, Wang J, Wang J and Hu J 2011 *Adv. Mater.* **23** 3542–47

[54] Huang X, Tang S, Mu X, Dai Y, Chen G, Zhou Z, Ruan F, Yang Z and Zheng N 2011 *Nat. Nanotechnol.* **6** 28–32

[55] Dykman L and Khlebtsov N 2012 *Chem. Soc. Rev.* **41** 2256–82

[56] Kim J W, Galanzha E I, Shashkov E V, Moon H M and Zharov V P 2009 *Nat. Nanotechnol.* **4** 688–94

[57] Cheng L *et al* 2014 *Adv. Mater.* **26** 1886–93

[58] Liu T, Wang C, Gu X, Gong H, Cheng L, Shi X, Feng L, Sun B and Liu Z 2014 *Adv. Mater.* **26** 3433–40

[59] Yang K, Xu H, Cheng L, Sun C, Wang J and Liu Z 2012 *Adv. Mater.* **24** 5586–92

[60] Song X, Chen Q and Liu Z 2015 *Nano Res.* **8** 340–54

[61] Hirsch L R, Stafford R J, Bankson J A, Sershen S R, Rivera B, Price R E, Hazle J D, Halas N J and West J L 2003 *Proc. Natl. Acad. Sci.* **100** 13549–54

[62] Riley R, O'Sullivan R, Potocny A, Rosenthal J and Day E 2018 *Nanomaterials* **8** 658

[63] Liu L, Xie H J, Mu L M, Liu R, Su Z B, Cui Y N, Xie Y and Lu W L 2018 *Int. J. Nanomed.* **13** 8119–35

[64] Pang B *et al* 2016 *ACS Nano* **10** 3121–31

[65] Zeng L, Pan Y, Tian Y, Wang X, Ren W, Wang S, Lu G and Wu A 2015 *Biomaterials* **57** 93–106

[66] Wang B, Yuan H, Liu Z, Nie C, Liu L, Lv F, Wang Y and Wang S 2014 *Adv. Mater.* **26** 5986–90

[67] Nomoto T and Nishiyama N 2018 Photodynamic therapy *Photochemistry for Biomedical Applications: From Device Fabrication to Diagnosis and Therapy* ed Y Ito (Singapore: Springer) pp 90 301–13

[68] Dos Santos A F, De Almeida D R Q, Terra L F, Baptista M S and Labriola L 2011 *J. Cancer Metastasis Treat.* **10** 324–32

[69] Lovell J F, Jin C S, Huynh E, Jin H, Kim C, Rubinstein J L, Chan W C W, Cao W, Wang L V and Zheng G 2011 *Nat. Mater.* **10** 324–32

[70] Jin C S, Cui L, Wang F, Chen J and Zheng G 2014 *Adv. Healthc. Mater.* **3** 1240–9

[71] Bharathiraja S, Manivasagan P, Santha Moorthy M, Bui N Q, Jang B, Phan T T V, Jung W K, Kim Y M, Lee K D and Oh J 2018 *Eur. J. Pharm. Biopharm.* **123** 20–30

[72] Liu J, Wang C, Wang X, Wang X, Cheng L, Li Y and Liu Z 2015 *Adv. Funct. Mater.* **25** 384–92

[73] Fang W, Tang S, Liu P, Fang X, Gong J and Zheng N 2012 *Small* **8** 3816–22

[74] Shen S, Tang H, Zhang X, Ren J, Pang Z, Wang D, Gao H, Qian Y, Jiang X and Yang W 2013 *Biomaterials* **34** 3150–8

[75] Khatun Z, Nurunnabi M, Nafiujjaman M, Reeck G R, Khan H A, Cho K J and Lee Y K 2015 *Nanoscale* **7** 10680–9

[76] Meng Z, Wei F, Wang R, Xia M, Chen Z, Wang H and Zhu M 2016 *Adv. Mater.* **28** 245–53

[77] Li W, Peng J, Tan L, Wu J, Shi K, Qu Y, Wei X and Qian Z 2016 *Biomaterials* **106** 119–33

[78] Link S and El-Sayed M A 2000 *Int. Rev. Phys. Chem.* **19** 409–53

[79] Salkho N M, Turki R Z, Guessoum O, Martins A M, Vitor R F and Husseini G A 2018 *Curr. Mol. Med.* **17** 668–88

[80] Elkhodiry M A, Momah C C, Suwaidi S R, Gadalla D, Martins A M, Vitor R F and Husseini G A 2016 *J. Nanosci. Nanotechnol.* **16** 1–18
[81] Kwon I K, Lee S C, Han B and Park K 2012 *J. Control. Release* **164** 108–14
[82] Liu Y, Miyoshi H and Nakamura M 2006 *J. Control. Release* **114** 89–99
[83] Azagury A, Khoury L, Enden G and Kost J 2014 *Adv. Drug Deliv. Rev.* **72** 127–43
[84] La Rosa M A D D, Husseini G A and Pitt W G 2013 *Ultrasonics* **53** 377–86
[85] Ueda H, Mutoh M, Seki T, Kobayashi D and Morimoto Y 2009 *Biol. Pharm. Bull.* **32** 916–20
[86] Li Y, An H, Wang X, Wang P, Qu F, Jiao Y, Zhang K and Liu Q 2018 *Nano Res.* **11** 1038–56
[87] Baghbani F, Chegeni M, Moztarzadeh F, Mohandesi J A and Mokhtari-Dizaji M 2017 *Mater. Sci. Eng. C* **77** 698–707
[88] Kiessling F, Liu Z and Gätjens J 2010 *J. Nanomater.* **2010** 1–15
[89] Lensen D, Gelderblom E C, Vriezema D M, Marmottant P, Verdonschot N, Versluis M, De Jong N and Van Hest J C M 2011 *Soft Matter* **7** 5417–22
[90] Rapoport N Y, Kennedy A M, Shea J E, Scaife C L and Nam K H 2009 *J. Control. Release* **138** 268–76
[91] Miller D L, Smith N B, Bailey M R, Czarnota G J, Hynynen K and Makin I R S 2012 *J. Ultrasound Med.* **31** 623–34
[92] Thiesen B and Jordan A 2008 *Int. J. Hyperthermia* **24** 467–74
[93] Luo S, Wang L F, Ding W J, Wang H, Zhou J M, Jin H K, Su S F and Ouyang W W 2014 *OA Cancer* **18** 2
[94] Tucker R D 2003 *J. Endourol.* **17** 601–7
[95] Deger S, Böhmer D, Türk I, Franke M, Roigas J, Budach V and Loening S A 2001 *Urologe A* **40** 195–8
[96] Thorat N D, Otari S V, Patil R M, Bohara R A, Yadav H M, Koli V B, Chaurasia A K and Ningthoujam R S 2014 *Dalton Trans.* **43** 17343–51
[97] Thorat N D, Patil R M, Khot V M, Salunkhe A B, Prasad A I, Barick K C, Ningthoujam R S and Pawar S H 2013 *New J. Chem.* **37** 2733–42
[98] Laurent S, Dutz S, Häfeli U O and Mahmoudi M 2011 *Adv. Colloid Interface Sci.* **166** 8–23
[99] Patil R M, Thorat N D, Shete P B, Otari S V, Tiwale B M and Pawar S H 2016 *Mater. Sci. Eng. C* **59** 702–9
[100] Kossatz S *et al* 2015 *Breast Cancer Res.* **17** 66
[101] Frei E H 1969 *J. Appl. Phys.* **40** 955–7
[102] Lübbe A S *et al* 1996 *Cancer Res.* **56** 4686–93
[103] Lee J-H *et al* 2013 *Angew. Chemie Int. Ed.* 52 4384–8
[104] Hu F X, Neoh K G and Kang E T 2006 *Biomaterials* **27** 5725–33
[105] Jain T K, Morales M A, Sahoo S K, Leslie-Pelecky D L and Labhasetwar V 2005 *Mol. Pharm.* **2** 194–205
[106] Rajesh Kumar S, Priyatharshni S, Babu V N, Mangalaraj D, Viswanathan C, Kannan S and Ponpandian N 2014 *J. Colloid Interface Sci.* **436** 234–42
[107] Dorniani D, Kura A U, Hussein-Al-Ali S H, Bin Hussein M Z, Fakurazi S, Shaari A H and Ahmad Z 2014 *Sci. World J.* **2014** 416354
[108] Carneiro M L *et al* 2011 *J. Nanobiotechnol.* **9** 11
[109] Peixoto R C A, Miranda-Vilela A L, de J, Filho S, Carneiro M L B, Oliveira R G S, da Silva M O, de Souza A R and Báo S N 2015 *Tumor Biol.* **36** 3325–36

[110] Bassiony H, Sabet S, El-Din T A S, Mohamed M M and El-Ghor A A 2014 *PLoS One* **9** e111960

[111] Zhang L, Zhou H, Belzile O, Thorpe P and Zhao D 2014 *J. Control. Release* **183** 114–23

[112] Al Faraj A, Shaik A P and Shaik A S 2015 *Int. J. Nanomed.* **10** 157

[113] Zhang Y, Zhang B, Luo J, Bai J and Liu F 2013 *Int. J. Nanomed.* **9** 33

[114] Zhang S, Gong M, Zhang D, Yang H, Gao F and Zou L 2014 *Eur. J. Radiol.* **83** 1190–8

[115] Jafari A, Salouti M, Shayesteh S F, Heidari Z, Rajabi A B, Boustani K and Nahardani A 2015 *Nanotechnology* **26** 075101

[116] Kato Y, Zhu W, Backer M V, Neoh C C, Hapuarachchige S, Sarkar S K, Backer J M and Artemov D 2015 *Pharm. Res.* **32** 3746–55

[117] van Ballegooie C, Man A, Win M and Yapp D 2019 *Pharmaceutics* **11** 125

[118] Zhang X-D, Wu D, Shen X, Chen J, Sun Y-M, Liu P-X and Liang X-J 2012 *Biomaterials* **33** 6408–19

[119] Lai C-W, Wang Y-H, Lai C-H, Yang M-J, Chen C-Y, Chou P-T, Chan C-S, Chi Y, Chen Y-C and Hsiao J-K 2008 *Small* **4** 218–24

[120] Liu Y, Chen W, Wang S and Joly A G 2008 *Appl. Phys. Lett.* **92** 043901

[121] Sicard-Roselli C, Brun E, Gilles M, Baldacchino G, Kelsey C, McQuaid H, Polin C, Wardlow N and Currell F 2014 *Small* **10** 3338–46

[122] Tagami T, Ernsting M J and Li S D 2011 *J. Control. Release* **154** 290–7

[123] Chen K J *et al* 2014 *ACS Nano* **8** 5105–15

[124] Nardecchia S, Sánchez-Moreno P, de Vicente J, Marchal J A, Boulaiz H, Nardecchia S, Sánchez-Moreno P, de Vicente J, Marchal J A and Boulaiz H 2019 *Nanomaterials* **9** 191

[125] Zagar T M *et al* 2014 *Int. J. Hyperthermia* **30** 285–94

[126] Raoof M, Corr S J, Kaluarachchi W D, Massey K L, Briggs K, Zhu C, Cheney M A, Wilson L J and Curley S A 2012 *Nanomedicine: Nanotechnol. Biol. Med.* **8** 1096–105

[127] Gannon C J *et al* 2007 *Cancer* **110** 2654–65

[128] Kulkarni V M, Bodas D, Dhoble D, Ghormade V and Paknikar K 2016 *Colloids Surf. B Biointerfaces* **145** 878–90

[129] Rejinold N S, Thomas R G, Muthiah M, Chennazhi K P, Park I-K, Jeong Y Y, Manzoor K and Jayakumar R 2014 *RSC Adv.* **4** 39408–27

IOP Publishing

External Field and Radiation Stimulated Breast Cancer Nanotheranostics

Nanasaheb D Thorat and Joanna Bauer

Chapter 10

Pharmacokinetics of nanomedicine for breast cancer

Vanishree Rao, Gautam Kumar and Nitesh Kumar

The introduction of nanomedicine in the treatment of breast cancer is intended to enhance the targeted drug delivery of conventional chemotherapeutic agents. Nanomedicine, being a trusted alternative is extensively used as an adjuvant therapy with favorable clinical outcomes. Nanomaterials (organic/inorganic) used in the preparation of nanomedicine vary widely in their physicochemical properties. Physicochemical properties like shape, size, surface area, surface charge, surface roughness, chemical composition, crystal structure and physicochemical stability play an important role in determining the pharmacokinetic advantages of each nanomaterial. Desired pharmacokinetic features such as improved rate of drug absorption, enhanced bioavailability of oral drugs, site-specific drug distribution, the modified pattern of drug metabolisms, prolonged time of drug residence in blood circulation and decreased or delayed renal excretion are achieved using the appropriate nanomaterial. This chapter reviews various types of nanoformulations/nanomedicines approved or in the pipeline for the treatment of breast cancer along with the detailed description on their improved pharmacokinetic properties which resulted from unique physicochemical features of nanomaterials used in the formulation. In addition, selection criteria for nano drug delivery systems, novel targeting approaches for improved pharmacokinetic and pharmacodynamic features, advantages and potential pharmacokinetic benefits of nanomedicines in the treatment of breast cancer are discussed. In conclusion, nanomedicines have an important role in improving the status and effectiveness of conventional breast cancer chemotherapy, since nanoformulations nullify the pharmacokinetic disadvantages and reduce the systemic cytotoxicity of conventional chemotherapy in breast cancer.

doi:10.1088/2053-2563/ab2907ch10

10.1 Introduction

Nanobiotechnology is used to develop tools which can serve as probes, sensors and vehicles to enhance the delivery of various biomolecules in different cellular systems [1]. Nanomaterials are of few nanometers in length, and they allow unique interactions with various biological systems at the molecular level [2]. Nano-oncology is an upcoming area which makes use of multiple nanomaterials for the detection, diagnosis and treatment of human cancers [3]. Breast cancer is one of the most commonly occurring cancers worldwide. Current treatment options mainly depend on the tumor characteristics like tumor size, metastasis, biomarkers, antigens or expression of endocrine receptors [4, 5]. The first line of treatment involves surgical removal of the tumor combined with radiation and appropriate chemo-therapy. Many efforts have been made to improve the chemotherapeutics in breast cancer but failure of drugs in reaching the target site at effective doses, increase in systemic toxicity, poor pharmacokinetic properties stand as hurdles, which ulti-mately results in decreased survival rates [6]. A major limitation in the therapy is less selective cytotoxicity. Using a nanotechnological approach for site-specific drug delivery can result in more effective and less toxic chemotherapy. This can improve the therapeutic index and prognosis [7]. Breast cancer is a solid tumor type, which is associated with increased interstitial fluid pressure (IFP); this behaves like a strong barrier for the trans-capillary transport of the anticancer agents [8]. Nanomedicine can easily surpass this disadvantage and provide good bio-distribution of the drug along with increased drug penetration at the site. Since distribution of drugs is also dependent on the half-life, high circulation time achieved by drug conjugate with nanocarrier will increase the half-life, which results in uniform distribution of the drug [1]. The nanomedicine paves the way for the targeted treatment approach. Knowing that accumulation of conventional chemotherapeutic agents are 10–20-fold higher in normal viscera, than in breast cancer site, involvement of nanocarriers becomes essential in targeted drug delivery. Nanoparticles are also used in imaging of the cancer *in vivo* and also for developing cancer biomarkers [9].

The use of nanoparticulates in breast cancer therapy allows us to modify the pharmacokinetic features of the conventional chemotherapeutic agents without having to alter the active compound [10]. Nanomedicine is more desirable than a drug alone formulation, since uptake of drugs by breast cancer cells is more efficient, the dosage required is lowered, hence side effects will be minimized and drugs are released in a controlled manner [8]. So, to improve the status and effectiveness of chemotherapy in breast cancer, nanotechnology can be used.

10.2 Nanobiotechnology-based platforms for breast cancer therapy

Nanomedicines can be defined as carrier systems in the nanosize range (preferably 10–100 nm), containing encapsulated, dispersed, adsorbed, or conjugated drugs and imaging agents. The delivery of anticancer agents is associated with many challenges which include poor absorption, low penetration into the target site, and non-specific dissemination in every part of the body [11]. Nanomedicines have the capacity to improve pharmacokinetics of drugs, but at the same time they may influence efficacy and toxicity of the anticancer drugs. Several nanomedicine strategies have become prominent approaches to improve drug delivery and pharmacodynamics in many diseases. Nanoparticles can be prepared by forming a solid polymer matrix to encapsulate drugs or by preparing vesicles like micelles, block copolymer, liposomes and nanoemulsions etc [12].

Earlier, researchers focused on improving the molecular properties of drugs which are already available as therapeutic and diagnostic agents, but nowadays nanotechnology has become a more prominent approach in novel therapeutic and diagnostic strategies to improve sensitivity and specificity [13]. The main objectives in the development of nanomedicines are:
- specific targeting and drug delivery,
- improved safety and biocompatibility,
- faster development of novel and safer medicines,
- improved pharmacokinetics of drugs.

10.3 Types of nanoformulations (nanomedicines) for breast cancer therapy

Currently, research has been focused on developing nanomedicines for breast cancer therapy. There are multiple types of nanoformulations used in the design of nanomedicine with improved targeting efficiency, safer, improved doses, reduced metabolic activities, increased circulation time, improved tumor uptake and effectiveness for breast cancer therapy. A variety of nanomedicines has been approved for the treatments of breast cancer and are described in table 10.1 [14, 15]. Nanomedicines can be made of organic or inorganic materials [16]. These can be classified as shown in figures 10.1 and 10.2.

10.4 Physicochemical properties of nanomedicines and their effects in pharmacokinetics and pharmacodynamics

Nanomedicines have unique properties that can alter pharmacokinetic and pharmacodynamic characteristic of therapeutic drugs. There are different properties of nanomedicines which are described below along with their effects in pharmacokinetic and pharmacodynamic characteristics of drugs.

10.4.1 Particle surface area and size of the nanomedicines

The size of the nanoparticle and surface area, plays an important part in drug interaction with biological systems. Lowering the size of nanocarrier used in the nanomedicines results in increased surface area, greater absorption and penetration, cellular uptake, endocytosis, and efficiency of the drugs [17, 18]. Reduction in size (50–200 nm) may also increase toxicity by producing reactive oxygen species which can damage DNA in normal biological systems such as liver, spleen and respiratory systems [19–21].

Table 10.1. Approved nanomedicines available in market for the treatment of breast cancer.

Brand/trade name	Type of nanoformulation	Therapeutic agent	Pharmacokinetic benefit	Approval/Indication	Company
DOXIL	PEGylated liposomes	Doxorubicin	Extended circulation time, reduced volume of distribution, increased tumor uptake and minimized cardiotoxicity	FDA 1995 Breast cancer, Ovarian, Kaposi's sarcoma	Sequus Pharmaceuticals Ltd
MYOCET	Non-PEGylated liposomes	Doxorubicin	Minimized cardiotoxicity and improved tumor uptake	Europe 2000 Breast cancer (IV)	Elan Pharmaceuticals
Genoxol-PM	Polymeric nanomicelles	Paclitaxel	Improved dose with minimal toxicity	Korea 2001 Breast cancer (IV)	Samyang biopharm
Abraxane	Nanoparticles	Albumin-bound paclitaxel	Shorter infusion time, safer and minimal side effects	2005, FDA Breast cancer, lung cancer (IV)	Abraxis BioScience, LLC
Nanoxel	Polymeric nanoparticles	Docetaxel	Improved efficacy and less toxic	DCGI, India 2006 Metastatic breast cancer	Dabur Pharma Ltd
Rexin-G	Protein tagged nanoparticles	micro-RNA-122	Tumor targeted delivery and improved efficacy	FDA 2007	Epeius Biotechnologies Corporation
Kadcyla	Immunoconjugate	Ado-trastuzumab emtansine	Tumor targeted delivery, less toxic and improved efficacy	FDA 2013 Metastatic breast cancer	Genentech, Inc.

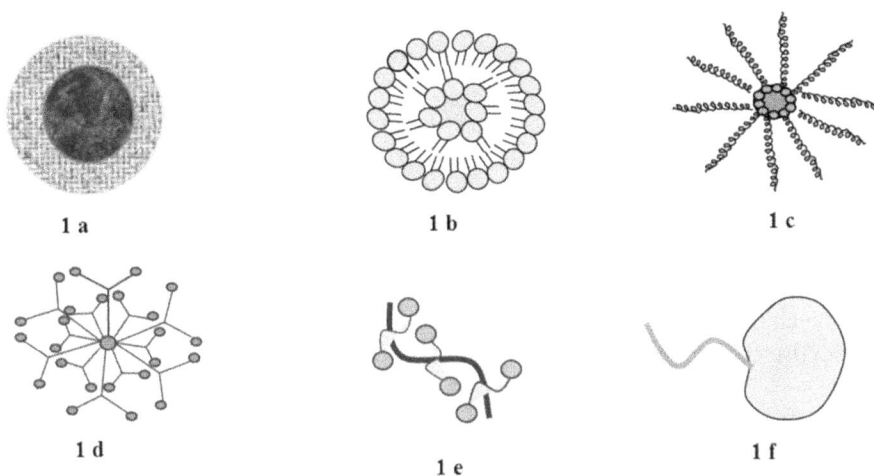

Figure 10.1. Organic nanomedicines: (a) polymeric nanoparticles, (b) liposomes, (c) polymeric micelles, (d) dendrimers, (e) drug–polymer conjugates, (f) protein–polymer conjugates.

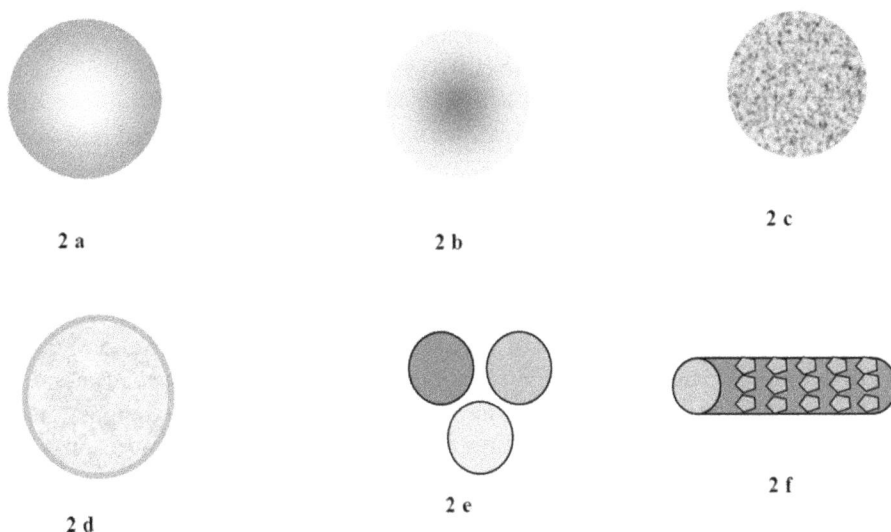

Figure 10.2. Inorganic nanomedicines: (a) gold nanoparticles, (b) iron oxide nanoparticles, (c) silica nanoparticles, (d) hollow/porous nanoparticles, (e) quantum dots, (f) carbon nanotubes.

10.4.2 Shape and aspect ratio of nanomedicines

The shape and aspect ratio of the nanomedicines play an important role in influencing membrane folding in biological systems during phagocytosis/endocytosis.

Spherical nanoparticles easily go for phagocytosis/endocytosis process as compared to fiber-like or rod-shaped nanoparticles and are less toxic [22, 23].

Rod-shaped nanoparticles are more able to block ion channels such as potassium channels, and increased uptake of nanorods is seen, which reaches maximum if the aspect ratio approaches unity [24].

Gold nanorods show less and slower uptake than spherical particles and TiO_2 fibers (15 mm size) have greater toxic potential than smaller size (5 mm) or spherical particles. Moreover, it is also known that the higher the aspect ratio, the more the toxicity and shelf life and reduced clearance of nanoparticles [25–27]. For example, fibers of asbestos longer than 10 μ cause lung cancer, fibers greater than 5 μ cause mesothelioma and fibers greater than 2 μ cause asbestosis because longer fibers cannot be cleared from the respiratory tract due to the inability of macrophages to phaogocytise the larger fibers [28].

10.4.3 Surface charge of nanomedicines

The role of surface charge is also important in nanomedicines since it describes the interaction of these nanocarriers with the biological systems. Multiple aspects of nanomedicines like blood–brain barrier integrity, colloidal behavior, transmembrane permeability, protein binding, selective adsorption of nanoparticles are mainly dependent on surface charge of nanoparticles [29–31]. Positively charged nanoparticles have greater potential for cellular uptake and phagocytosis compared to negatively charged or neutral nanoparticles. They can induce platelet aggregation and hemolysis which causes severe toxicity to the system [32].

Negatively charged nanoparticles in the size range of 50–500 nm can easily permeate skin after dermal administration, whereas positively charged and neutral particles cannot [33].

10.4.4 Composition and crystalline structure of nanomedicines

It is known that particle size plays an important role in deciding toxicity of nanoparticles, but we cannot simply ignore the comparable toxicity studies of different nanoparticles chemistries of the same dimensions. It was found that silver and copper nanoparticles resulted in toxicity of many tested biological systems, but TiO_2 from the similar dimensions never produced any toxicity signs thereby emphasizing the role of the composition of nanoparticles in lowering toxicity [34].

It was also found that the rutile structure of TiO_2 nanoparticles produced oxidative damage to DNA, formation of micronuclei in the absence of light and lipid peroxidation, while the anatase structure of nanoparticles of the same size and chemical composition did not, thus emphasizing the role of the crystal structure of nanoparticles in determining toxicity [35].

10.4.5 Aggregation and concentration of nanomedicines

The aggregation states and concentration of nanoparticles also plays an important role in the pharmacokinetic and pharmacodynamic properties of nanoparticles. Generally, the aggregation states of nanoparticles depend on surface charge, size and composition. It was seen that nanotubes of carbon, mainly accumulate in the spleen, liver and lungs without causing acute toxicity but produce cytotoxicity when retained for prolonged periods. More side effects are seen with agglomerated carbon nanotubes than carbon nanotubes that are well-dispersed and enhance the pulmonary interstitial fibrosis [36, 37].

10.4.6 Surface properties of nanomedicines

The surface properties of nanoparticles (coating and surface roughness) also play an important role in pharmacokinetic and pharmacodynamic properties of nanoparticles as they are involved in deciding the final results of their interaction with the biological entities or cells [38, 39]. Surface coating may affect the physicochemical characteristics of nanoparticles, like optical, magnetic, electric properties and chemical reactivity, and may change the pharmacokinetics, accumulation, distribution and toxicity of nanocarriers. It is also known that the presence of ozone, oxygen, oxygen radicals and transition metals on the surface of nanocarriers may lead to the production of reactive oxygen species and the initiation of inflammatory response; these can influence their associated toxicity issues. For example, the presence of oxygen radicals on the surface of silica nanoparticles leads to more specific cytotoxicity potential [40, 41]. On the other hand, surface coating may also be used to minimize the toxicity issues and to improve stability of the nanoparticles [42]. However, care should be taken when choosing an appropriate coating material, because weaker surface coatings can cause photolytic or oxidative degradation resulting in the exposure of the metallic core, which can be toxic or may cause hazardous reactions inside the body. Examples of surface coating agents include, polyethylene glycol, poloxamers and polyethylene etc [43].

Moreover, during the selection of suitable coating material, the charge of the coating material might be considered. Coating with negatively charged serum albumin shows a higher uptake by liver and faster clearance from blood [44, 45]. The attributes of nanocarriers like hydrophobicity, surface roughness and charge of nanoparticles increases the process of cellular uptake of nanocarriers, these can also alter the pharmacokinetic and pharmacodynamic potential of nanocarriers. Surface roughness defines the strength of nanocarrier–cell interactions and promotes cell adhesion. The structure of the pore is critical in cell–nanocarrier interactions. It was found that long-range ordered mesoporous silica nanoparticles cause size dependent hemolysis. Nanoporous silicon nanoparticles or luminescent porous silicon nanoparticles with size of the pore about 2 nm will not have any toxicity in mouse-models [46–49].

10.4.7 Solvents/media in nanomedicines

Solvents/media are also said to affect particle agglomeration and dispersion state of nanocarriers, which can alter their particle size, thereby altering pharmacokinetic and pharmacodynamic potential of nanoparticles. It was found that particles of ZnO, TiO_2, and carbon black have significantly greater size in PBS than in water; moreover, it is also known that nanoparticles show different sizes in biological fluids [50, 51]. Accordingly, the effect of nanocarriers varies based on the composition of the medium in which the nanocarriers are suspended; on the other hand, the same

Table 10.2. Selection criteria for nano drug delivery system.

Desired characteristics	Comments
Inherently non-toxic and biocompatible materials	Non-toxic and biocompatible materials should be used
Small size (10–200 nm)	Best particle size proven with *in vivo* studies with effective drug delivery
Encapsulation of active molecules at therapeutic levels	Nanoparticle vehicle must encapsulate a high percent of the agent (>50 mol%) to achieve therapeutic dosage
Stability of colloidal form in physiological environment	The nanocarriers and surface functionalization should be resistant to the agglomeration
Reasonable circulation time	Reasonable circulation time in the circulatory system for the nanoparticle to finish its task
Targeting to cell or tissue of choice	Greatest uptake concentration of chemotherapeutics within the desired lesions and the least side effects with healthy tissues
In vivo controlled release of active molecules	Should have trigger mechanism like acidic pH inside the tumor cells or during maturation process of endosome for drug penetration
Clearance	To reduce the cumulative and/or long-term systemic side effects a proper clearance mechanism should be present

nanocarriers exhibit various pharmacokinetic and pharmacodynamic properties when dissolved in different mediums [52, 53].

10.5 Selection criteria for nano drug delivery system

The size of the nanocarriers, charge and shape are the major parameters in nanocarrier systems that determine the *in vivo* tissue distribution, specific targeting potential and biological delivery of nanoparticles. Taking into account that each active agent is expected to act via different pathways and follow different pharmacokinetic patterns, selection of nanocarriers for each desired effect or each different active compound/conventional chemotherapeutic agents becomes very important. Based on the expected criteria that are required in a nanocarrier the type and size of nanocarriers are chosen [54, 55]. Various selection criteria are described in table 10.2.

The nanocarriers are expected to enhance the pharmacokinetic features of conventional chemotherapeutic agents which lack the specific target recognition features, with poor ADME profile [10, 56]. The advantages of using nanocarrier systems are listed below:

- protection of drugs from early degradation by many physiological factors;
- enhances stability of the drug;
- incorporation of both hydrophilic and hydrophobic drugs is feasible;

- reduction in dosing frequency;
- enhances target specific drug absorption;
- controls the tissue distribution and other pharmacokinetic parameters of the drug;
- allows enhanced penetration into the tumor cells;
- prevention of any possible drug interactions with biological system;
- longer shelf life;
- reduced systemic toxicity.

10.6 Arsenal for drug delivery

To prepare a suitable nanocarrier for immediate and effective clinical benefit, some important properties are required to be considered. An ideal nanocarrier should be biodegradable, easily soluble, biocompatible, and non-toxic in nature. These features are the most important ones since this decides the primary selection of any nanocarrier [57, 58]. Along with the features mentioned above, some other important features are mentioned in figure 10.3.

10.7 Importance of nanomedicines in pharmacokinetics of breast cancer therapy

All the FDA approved anticancer agents fail in differentiating between cancerous and normal surrounding cells for their action. This non-specific cytotoxicity leads to systemic toxicity and decrease in anticancer efficacy. This can be overcome with target specific drug delivery and action. This is exactly where nanocarriers will come into the picture in the treatment of breast cancer. Nanotechnology offers a target specific approach and could therefore decrease the systemic toxicity to provide significant benefits to cancer patients. This can successfully reverse the limitations of conventional chemotherapy. More attention has been focused by the researchers in pharmacokinetic studies because it is an important step in designing novel drug

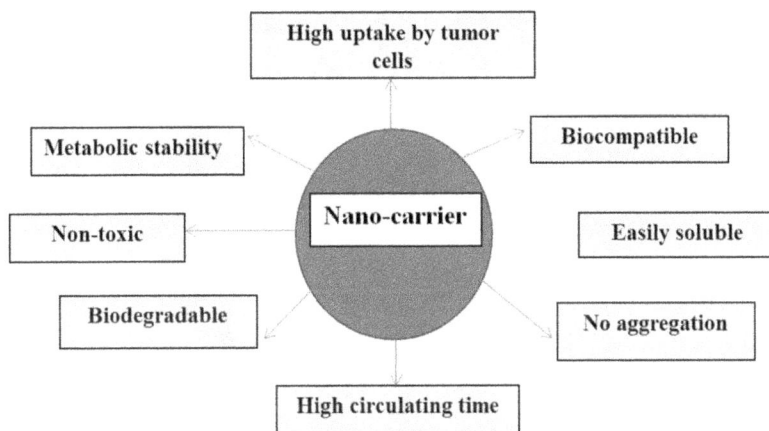

Figure 10.3. Nanocarrier.

delivery systems which requires ADME profile (absorption, distribution, metabolism and excretion) from the early stage of the development to the final clinical investigations [59, 60].

The main objectives of nanomedicines from a pharmacokinetic point of view include:

- an enhanced drug release profile in biological systems;
- improved rate of drug absorption;
- site-specific drug distribution;
- modified pattern of drug metabolisms;
- prolonged time of drug residence in the body (in blood circulation);
- decreased or delayed renal excretion of the drugs.

10.8 Pharmacokinetics of nanomedicines for breast cancer therapy

Taking into account the benefits of nanomedicine in breast cancer therapy in comparison with available conventional chemotherapy, the FDA has approved a few nanocarrier based chemotherapeutic delivery platforms between 1995 and 2005. The pharmacokinetic details of nanoformulations based on liposomes, polymeric nanoparticles and approved formulations like Doxil and Abraxane are discussed below along with a few newly approved nanomedicines in the pipeline (figure 10.4).

Liposomes are one of the popular types of nanocarriers which are made up of phospholipid vesicles. They are mainly used for controlled and targeted drug delivery. Nanopores are aerogels which are produced by sol–gel chemistry for controlled release of the conventional chemotherapeutics. Nanowires or carbon nanotubes are made of semiconducting metals or carbon, for drug delivery into

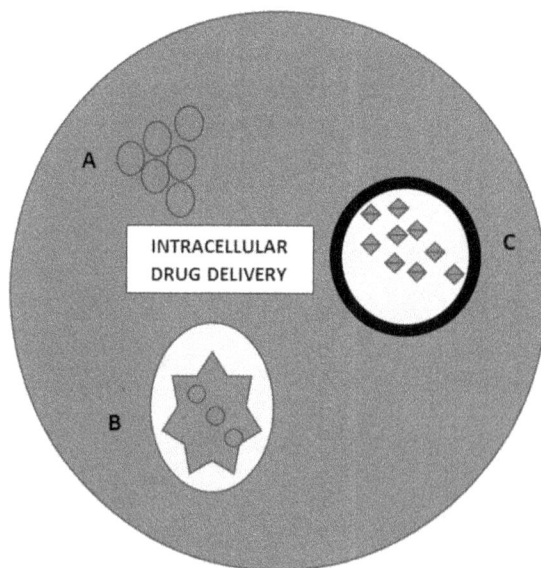

Figure 10.4. (A) Liposomes. (B) Polymeric nanoparticles including polymeric micelles. (C) Albumin-bound nanoparticles.

genes or specific DNA in breast cancer. Nanoshells are typically made of gold sulfide or silica, with dielectric core and metal shell for enhanced tumor targeting. Ceramic nanocarriers are designed for selective drug targeting and bio-molecular delivery in breast cancer, made up of silica or alumina. Polymeric micelles are amphiphilic block polymers used for delivery of water insoluble drugs. Polymeric nanocarriers are biodegradable and used for the controlled and target specific drug delivery in breast cancer [16].

10.8.1 Pharmacokinetics of liposomes

Liposomes are a group of self-assembling closed round shaped vehicles which are made of a membrane consisting of a lipid bilayer enclosing an aqueous core. This structure was first explained by Bangham and Horne in 1964 [61]. Later the ability of liposomes, typically of 50–100 nm, to encapsulate drug particles within the core was described by Gregoriadis *et al* in 1971 [62]. liposomes became the first ever nanocarriers to be approved by the FDA (1996), to carry various conventional chemotherapeutic agents. Various hydrophilic drugs, genes, small interfering RNA, can be carried by the liposomes, to some extent even hydrophobic drugs also can be encapsulated within the lipid bilayer [63].

The half-life of liposomes used in breast cancer is improved by the addition of inert polymers like polyethylene glycol (PEG). This forms protection around each liposome, thereby decreasing the possibility of recognition by the reticuloendothelial system, which usually decreases the circulation time. Subsequent clearance is also decreased by this addition. To improve oral bioavailability in breast cancer patients, liposomes are employed. The lipid bilayer enhances the adherence to bio-membranes and helps in the formation of mixed-micelle structures with bile salts to increase the solubility [64]. The membranes of liposomes contain amphiphilic compounds, like glycolipids and phospholipids, which makes them biodegradable. Moreover, the localization of liposomal preparations into the tumor in breast cancer is achieved by increased permeability, enhanced retention at the tumor site and bioconjugation of the liposome-based drug delivery system with specific antibodies, peptides and aptamers which are specific to the breast cancer [65].

10.8.2 Pharmacokinetics of polymeric nanoparticle system

These are biodegradable polymers. Dendrimers, polymeric micelles and polymer–drug conjugates are included in this system. Poly D L-lactic-*co*-glycolic acid (PLGA), poly D L-lactic acid (PLA), and polyethylene glycol (PEG) are used to produce polymeric nanoparticles. Chitosan, alginate, and pectin are polysaccharides helpful in encapsulating these nanostructures. Block copolymers are used with different hydrophilicity to formulate self-assembled nanoparticles and usually consist of two or more polymer chains [66]. Either hydrophilic or hydrophobic drugs are encapsulated by polymeric nanoparticles. This improves pH dependent controlled release and surface related modifications. This can be explored for controlled and site-specific drug delivery. Dendrimers consist of various arms extending from a center. Three main components include a central core, peripheral functional groups dendrimer, peripheral groups with charged hydrophilic and lipophilic function. Dendrimers can vary in size, chemical compositions and molecular weights. Targeting molecules can be simply tagged with dendrimers to achieve target specific delivery of drugs [67].

Paclitaxel was isolated from the bark of the yew tree by Wall and Wani in 1967 [68]. Paclitaxel, an antimitotic drug acts by inhibiting the depolymerization of the microtubule leading to cell cycle arrest and mitosis inhibition. Genexol-PM is a polymeric based nanoparticle preparation of paclitaxel. It has been proven that Genexol-PM is able to deliver a higher dose of paclitaxel when compared to conventional paclitaxel. Genexol-PM was used in a dose of 300 mg m^{-2} as against 175 mg m^{-2} of conventional paclitaxel. Systemic side effects were very low with the nanoparticle preparation of paclitaxel when compared to conventional paclitaxel [69, 70]. The circulating half-life was 1.8 times higher for the polymeric form of paclitaxel in comparison with the conventional one. The overall survival rates were also increased with the polymeric form and the reason for the lower systemic toxicity was mainly due to the absence of castor oil which was added in conventional preparation to increase solubility. Conventional preparation was mainly associated with increased hypersensitivity and neuropathy. Abraxane is a castor oil free colloid suspension of unmodified paclitaxel stabilized with albumin by homogenization under high pressure. In Abraxane, albumin and paclitaxel are linked via reversible noncovalent binding. This allows safe administration of paclitaxel with a much higher dose and a significant shorter injection time period. A 9.9 fold increase in endothelial cell binding and a 4.2 fold increment endothelial transcytosis was observed with Abraxane when compared with conventional paclitaxel [71].

10.8.3 Pharmacokinetics of Doxil

Doxil is the trade name given to a nanoformulation containing doxorubicin hydrochloride encapsulated by PEGylated liposome for intravenous use in breast cancer patients. This was the first FDA approved nanomedicine [72]. It is currently used as adjuvant chemotherapy along with other drugs before the surgical removal of the tumor to shrink the tumor size and to treat advanced stages of breast cancer. The active ingredient, i.e. doxorubicin produces cytotoxicity in cancer cells by intercalating between the base pairs present in the DNA, thereby inhibiting DNA synthesis and transcription [73]. Doxorubicin was known to cause cardiotoxicity on cumulative dose greater than 600 mg m^{-2}.

Doxil was mainly introduced to reduce the systemic side effects of doxorubicin formulation without impairing its anticancer effect. Extended circulation time was achieved with Doxil by avoiding premature elimination of nanoparticles by the reticuloendothelial system due to PEGylation. It also has the capacity to achieve intramural drug levels by passive targeting effect [55]. Doxil improved survival rates and three-fold reduction of cardiotoxicity. Doxil and doxorubicin were comparable with respect to overall as well as progression-free survival in a Phase III trial of 22 509 women with metastatic breast cancer. The circulating half-life of Doxil was 74 h compared to 5 min for free doxorubicin. Plasma clearance rates of

Doxil were also much better than free doxorubicin. 50 mg m^{-2} of Doxil given every four weeks was as effective as 60 mg m^{-2} of doxorubicin given once in three weeks. In a safety study, only 10 patients out of 254 patients developed cardiotoxicity with Doxil, whereas 48 out of 255 patients treated with free doxorubicin developed severe cardiotoxicity [74].

Few disadvantages have been encountered in the pharmacokinetic profile of the liposomal nanomedicine. In many studies it was found that 50%–80% of liposomes were absorbed mainly by liver cells (Kupffer cells) and reticuloendothelial system (RES) within the 15–30 min of intravenous administration [75]. Stability, poor reproducibility and difficulty in sterilization are the other problems with liposomal formulation.

10.8.4 Pharmacokinetics of Myocet

Myocet is a recently approved nanomedicine containing doxorubicin as an active agent in non-PEGylated liposome formulation. This was designed to minimize the cardiotoxicity and to improve the tumor uptake [76]. Conventional doxorubicin has fast distribution and slow elimination phase. Myocet (60 mg m^{-2}) use is estimated to have five-fold less clearance than doxorubicin alone (60 mg m^{-2}). Prolonged terminal half-life was observed in the case of patients receiving Myocet (16.4 h), when compared to conventional doxorubicin (10.4 h), along with reduced volume of distribution. Therapeutic index is also improved in the case of Myocet since it can only escape from leaky vasculature like in areas of inflammation, which supports the hypothesis that distribution in the tumor area is very high compared to healthy tissues like the heart [77].

10.8.5 Pharmacokinetics of Genoxol-PM

Genoxol-PM is nanomedicine containing paclitaxel as an active ingredient encapsulated with polymeric nanomicelles. This is a cremophor EL free formulation which decreased the incidence of hypersensitivity reaction and neuropathy. Nonlinear pharmacokinetic behavior of paclitaxel was the major disadvantage, but here since the drug is trapped inside the nanomicelles it was less available for tissue distribution, metabolism and biliary excretion. Genoxol-PM permits higher drug delivery dose of paclitaxel without any additional toxicity [70, 78].

10.8.6 Pharmacokinetics of nanoxel

Nanoxel is a nanomedicine containing docetaxel as an active compound encapsulated with polymeric nanoparticles. This formulation reduced the toxicity and improved the efficacy of the drug. Nanoxel showed three-fold higher tumor cell uptake when compared to conventional docetaxel. Release of docetaxel was very fast at lower pH, which resulted in high tubulin stabilization and induction of apoptosis in breast cancer cells when compared to conventional drugs [79, 80].

10.8.7 Pharmacokinetics of Rexin-G

Rexin-G is an FDA approved formulation of micro-RNA 122 trapped in protein tagged nanoparticles, developed for enhanced tumor target and improved efficacy. Peak blood levels after the administration were achieved within 5 min, and no circulating vectors were found after an hour, this contributes to the enhanced tumor targeting ability. Accumulation of this nanomedicine was specific to the tumor site with no signs of accumulation in other healthy tissue [81, 82].

10.8.8 Pharmacokinetics of Kadcyla

Kadcyla is an immunoconjugate of ado-trastuzumab emtansine. This particular conjugated formulation allows intracellular drug delivery of trastuzumab in cells overexpressing human epidermal growth factor receptor 2, making it exclusively toxic to breast cancer cells. Peak concentrations were observed immediately after the infusion, with 3.13 L of volume of distribution. The half-life of this formulation is about four days, with no observed toxicity to healthy cells when compared to conventional trastuzumab [83].

10.8.9 Pharmacokinetics of Abraxane

Abraxane is a castor oil free colloid suspension of unmodified paclitaxel stabilized with albumin by homogenization under high pressure. In Abraxane, albumin and paclitaxel are linked via reversible noncovalent binding. This allows safe administration of paclitaxel with a much higher dose and a significantly shorter injection time period. A 9.9 fold increase in endothelial cell binding and a 4.2 fold increment endothelial transcytosis was observed with Abraxane when compared with conventional paclitaxel [71, 84]

10.9 Novel targeting approaches for improved pharmacokinetic and pharmacodynamic features for breast cancer therapy

10.9.1 Her2 targeting approach for her2 positive breast cancer

Strategy: Target overexpressed her2 transmembrane receptor to inhibit tyrosine kinase activity and her2 homo- and/or heterodimerization into tumor promoting combinations such as the her2/her3 [85].
  Active agents: Trastuzumab, pertuzumab and lapatinib.

10.9.2 AAV2 receptor targeting approach for triple negative breast cancer

Strategy: Induce necrotic tumor cell death to inhibit tumor growth. Method can be further improved by encapsulation in targeting nanocarriers [86].
  Active agents: AAV2.

10.9.3 Gene targeting approach for breast cancer

Strategy: The active agent can be encapsulated in a targeting nanocarrier such as liposomes or form biocompatible complexes with atelocollagen to improve uptake [87].

Active agents: siRNA, siRNA–atelocollagen complex.

10.9.4 Photothermal ablation (PTA) for breast cancer

Strategy: Targeted nanoparticle-mediated PTA that can absorb NIR irradiation and release heat energy above the thermal ablation threshold (50 °C) to damage tumor cells [88].

Active agents: Metallic-based nanomaterials (gold or silver).

10.10 Advantages of nanomedicine in breast cancer therapy

(a) Active and passive targeting by nanomedicine increases tumor drug concentration and reduced uptake by noncancerous cells.

(b) A variety of nanomedicines potentially improve tumor permeability.

(c) Use of strategies such as PEGlyation to extend the circulation time.

(d) Enhanced tumor specificity.

(e) Active and passive targeting may increase phagocytosis/endocytosis; inhibit drug efflux mechanisms; co-delivery of active agents that can overcome drug resistance mechanisms.

(f) Targeting of tumor microenvironment can also be done by stimulus-responsive nanocarriers such as pH-sensitive nanodevices.

(g) Targeting cancer stem cells.

(h) Several nanoformulations can improve drug solubilization and protection of unstable drugs.

(i) A variety of advantage and disadvantages of nanomedicines are described in table 10.3.

10.11 Potential pharmacokinetic benefits of nanomedicine

A common practice in the development of nanodrugs is to conjugate or encapsulate a therapeutically active agent to an NP to alter its PK. Nanoformulations can be used to overcome many of the limitations of conventional chemotherapeutic agents by enhancing the pharmacokinetics and distribution, without altering the molecular structure of the active compound. They can be designed to enable a medicine to reach deeper areas, to prolong the time of circulation to have greater accumulation and to target a specific tumor site [1, 6]. The incorporation of NPs in a pharmaceutical formulation can also alter the concentration–time profile of a drug, enabling its release (and exposure to diseased and/or healthy tissues) in a controlled and sustained manner. Currently, most nanodrugs are previously existing drugs conjugated to NPs to improve PK and pharmacodynamic properties. In many cases, the drug–nanoparticle conjugates use 'passive targeting,' that involves non-specific accumulation in diseased tissue, often tumors [13]. However,

Table 10.3. Advantages and disadvantages of nanomedicines.

Nanoparticle delivery system	Advantages	Disadvantages
Solid–lipid nanoparticles	Better pharmacokinetics with improvement in absorption	Drug loading capacity is lowered
Liposome	Reduce toxicity and increased targeted release	Toxicity by cationic lipids Rapid degradation of the nanocarriers by MPS
Polymeric	Chemical composition is very novel	Degradation of the carrier
Magnetic nanoparticle	Magnetic field influence	Potential material toxicity
Carbon nanotubes	Better penetration and localization	Material toxicity
Quantum dots	Fluorescent properties for imaging and drug tracking	Material toxicity

'active targeting' can be done by attaching ligands/molecules (e.g. proteins, antibodies, or small molecules) to the surface of the drug–NP conjugate that are designed to attach to receptors on specific cells. Active targeting can result in an increase in intracellular drug accumulation and uptake by the cells of the targeted tissue. Preclinical and clinical studies are necessary to characterize the PK, PD, bio-distribution, efficacy, and toxicity of nanopharmaceuticals to understand how they differ from conventional dosage forms [8]. These studies are needed because drugs formulated with nanoparticles can dramatically alter PK. For example, administration of a 50 mg m^{-2} dose of liposomal doxorubicin in humans was found to increase the area under the curve (AUC) by 300 fold and reduce clearance 250 fold compared to free drug [53, 57].

10.12 Conclusion

Breast cancer is the leading cause of death among women worldwide. Currently available treatment modalities have many limitations like poor bioavailability, lack of target specificity, low tumor permeability, higher volume of distribution, higher metabolic and elimination rate, owing to increased cytotoxicity in normal/healthy biological systems and development of resistance that ultimately leads to failure of therapy. Nanomedicines can achieve site-specific drug delivery which will result in more effective and less toxic chemotherapy. Breast cancer is a solid tumor type, which is associated with increased interstitial fluid pressure (IFP), this behaves like a strong barrier for the trans-capillary transport of the anticancer agents. Nanomedicine can easily surpass this disadvantage and provide good bio-distribution of the drug along with increased drug penetration at the tumor site. Since distribution of drugs is also dependent on the half-life, high circulation times achieved by drug conjugate with nanocarrier will increase the half-life, which results in uniform distribution of the drug. Knowing that accumulation of conventional chemotherapeutic agents are 10–20 fold higher in normal viscera than in breast cancer sites, involvement of nanomedicines becomes essential in targeted drug delivery. Nanoparticles are also used in imaging of the various tumors including breast cancer *in vivo* and also for developing cancer biomarkers. Nanomedicines in breast cancer therapy allow us to modify the pharmacokinetic and pharmacodynamic features of the conventional chemotherapeutic agents without altering the active pharmaceutical entity. Nanomedicine is more desirable than conventional medicines, since uptake of drug by breast cancer cells is more efficient, the dosage required is lowered, hence side effects will be minimized and drugs are released in a controlled manner. So, nanomedicines play an important role in improving the status and effectiveness of conventional breast cancer chemotherapy.

References

[1] Patra J K, Das G, Fraceto L F, Campos E V R, del Pilar Rodriguez-Torres M and Acosta-Torres L S *et al* 2018 Nano based drug delivery systems: recent developments and future prospects *J. Nanobiotechnol.* **16** 71

[2] Yezhelyev M V, Gao X, Xing Y, Al-Hajj A, Nie S and O'Regan R M 2006 Emerging use of nanoparticles in diagnosis and treatment of breast cancer *Lancet Oncol.* **7** 657–67

[3] Portney N G and Ozkan M 2006 Nano-oncology: drug delivery, imaging, and sensing *Anal. Bioanal. Chem.* **384** 620–30

[4] Waks A G and Winer E P 2019 Breast cancer treatment: a review *J. Am. Med. Assoc.* **321** 288–300

[5] Bowman K M and Kumthekar P 2018 Medical management of brain metastases and leptomeningeal disease in patients with breast carcinoma *Future Oncol.* **14** 391–407

[6] Sledge G W, Mamounas E P, Hortobagyi G N, Burstein H J, Goodwin P J and Wolff A C 2014 Past, present, and future challenges in breast cancer treatment *J. Clin. Oncol.* **32** 1979

[7] Baghaee P T and Donya A 2018 Review of studies in field of the effects of nanotechnology on breast cancer arXiv:181207494

[8] Wu D, Si M, Xue H-Y and Wong H-L 2017 Nanomedicine applications in the treatment of breast cancer: current state of the art *Int. J. Nanomed.* **12** 5879

[9] Franco M S, Roque M C and Oliveira M C Active targeting of breast cancer cells using nanocarriers *Mod. Applic. Pharm. Pharmacol.* **1** 2

[10] Sebastian R 2017 Nanomedicine-the future of cancer treatment: A review *J. Cancer Prev. Curr. Res.* **8** 00265

[11] Kaddi C D, Phan J H and Wang M D 2013 Computational nanomedicine: modeling of nanoparticle-mediated hyperthermal cancer therapy *Nanomedicine* **8** 1323–33

[12] Moss D M and Siccardi M 2014 Optimizing nanomedicine pharmacokinetics using physiologically based pharmacokinetics modelling *Br. J. Pharmacol.* **171** 3963–79

[13] De Jong W H and Borm P J 2008 Drug delivery and nanoparticles: applications and hazards *Int. J. Nanomed.* **3** 133

[14] Tang X, Loc W S, Dong C, Matters G L, Butler P J and Kester M *et al* 2017 The use of nanoparticulates to treat breast cancer *Nanomedicine* **12** 2367–88

[15] Ventola C L 2017 Progress in nanomedicine: approved and investigational nanodrugs *Pharm. Ther.* **42** 742

[16] Khan I, Saeed K and Khan I 2017 Nanoparticles: Properties, applications and toxicities *Arabian J. Chem.* at press

[17] Nel A, Xia T, Mädler L and Li N 2006 Toxic potential of materials at the nanolevel *Science* **311** 622–7

[18] Aillon K L, Xie Y, El-Gendy N, Berkland C J and Forrest M L 2009 Effects of nanomaterial physicochemical properties on *in vivo* toxicity *Adv. Drug Deliv. Rev.* **61** 457–66

[19] De Jong W H, Hagens W I, Krystek P, Burger M C, Sips A J and Geertsma R E 2008 Particle size-dependent organ distribution of gold nanoparticles after intravenous administration *Biomaterials* **29** 1912–9

[20] Alarifi S, Ali D and Alkahtani S 2017 Oxidative stress-Induced DNA damage by manganese dioxide nanoparticles in human neuronal cells *BioMed Res. Int.* **2017** 5478790

[21] Donaldson K, Stone V, Gilmour P, Brown D and MacNee W 2000 Ultrafine particles: mechanisms of lung injury *Philos. Trans. R. Soc. London, Ser.* A **358** 2741–9

[22] Verma A and Stellacci F 2010 Effect of surface properties on nanoparticle–cell interactions *Small* **6** 12–21

[23] Champion J A and Mitragotri S 2006 Role of target geometry in phagocytosis *Proc. Natl. Acad. Sci.* **103** 4930–4

[24] Park K H, Chhowalla M, Iqbal Z and Sesti F 2003 Single-walled carbon nanotubes are a new class of ion channel blockers *J. Biol. Chem.* **278** 50212–6

[25] Chen X, Zhang M, Gan H, Wang H, Lee J-H and Fang D *et al* 2018 A novel enhancer regulates MGMT expression and promotes temozolomide resistance in glioblastoma *Nat. Commun.* **9** 2949

[26] Hsiao I-L and Huang Y-J 2011 Effects of various physicochemical characteristics on the toxicities of ZnO and TiO_2 nanoparticles toward human lung epithelial cells *Sci. Total Environ.* **409** 1219–28

[27] Chithrani B D, Ghazani A A and Chan W C 2006 Determining the size and shape dependence of gold nanoparticle uptake into mammalian cells *Nano Lett.* **6** 662–8

[28] Lippmann M 1990 Effects of fiber characteristics on lung deposition, retention, and disease *Environ. Health Perspect.* **88** 311–7

[29] Hoshino A, Fujioka K, Oku T, Suga M, Sasaki Y F and Ohta T *et al* 2004 Physicochemical properties and cellular toxicity of nanocrystal quantum dots depend on their surface modification *Nano Lett.* **4** 2163–9

[30] Pietroiusti A, Massimiani M, Fenoglio I, Colonna M, Valentini F and Palleschi G *et al* 2011 Low doses of pristine and oxidized single-wall carbon nanotubes affect mammalian embryonic development *ACS Nano* **5** 4624–33

[31] Georgieva J V, Kalicharan D, Couraud P-O, Romero I A, Weksler B and Hoekstra D *et al* 2011 Surface characteristics of nanoparticles determine their intracellular fate in and processing by human blood–brain barrier endothelial cells *in vitro Mol. Ther.* **19** 318–25

[32] Fallahi F 2013 Bioaccumulation and neuroinflammation of gold nanoparticles in the central nervous system *MS Thesis* Wright State University

[33] Kohli A and Alpar H 2004 Potential use of nanoparticles for transcutaneous vaccine delivery: effect of particle size and charge *Int. J. Pharm.* **275** 13–7

[34] Griffitt R J, Luo J, Gao J, Bonzongo J C and Barber D S 2008 Effects of particle composition and species on toxicity of metallic nanomaterials in aquatic organisms *Environ. Toxicol. Chem.* **27** 1972–8

[35] Gurr J-R, Wang A S, Chen C-H and Jan K-Y 2005 Ultrafine titanium dioxide particles in the absence of photoactivation can induce oxidative damage to human bronchial epithelial cells *Toxicology* **213** 66–73

[36] Yang S-T, Wang X, Jia G, Gu Y, Wang T and Nie H *et al* 2008 Long-term accumulation and low toxicity of single-walled carbon nanotubes in intravenously exposed mice *Toxicol. Lett.* **181** 182–9

[37] Wick P, Manser P, Limbach L K, Dettlaff-Weglikowska U, Krumeich F and Roth S *et al* 2007 The degree and kind of agglomeration affect carbon nanotube cytotoxicity *Toxicol. Lett.* **168** 121–31

[38] Gupta A K and Gupta M 2005 Cytotoxicity suppression and cellular uptake enhancement of surface modified magnetic nanoparticles *Biomaterials* **26** 1565–73

[39] Yin H, Too H and Chow G 2005 The effects of particle size and surface coating on the cytotoxicity of nickel ferrite *Biomaterials* **26** 5818–26

[40] Sayes C M, Fortner J D, Guo W, Lyon D, Boyd A M and Ausman K D *et al* 2004 The differential cytotoxicity of water-soluble fullerenes *Nano Lett.* **4** 1881–7

[41] Risom L, Møller P and Loft S 2005 Oxidative stress-induced DNA damage by particulate air pollution *Mutat. Res.* **592** 119–37

[42] Kirchner C, Liedl T, Kudera S, Pellegrino T, Muñoz Javier A and Gaub H E *et al* 2005 Cytotoxicity of colloidal CdSe and CdSe/ZnS nanoparticles *Nano Lett.* **5** 331–8

[43] Mancini M C, Kairdolf B A, Smith A M and Nie S 2008 Oxidative quenching and degradation of polymer-encapsulated quantum dots: new insights into the long-term fate and toxicity of nanocrystals *in vivo* *JACS* **130** 10836–7

[44] Bang J, Chon B, Won N, Nam J, Joo T and Kim S 2009 Spectral switching of type-II quantum dots by charging *J. Phys. Chem.* C **113** 6320–3

[45] Dorfs D, Franzl T, Osovsky R, Brumer M, Lifshitz E and Klar T A *et al* 2008 Type-I and type-II nanoscale heterostructures based on cdte nanocrystals: a comparative study *Small* **4** 1148–52

[46] Nel A E, Mädler L, Velegol D, Xia T, Hoek E M and Somasundaran P *et al* 2009 Understanding biophysicochemical interactions at the nano–bio interface *Nat. Mater.* **8** 543

[47] Lin Y-S and Haynes C L 2010 Impacts of mesoporous silica nanoparticle size, pore ordering, and pore integrity on hemolytic activity *JACS* **132** 4834–42

[48] De Angelis F, Pujia A, Falcone C, Iaccino E, Palmieri C and Liberale C *et al* 2010 Water soluble nanoporous nanoparticle for *in vivo* targeted drug delivery and controlled release in B cells tumor context *Nanoscale* **2** 2230–6

[49] Park J-H, Gu L, Von Maltzahn G, Ruoslahti E, Bhatia S N and Sailor M J 2009 Biodegradable luminescent porous silicon nanoparticles for *in vivo* applications *Nat. Mater.* **8** 331

[50] Sager T M, Porter D W, Robinson V A, Lindsley W G, Schwegler-Berry D E and Castranova V 2007 Improved method to disperse nanoparticles for *in vitro* and *in vivo* investigation of toxicity *Nanotoxicology* **1** 118–29

[51] Jiang J, Oberdörster G and Biswas P 2009 Characterization of size, surface charge, and agglomeration state of nanoparticle dispersions for toxicological studies *J. Nanopart. Res.* **11** 77–89

[52] Zhang L-q, Zhang Y-k, Lin X-c, Yang K and Lin D-h 2014 The role of humic acid in stabilizing fullerene (C 60) suspensions *J. Zhejiang Univ. Sci.* A **15** 634–42

[53] Hou W-C, Westerhoff P and Posner J D 2013 Biological accumulation of engineered nanomaterials: a review of current knowledge *Environ. Sci.: Process. Impacts* **15** 103–22

[54] Hare J I, Lammers T, Ashford M B, Puri S, Storm G and Barry S T 2017 Challenges and strategies in anti-cancer nanomedicine development: an industry perspective *Adv. Drug Deliv. Rev.* **108** 25–38

[55] Lammers T, Kiessling F, Ashford M, Hennink W, Crommelin D and Storm G 2016 Cancer nanomedicine: is targeting our target *Nat. Rev. Mater.* **1**

[56] Skandalis S S, Gialeli C, Theocharis A D and Karamanos N K 2014 Advances and advantages of nanomedicine in the pharmacological targeting of hyaluronan-CD44 interactions and signaling in cancer *Adv. Cancer Res.* **123** 277–317

[57] Dawidczyk C M, Russell L M and Searson P C 2014 Nanomedicines for cancer therapy: state-of-the-art and limitations to pre-clinical studies that hinder future developments *Front. Chem.* **2** 69

[58] Wang R, Billone P S and Mullett W M 2013 Nanomedicine in action: an overview of cancer nanomedicine on the market and in clinical trials *J. Nanomater.* **2013** 1

[59] Hamidi M, Azadi A, Rafiei P and Ashrafi H 2013 A pharmacokinetic overview of nanotechnology-based drug delivery systems: an ADME-oriented approach *Crit. Rev. Ther. Drug Carrier Syst.* **30** 435

[60] Samarasinghe R M, Kanwar R K and Kanwar J R 2012 The role of nanomedicine in cell based therapeutics in cancer and inflammation *Int. J. Mol. Cell. Med.* **1** 133

[61] Bangham A D and Horne R W 1964 Negative staining of phospholipids and their structural modification by surface-active agents as observed in the electron microscope *J. Mol. Biol.* **8** 660–IN10

[62] Gregoriadis G, Leathwood P D and Ryman B E 1971 Enzyme entrapment in liposomes *FEBS Lett.* **14** 95–9

[63] Paliwal S R, Paliwal R, Agrawal G P and Vyas S P 2011 Liposomal nanomedicine for breast cancer therapy *Nanomedicine* **6** 1085–100

[64] Brown S and Khan D R 2012 The treatment of breast cancer using liposome technology *J. Drug Deliv.* **2012** 212965

[65] Rocha M, Chaves N and Báo S 2017 Nanobiotechnology for breast cancer treatment *Breast Cancer-From Biology to Medicine* (Books on Demand)

[66] Farias Azevedo L, da Fonseca Silva P, Porfírio Brandão M, Guerra da Rocha L, CFS A and da Silva S A S *et al* 2018 Polymeric nanoparticle systems loaded with red propolis extract: a comparative study of the encapsulating systems, PCL-Pluronic versus Eudragit® E100-Pluronic *J. Apic. Res.* **57** 255–70

[67] Stanczyk M, Dziki A and Morawiec Z 2012 Dendrimers in therapy for breast and colorectal cancer *Curr. Med. Chem.* **199** 4896–902

[68] Wall M E and Wani M C 1967 Recent progress in plant anti-tumor agents (153rd National Meeting of the American Chemical Society)

[69] Perez E A 1998 Paclitaxel in breast cancer *Oncologist* **3** 373–89

[70] Kim T-Y, Kim D-W, Chung J-Y, Shin S G, Kim S-C and Heo D S *et al* 2004 Phase I and pharmacokinetic study of Genexol-PM, a cremophor-free, polymeric micelle-formulated paclitaxel, in patients with advanced malignancies *Clin. Cancer Res.* **10** 3708–16

[71] Miele E, Spinelli G P, Miele E, Tomao F and Tomao S 2009 Albumin-bound formulation of paclitaxel (Abraxane® ABI-007) in the treatment of breast cancer *Int. J. Nanomed.* **4** 99

[72] Gabizon A, Shmeeda H and Barenholz Y 2003 Pharmacokinetics of pegylated liposomal doxorubicin *Clin. Pharmacokinet.* **42** 419–36

[73] Soundararajan A, Bao A, Phillips W T, Perez R III and Goins B A 2009 [186Re] Liposomal doxorubicin (Doxil): *in vitro* stability, pharmacokinetics, imaging and biodistribution in a head and neck squamous cell carcinoma xenograft model *Nucl. Med. Biol.* **36** 515–24

[74] Duggan S T and Keating G M 2011 Pegylated liposomal doxorubicin *Drugs* **71** 2531–58

[75] Bozzuto G and Molinari A 2015 Liposomes as nanomedical devices *Int. J. Nanomed.* **10** 975

[76] Batist G, Barton J, Chaikin P, Swenson C and Welles L 2002 Myocet (liposome-encapsulated doxorubicin citrate): a new approach in breast cancer therapy *Expert Opin. Pharmacother.* **3** 1739–51

[77] Marty M 2001 Liposomal doxorubicin (Myocet™) and conventional anthracyclines: a comparison *Breast* **10** 28–33

[78] Werner M E, Cummings N D, Sethi M, Wang E C, Sukumar R and Moore D T *et al* 2013 Preclinical evaluation of Genexol-PM, a nanoparticle formulation of paclitaxel, as a novel radiosensitizer for the treatment of non-small cell lung cancer *Int. J. Radiat. Oncol. Biol. Phys.* **86** 463–8

[79] Ranade A, Joshi D, Phadke G, Patil P, Kasbekar R and Apte T *et al* 2013 Clinical and economic implications of the use of nanoparticle paclitaxel (Nanoxel) in India *Ann. Oncol.* **24** v6–12

[80] Nigam A, Sharma A and Singh S K 2012 Nanoparticle paclitaxel (Nanoxel) as a safe and cost-effective radio-sensitizer in locally advanced head and neck carcinoma *J. Cancer Ther.* **3** 44

[81] Chawla S P, Bruckner H, Morse M A, Assudani N, Hall F L and Gordon E M 2019 A phase I-II study using rexin-g tumor-targeted retrovector encoding a dominant-negative cyclin G1 inhibitor for advanced pancreatic cancer *Mol. Ther. Oncolytics* **12** 56–67

[82] Gordon E M and Hall F L 2010 Rexin-G, a targeted genetic medicine for cancer *Expert Opin. Biol. Ther.* **10** 819–32

[83] von Minckwitz G, Huang C-S, Mano M S, Loibl S, Mamounas E P and Untch M *et al* 2019 Trastuzumab emtansine for residual invasive HER2-positive breast cancer *N. Engl. J. Med.* **380** 617–28

[84] Green M, Manikhas G, Orlov S, Afanasyev B, Makhson A and Bhar P *et al* 2006 Abraxane®, a novel Cremophor®-free, albumin-bound particle form of paclitaxel for the treatment of advanced non-small-cell lung cancer *Ann. Oncol.* **17** 1263–8

[85] Cirstoiu-Hapca A, Buchegger F, Bossy L, Kosinski M, Gurny R and Delie F 2009 Nanomedicines for active targeting: physico-chemical characterization of paclitaxel-loaded anti-HER2 immunonanoparticles and *in vitro* functional studies on target cells *Eur. J. Pharm. Sci.* **38** 230–7

[86] Tekedereli I, Alpay S N, Akar U, Yuca E, Ayugo-Rodriguez C and Han H-D *et al* 2013 Therapeutic silencing of Bcl-2 by systemically administered siRNA nanotherapeutics inhibits tumor growth by autophagy and apoptosis and enhances the efficacy of chemotherapy in orthotopic xenograft models of ER (−) and ER (+) breast cancer *Mol. Ther. Nucl. Acids* **2** e121

[87] Minakuchi Y, Takeshita F, Kosaka N, Sasaki H, Yamamoto Y and Kouno M *et al* 2004 Atelocollagen-mediated synthetic small interfering RNA delivery for effective gene silencing *in vitro* and *in vivo* *Nucl. Acids Res.* **32** e109

[88] Song G, Han L, Zou W, Xiao Z, Huang X and Qin Z *et al* 2014 A novel photothermal nanocrystals of Cu 7s 4 hollow structure for efficient ablation of cancer cells *Nano-Micro Lett.* **6** 169–77

IOP Publishing

External Field and Radiation Stimulated Breast Cancer Nanotheranostics

Nanasaheb D Thorat and Joanna Bauer

Chapter 11

Clinical and preclinical trials of breast cancer

S S Rohiwal and A P Tiwari

Breast cancer (BC) is the second-most common cause of cancer-related deaths and it is caused by metastatic BC (MBC). This chapter stresses the inadequate success of traditional therapies as well as the use of nanomedicine for treating BC. An overview of nanomedicines applied in BC therapy is summarized with their present status, which is attaining attention in the pre-clinically as well as clinically applied landscape. Here we highlight the biology of MBC, the nanomaterial used for BC. Furthermore, the types, strategies and challenges in preclinical and clinical trials are explored in detail.

11.1 Introduction

Breast cancer is a highly heterogeneous disease. Consequently, the classification of breast cancer is very complicated to understand [1]. With the advancement of the disease, the primary tumor within the breast (stage 1) frequently spreads to the tissues and lymph nodes nearby (stage 2–3) or the distant organs (distant metastasis, stage 4). Lung, bone, liver and brain are the frequent sites of breast cancer metastasis [2]. Regardless, once cancer metastasizes, the value of many standard treatment options will considerably diminish as they are either not suitable for systemic use or their effectiveness against metastasized, high-grade cancer is far from optimal [3]. The need for a more robust, accurate, cost-effective, and rapid drug development mechanism is clearly evident. The value of many new drugs (in the broadest context) with respect to increasing life expectancy remains somewhat controversial [4]. Moreover, the emergence of substantially more effective and less toxic new breast cancer therapies has been slow. To some degree, this may reflect the complexity of biological signaling in cancer cells [5]. Rapid growth in nanotechnology towards the development of nanomedicine product shows great promise to improve therapeutic strategies against cancer. Nanomedicine products represent an opportunity to

doi:10.1088/2053-2563/ab2907ch11

achieve sophisticated targeting strategies and multi-functionality. They can improve the pharmacokinetic and pharmacodynamic profiles of conventional therapeutics and may thus optimize the efficacy of existing anticancer compounds [6]. Preclinical studies provide the opportunity to measure various pharmacokinetic/pharmacodynamic properties of a drug (absorption, distribution, metabolism, elimination, toxicity). For molecularly targeted therapies, measuring tissue concentrations and whether the test drug successfully modulates the target (or a surrogate biomarker predictive of drug action) are important. These data can also be used to guide their incorporation into the first-in-human studies, a major goal in phase 0 trials for molecular proof of concept [7]. Once a drug is approved for a specific disease, it may be used for other indications. This off-label usage is increasing and is potentially problematic [8], particularly if applied without adequate prior experimentation. The failure (poor activity, unpredictable toxicity) of some new combinations may reflect a lack of rigorous preclinical investigation [9]. Since there is no *a priori* requirement for animal efficacy studies, clinical trials may be designed primarily from *in vitro* data where adverse pharmacokinetic, pharmacodynamic, toxicologic, and/or molecular feedback signaling interactions may be inadequately modeled. Adverse interactions could be missed, such as increased toxicity and/or reduced therapeutic efficacy [10].

This chapter focuses on common nanomaterials having potential in cancer therapeutics. Also, the need of preclinical and clinical studies towards development of effective breast cancer therapy is covered. The existing challenges with the preclinical and clinical studies are briefly discussed. A proper understanding of the origin of breast cancer, nanocarriers employed for the therapy and their effective clinical application is needed for the researchers for successful translation of research into medicine for effective treatment of breast cancer.

11.2 Biology of breast cancer metastasis

Metastasis is a pathogenic agent's spread from an initial site to a different site within the host's body spread as such by a cancerous tumor, the newly pathological sites are then metastases [11, 12]. Metastatic cells are a specified subset of tumor cells within a primary tumor mass that have acquired the ability to complete the multistep metastatic cascade. Breast cancer is the most common cancer in females worldwide, with an incidence rate of over 1.6 million cases per year [13]. The process of metastasis to distant organs from the initial site is not random; relatively metastasis is highly reliant on the nature of the disseminated tumor cells and microenvironment of the secondary organ [14]. It is a complex and heterogeneous type of cancer, characterized by the occurrence of multiple sequential molecular changes, which can be used as diagnostic and prognostic markers of the disease [15]. These steps are instigated with dysregulation of normal epithelial cell proliferation to different hyperplasia, carcinoma *in situ*, invasive carcinoma and finally to metastatic disease (figure 11.1) [16, 17]. The most common reason of hereditary breast cancer is an inherited mutation in theBRCA1 or BRCA2 gene, which are tumor suppressor genes involved in such crucial functions as DNA repair [16, 17].

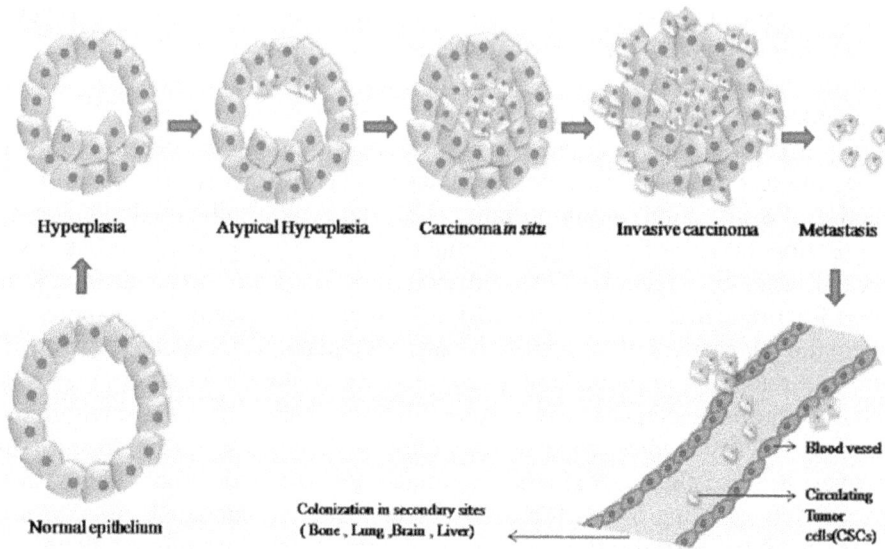

Figure 11.1. Progression of breast cancer through various stages and its metastasis [18], reproduced with permission.

Concisely, cancer cells in solid tumors recruit inflammatory cells leading to inflammatory stroma. This in turn releases epithelial to mesenchymal transition (EMT) inducing growth factor signaling, such as TGF-β, STAT, hypoxia, and tumor-stromal cell communication. These complex sets of events disrupt mainly the tight junctions, desmosomes and promote the transition of epithelial to transient mesenchymal cells via type 3 EMT [19]. The invasiveness, motility, and anoikis resistance of the metastatic breast cancer cells are derivative from oncogenic epithelial to mesenchymal transition (EMT) through the loss of E-cadherin expression with the increase in fibroid morphology. Several biological processes and genetic mutations lead to the appearance of breast cancer, and sensitivity to various drugs such as hormone receptor, vascular endothelial growth factor (VEGF), human epidermal growth factor receptor 2 (HER2), EGF, cyclin-dependent kinase 4/6 (CDK4/6) and mechanistic target of rapamycin (mTOR). Additionally, the non-genetic risk factors of breast cancer include race, ethnicity, benign breast conditions, lobular carcinoma *in situ* or lobular neoplasia, exposure to diethylstilbestrol, personal behavior-related risk factors of breast cancer lifestyle, birth control and contraceptives, significant overweight or obesity, excessive alcohol consumption, not breast feeding and lack of physical activity. With the mesenchymal to epithelial transition (MET) colonization of circulating tumor cells at the distant organs was initiated [19]. Though, it is still uncertain about the preferential colonization of breast cancer cells at distant organs.

11.3 Nanomaterials used for breast cancer

Due to the ability to tune to size, shape, composition and surface functionality, nanoparticles provide a platform as a significant resource for nanomedicine. The development of a wide-ranging nanoparticle platform can be broadly categorized as lipid based nanocarriers, polymeric, inorganic and hybrid. Furthermore, nanomedicine therapeutics is explored in phase 1 trials in patients with solid tumors, and specific cancer indications are also explored in advanced phases 2 and 3 clinical trials. The most used therapeutic measure for metastatic breast cancer is palliative in nature and less towards the existence rate of the patient. The conventional therapies such as rapid clearance from the circulatory system, unfavorable distribution and insufficient dosage at tumor site have many drawbacks. These drawbacks can be overcome by nanoparticle based drug delivery systems, which targets the tumor cells either by passive (EPR effect) or active (ligand) targeting [20].

11.3.1 Lipid based nanocarrier

Liposomes are one of the most popular and diversified, versatile candidates that have their own unique properties. Liposomes are unilamellar lipid bilayers with an aqueous core. These have numerous advantages, namely the ability to incorporate hydrophilic and hydrophobic compounds, biocompatibility, and biodegradability in terms of their main constituents and low toxicity. Numerous lipid based nanosytems are available in the market for clinical trials, they have also been explored as delivery systems of great diagnostic agents including 14C isotope, quantum dots, 64Cu, gadolinium (Gd) based contrast agents, fluorescent probes and SPIONS [20–26]. Because of these factors, liposomes emerge as a highly promising theranostic tool, with a wide spectrum of clinical applications in cancer management. Several researchers, have studied different animal models and tried different liposomal based nanosystems to cure breast cancer. Feng *et al* in 2017 synthesized long-circulating liposomes, co-encapsulating a hypoxia-activated prodrug, AQ4N and the photosensitizer hexadecylamine conjugated chlorine6 (hCe6). After chelating with 64Cu isotope, liposomes displayed properties suitable for positron emission tomography to treat solid tumor microenvironment particularly for hypoxia condition [27]. These results suggested that these nanosystems have the ability to accumulate at the tumor site which remarkably enhances the inhibition of tumor growth via a synergistic effect attained by sequential PDT and hypoxia-activated chemotherapy [27]. He *et al* synthesized gonadorelin-functionalized liposomes carrying both magnetic iron oxide nanoparticles (MIONs) with conventionally used chemotherapeutic mitoxantrone. These NPs effectively inhibit tumor progression, whereas providing a real-time and non-invasive visualization of the therapeutic protocol [28]. Ma *et al* developed doxorubicin loaded into folate receptor-targeted long-circulating liposomes embedding conjugated polymer dots. These are a class of fluorescent macromolecules that have photochemical and electroluminescence features including light harvesting properties, high brightness and good biocompatibility [29].

Similarly, Dai *et al* synthesized doxorubicin loaded liposomes which had been surface modified with a cyclic octapeptide further targeting integrin α3. This is a highly expressed receptor on the surface of the triple negative breast cancer cell line MDA-MB-231. Dia *et al* successfully accomplished tumor site by accumulating liposomes and non-invasive real-time monitoring [30].

11.3.2 Polymeric NPs

Polymeric based nanoparticle systems have been extensively used as nanocarriers for certain drugs. These polymeric particles are used because of their unique properties like increasing the absorption, distribution and bioavailability of active substances; they also have high active substance-loading capacity. Therefore, they enhance the intracellular distribution of the active substance and protect the active substance entrapped in the solid matrix without degradation [31]. These NPs have better penetration ability to colloids such as biological barriers, tissues and cellulosic membranes. Due to their small size they have to escape from phagocytes by virtue of their small size, which enhances circulation times with prolonged and controlled release of drug molecules. There are different types of natural polymers such as albumin, alginate, gelatin; alginate, PCL (polycaprolactone), PLGA (poly (lactic-*co*-glycolic acid)) and PVA (poly (vinyl alcohol)) are commonly used for drug delivery systems. In the treatment of breast cancer, many nanodrug delivery systems have been developed using polymeric biodegradable materials for controlled release [32]. Cyclodextrin (CD) is one of the most popular polymers for a gene delivery system reported for the delivery of plasmid DNA in 1999. Soon thereafter CD-mediated NPs advanced into clinical trials, which was the first targeted siRNA delivery system entering clinical trials for cancer treatment [33].

11.3.3 Inorganic nanoparticles

Inorganic NPs include both metal as well as metal oxides. Examples of metal oxide NPs includes silver (Ag), copper oxide (CuO), iron oxide (Fe_3O_4), zinc oxide (ZnO), and titanium oxide (TiO_2). Numerous inorganic NPs like iron oxide, gold nanoshell and hafnium oxide NPs have been tested on cancer patients. Iron oxide NPs or magnetic NPs are already commercialized in Europe for the treatment of glioblastoma. These NPs can be managed with the help of a magnetic field. These are established from chemical compounds from a magnetic material like nickel, cobalt and iron [34] used in magnetic resonance imaging (MRI), adhering to target tissue and drug release system technology [35]. In recent times, novel inorganic NPs like nano-diamond and grapheme have also been explored in cancer treatment [36, 37]. Furthermore, Semkina *et al* developed a new system based on vascular endothelial growth factor (VEGF) targeted magnetic NPs for cancer treatment. These VEGF-targeted NPs which were loaded with the anticancer agent doxorubicin displayed 50% enhancement in the 4T1 tumor-bearing mice survival rate. Gold NPs are effective and appropriate inorganic constructs for drug release, gene therapy and imaging applications. They are 130 nm in diameter and are biocompatible and non-toxic in nature [38]. These NPs are flexible and easily conjugated with different biomolecules and drugs making them more successful for dealing with molecular pathways involved in heterogeneous cancers including melanoma. Silva *et al* in 2016 designed gold NPs coated with hyaluronic and oleic acids (HAOA), conjugated with epidermal growth factor (EGF) for a photothermal strategy in cutaneous melanoma. This targeted therapy established a potential method to destroy melanoma cells at their initial stage by photo activation and thermoablation, considering a neoadjuvant setting.

These spherical NPs were actively internalized by tumor cells through EGFR-mediated endocytosis in *in vivo* assays [39]. Balakrishnan *et al* targeted gold NPs to EGFR/VEGFR2 signaling pathway for breast cancer. The results showed the suppression of multiple proteins which are snail, PI3K, slug, vimentin, N-cadherin, Akt and p-GSK3b of these cells by AuNPs–Qu-5 NPs [40]. These gold NPs were further coupled with cytotoxic abietane (i.e. 6,7-dehydroroyleanone) adapted for polymeric NPs loaded with parvifloron D resulting in targeting and reducing the growth of melanoma cells [41]. Lee *et al* synthesized folate-conjugated, doxorubicin (Dox) loaded poly-(lactic-*co*-glycolic acid) (PLGA) in gold half-shell NPs, showing effective elimination of tumors in target tissues through NIR-responsive manner. The gold nanorods are effective with an NIR photo-to-thermal transducer model which also denatures the DNA double helix using NIR light to increase the release of antitumor drug at a particular site for chemotherapy [42]. Briefly, to conclude, these NPs are flexible and adaptable for further surface modification by conjugating with photosensitizer to a triple-action nanotheranostic system (figure 11.2).

11.4 Concept of preclinical trials

11.4.1 Strategies and preclinical animal models

Many nanoformulation based chemotherapeutics are available in the market, but there is a search for new formulations that can target more specifically metastatic tumors. This is because cancerous cells have to invade the ECM of the tumor microenvironment, enter the blood circulation, reach the secondary site to form micrometastasis and finally proliferate to produce macro-metastasis to colonize a secondary site. Hence, numerous approaches have been exploited to target the molecules accountable for metastasis in the control of metastatic breast cancer. Basically there are three strategies to tackle control of metastatic breast cancer: invasion inhibition strategies, strategies against circulating tumor cells (CTCs) and strategies to dissuade colonization of CTCs.

- Invasion inhibition strategies:

    In an attempt to invade and enter into the blood circulation, tumor cells need matrix metalloproteinases (MMPs) particularly MMP-2 and MMP-9 to damage the surrounding ECM components and nanoformulations that aim for these MMPs to have inhibited tumor cell invasion and migration [43]. A plant derived component named Silybum marianum loaded lipid NPs decreases MMP-9 and Snail (EMT regulator) levels [34]. Gadolinium metallo-fullerenol NPs (f-NPs) down regulated the mRNA expression of both MMP-2 (~40%) and MMP-9 (90%) [44]. Likewise, both these nano-particle complexes decreased pulmonary metastasis in mammary tumor model [43, 45]. An enzyme lysyl oxidases (LOX) over expressed in the tumor

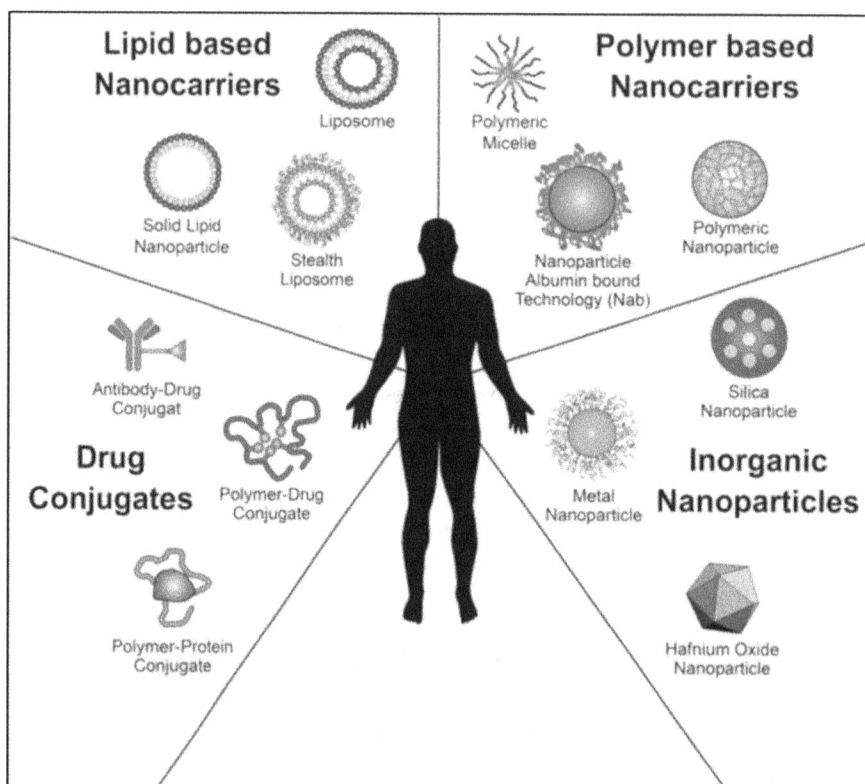

Figure 11.2. Schematic illustration of established nanotherapeutic platforms used in clinical cancer care [37], reproduced with permission.

microenvironment also effects tumor cell invasion and migration. But with disparate MMPs, this enzyme does not damage the tumor ECM, but increases the rigidity by cross-linking collagen and elastin, thus preferring the migration of tumor cells. Granchi *et al* 2009 reported the suppression of mammary tumors and its metastasis by inhibiting LOX activity by applying high doses of antibodies or chemical inhibitors. The use of LOX antagonistic substances reports side effects such as lathyrism a neurotoxic disease, which bounds its potential application [46]. Kanapathipillai *et al* synthesized LOX antibody coated PLGA NPs (LOXabNPs) which is five-fold more effective than soluble LOX antibody *in vitro* and also capable of inhibiting collagen cross-linking in 4T1 cell bearing orthotopic mammary tumor model at 5 mg kg^{-1} concentration [47]. Several EMT regulators such as Snail, Slug, Twist and Notch Metastasis can inhibit metastasis. Anti-metastasis therapy retaining RNA interference is found to be one of the promising strategies for treating breast cancer metastasis [44, 48]. Similarly, Zang *et al* have conjugated gold nanorods with PAR-1siRNA (AuNRs@PAR-1siRNA) and treated cells showed lesser PAR-1 mRNA (61.8%) and protein (56.8%) expression with very low level of migration. Herein, the degree of inhibition

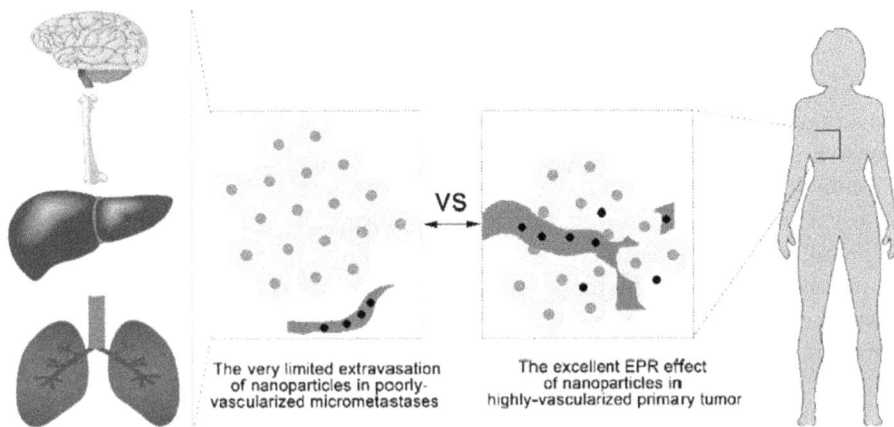

Figure 11.3. The limited enhanced permeation and retention (EPR) effect in micrometastases.

was proportional to the degree of gene silencing confirming the importance of PAR-1 expression in the metastatic ability of tumor cells [49] (figure 11.3).

- Strategies against circulating tumor cells (CTCs):

    CTCs and their detection in systemic circulation are associated with poor survival, which makes them a potential target for anti-metastatic therapy, Palange *et al* have synthesized lipid–polymer NPs encapsulating curcumin (NANOCurc) to reduce the adhesion (~70%) of CTCs (MDA-MB-231) to HUVECs by down-regulating the expression of ICAM-1 and MUC-1 as the adhesion of CTCs to vascular endothelial cells has been found to be critical in migration towards secondary sites [50].

    Xu *et al* used another strategy to control the metastasis by disturbing the microvasculature since tumor vasculature is critical for the CTCs to reach the secondary site [51]. Wang *et al* have used a combination of gold nanorods with doxorubicin (GNR@ DOX) against lung metastasis of breast cancer cells upon laser irradiation. The results showed that about 90% reduction in micro vessel density in metastatic tumor model were obtained due to the inherent property of gold nanorods to emit heat when subjected to NIR [52]. Furthermore, Carneiro *et al* treated citrate and rhodium(II) citrate coated maghemite nanoparticles (Magh-Rh2(H2cit)4) in orthotopic 4T1 cells bearing mouse model that were found to decrease CD31 expression (angiogenesis marker) levels at the tumor site, which was directly correlated with reduced microvascularization resulting in tumor regression *in vivo* [53].

- Strategies dissuade colonization of CTCs.

Another strategy to control the colonization of CTCs at secondary sites is to inhibit lymph node metastasis as CTCs reach the regional lymph nodes for migration towards the secondary organs like bone, lungs, brain and liver [54]. Liu *et al* synthesized expansile nanoparticle formulation of paclitaxel (Pax-eNP) for local administration in a murine model bearing antagonistic triple negative breast cancer cells, which has been observed to decrease the primary tumor load and prevent lymph node metastasis by the retention of the same in regional lymph nodes for several days [55]. Furthermore, Zubris *et al* and Lui *et al* have synthesized a pH responsive polymer to release its contents at acidic pH (pH b 5) and found it in endosome of tumor cells and in tumor microenvironment of some solid tumors [55, 56]. Murakami *et al* synthesized an adocetaxel formulation called Cellax which was capable of interacting with the α-SMA expressing stromal cells and eventually reducing them leading to increased tumor permeability, perfusion and reduced lung metastasis [57]. This micelle is 10–30 nm in diameter and exhibits maximal accumulation in the lymph nodes. Furthermore, it also inhibited liver and lung metastasis in animal models [58].

Various mouse models are widely used for research studies of disease progression and the development of new therapies [59]. For preclinical studies rat and rabbit models are most commonly used. Moreover, standard tumor models include subcutaneous xenografts of human cell lines or explants, orthotopic xenografts, and genetically engineered mouse models. Though these models are valuable for preclinical studies, the differences in physiology can lead to differences in circulation and tumor accumulation compared to humans [60]. Additionally, another innovative lipid-based approach in preclinical testing is the squalenoylation technology. Various anticancer drugs, e.g. doxorubicin, are chemically linked to the natural lipid squalene, these conjugates are self-organized in water to nanoassemblies with exclusive characteristics such as high drug cargo and prolonged systemic circulation. Squalene–doxorubicin nanoassemblies showed a 95% reduction of pancreatic tumors in xenograft mice models, compared to 29% achieved by free doxorubicin [61]. Furthermore, in numerous preclinical experiments, the surface of liposomes was coated with anti-EGFR, anti-HER2, anti-VEGFR2, or other antibodies [62]. In mouse models of human cancer, these nanocarriers were shown to transport cytotoxic compounds precisely to cells expressing the target antigen. Cellular uptake was determined by the interaction of the target antigen on the cell surface and the antibody coating the surface of the liposome. Tumors treated with anti-VEGFR2-targeted and doxorubicin-loaded ILs shrank to 1/6 of the size of those treated with an identical dose of non-targeted liposomal doxorubicin [62]. This demonstrates the potency of targeted lipid nanocarriers compared to their non-targeted counterparts.

11.4.2 Challenges

To achieve the beneficial goal, it is necessary to understand the complexity and interplay between the biology of metastasis of breast cancer (MCB) metastasis and the fabricated nano-system.

- Challenge 1—limited access to micrometastases: identification and delivery of therapeutic agents to metastatic lesions is a much bigger challenge compared to the eradication of primary tumors. There are three distinct kinds of niches that coexist within the metastatic lesion: single dormant cells, dormant micrometastases, and actively growing and vascularized metastases, once the cancer cells are dispersed into a secondary organ, which may be detected by their effects on vital organ function or by many imaging instruments [63]. So far, few studies have investigated the effects of drugs on dormant cancer cells. In models of BC metastases to the liver, cytotoxic chemotherapy effectively inhibited the development of metastases, but had little effect on the elimination of dormant cancer cells disseminated to the liver [64]. Also, no effect on late-developing BC metastases, apparently due to a dormant state of the cancer cells at the time of treatment has been proven in another study that chemotherapy treatment given at the early stage of metastasis had. Therefore, dormant cancer cells remain an elusive target for therapy using small-molecular drugs and provide a greater challenge for nanomedicine which uses a much larger size to transport therapeutic agents. When used to target micrometastases and poorly-vascularized metastases, nanotherapeutics also encounters the same problem. The primary tumor vasculature can have fenestrae up to 600 nm and tumor-associated lymphatic drainage is often poor, therefore, some solid tumors will exhibit an EPR effect for nano-vehicles [65]. Yet, the EPR effect is only present in tumors of more than ~100 mm^3 in volume, and fails in the unvascularized metastases [66]. This significantly hinders the application of nanotherapeutics in the treatment of MBC metastasis. Mostly, dispersion in the body, the smaller size, and the presence of less vasculature than primary tumors are unique physiological barriers of metastases, which make metastases less accessible to molecular and NPs agent.

- Challenge 2—the anti-apoptosis characteristic of cancer stem cells (CSCs): generally, undifferentiated CSCs seem to coexist with the fully differentiated end stage cells and partially differentiated transit amplifying cells simulta-neously in breast carcinoma [67]. One feature that helps to define a CSC is its unique ability to resist apoptosis. Metastasis is a truly inefficient process; despite the large number of cells migrating away from the primary tumor, only a few successfully colonize a secondary organ. One proposed reason is provided by the CSC model, which hypothesizes that colonization, can only be achieved by CSCs, which are rare within BC primary tumors [66]. As few as 100–200 breast CSCs have been shown to develop tumors when being inoculated into the mammary fat pad of immune compromised mice. Furthermore, CSCs are one of the major causes of treatment resistance in MBC patients. This resistance presumably stems from two sources: (1) CSCs often retreat from the activated G1/S phase into G0 phase; and (2) the close association between CSCs and the mesenchymal cancer cells. Mesenchymal cells have been shown to be the product of EMT and typically exhibit intrinsic drug resistance [68]. In support of this, various studies have identified that the drug-resistant cancer cell subtype often exhibits a more mesenchy-mal-like phenotype [69].

11.5 Concept of clinical trials

Over the years, breast cancer treatment has significantly improved due to lessons learned through clinical trials. Clinical trials test the safety and benefits of new treatments as well as new combinations (or new doses) of standard treatments. They can also study other parts of care including risk reduction, diagnosis and screening. To date, an abundance of studies implementing high-throughput molecular profiling techniques such as gene-expression profiling and next-generation sequencing (NGS) have been conducted in the setting of breast cancer, resulting in molecular fragmentation of this group of diseases [70]. Specifically, in order to facilitate the successful clinical development of these anticancer drugs within specific molecular niches of cancer diagnoses, there have been developed new, innovative trial designs that could be classified as follows: (1) longitudinal cohort studies that implement (or not) 'nested' downstream trials, (2) studies that assess the clinical utility of molecular profiling, (3) 'master' protocol trials, (4) 'basket' trials, (5) trials following an adaptive design

11.5.1 Types

Longitudinal cohort studies with or without downstream clinical trials
This is a study design corresponding to the enrollment of patients in a program of extensive molecular profiling coupled with prospective follow-up for the clinical outcome of the enrolled patients; they can be treated either with standard care, or be directed to downstream clinical trials [71].

Studies assessing the clinical utility of molecular profiling
The new powerful molecular profiling tools have provided much information about the molecular landscapes of breast cancer and they promise to guide patients to targeted treatment based on the molecular profile of their disease. However, their clinical utility still needs to be proven, with a newly emerged study design specifically trying to address this issue. In particular, this study design attempts primarily to answer the question whether there is clinical benefit for patients with cancer to receive targeted agents guided by molecular profiling as compared to conventional treatment. It must be noted that such studies are 'proof-of-concept' not assessing individual treatment options, but the whole concept of molecular profiling guiding treatment selection. Such a study has been reported by Tsimberidou *et al* through a non-randomized phase 1 clinical trial program with promising results conducted at the University of Texas MD Anderson Cancer Center. In the context of this program, 1144 patients with advanced solid tumors of several different histologies were enrolled, with their tumor tissue being molecularly profiled and then directed to one of several phase 1 trials assessing targeted compounds: patients that had one molecular aberration and received targeted treatment based on the molecular profile of their disease ($n = 175$) showed increased overall response rate (ORR, 27% versus 5%, $P < 0.0001$), longer time-to-treatment failure (TTF, 5.2 months versus 2.2 months, $P < 0.0001$) and longer overall survival (OS, 13.4 versus 9.0 months, $P = 0.017$) as compared to patients receiving conventional treatment ($n = 116$) [72].

Master protocol trials

This type of study design enables the assessment of several targeted agents in parallel. After one molecular prescreening step, patients are directed to one of several downstream treatment arms, receiving a targeted agent matching the molecular profiles of their disease. The main advantage of this innovative type of study design is the reduction in the percentage of screening failure rate, since patients undergoing the molecular prescreening can have more options for subsequent matched targeted treatment in one of the downstream trials. Additionally, there is increased efficiency in some of the important preparatory steps to activate such a trial, such as the need to have one common ethics committee approval that will allow the clinical assessment of several different investigational agents. An important initiative implementing study design, recently launched by the NCI (National Cancer Institute) in collaboration with SWOG, the Foundation for the National Institutes of Health, and the Friends of Cancer Research and the FDA, is the master protocol for second-line treatment of patients with squamous NSCLC; according to this protocol NGS of tumor samples will be performed using a panel of 250 cancer-related genes and patients will be guided to one of the five integrated study strata, with a total of 10 treatment arms. Within each one of these strata, a phase 2/3 study design has been incorporated, with predefined thresholds of efficacy that need to be demonstrated prior to the phase 3 component activation. In the setting of BC, BIG is currently designing such a master protocol trial, aiming to assess several molecularly targeted agents for patients with aggressive metastatic triple negative breast cancer (TNBC) [73].

Basket-trials

This is an innovative, histology-independent trial design, where patients with cancer diagnoses of different histologies can be enrolled in the study protocol based on the presence of a specific molecular aberration. There is a currently ongoing clinical trial that aims to develop a small molecule HER2 blocking agent within patients with ERBB2 mutated cancers that exemplifies this approach. The main disadvantage in this innovative design is a biology-driven one; in particular this is the issue of the potentially different functional outputs that a specific molecular aberration could have among different types of cancer. This has been reported in studies documenting lack of antitumor efficacy of vemurafenib, a BRAF small molecule inhibitor, in the setting of BRAF mutated metastatic colorectal cancer; these findings are in direct contradiction with the dramatic antitumor activity seen among patients with metastatic melanoma bearing the V600E BRAF mutation [73].

Adaptive-trials

Another innovative study design that has entered the arena of clinical trials in oncology are the so-called adaptive trials. This type of study design corresponds to trials that allow modifications in the study during its conduct, related among other parameters to the study population, or the statistical framework. The initial conceptual development of this study design dates back to the 1970s, when the concept of adaptive randomizations was firstly introduced. Adaptive trials can be conducted in different phases, namely phase 1 adaptive dose finding studies, or phase 1/2 adaptive seamless studies for early clinical development of experimental compounds, as well as phase 2/3 adaptive seamless trials for late clinical development. Such a study design has been recently exemplified by the BATTLE (Biomarker-integrated Approaches of Targeted Therapy for Lung Cancer Elimination)-1 and -2 clinical trials, conducted among patients with metastatic NSCLC as an effort to personalize treatment of patients with lung cancer, or the I-SPY (The Investigation of Serial Studies to Predict Your Therapeutic Response with Imaging and Biomarker Analysis)-1 and -2 trials conducted among patients with early-stage BC, in the neoadjuvant setting. Studies following an adaptive design are more laborious in the sense that clinical trials' simulations are needed, with different statistical scenarios needing to be developed by highly skilled biostatisticians and trialists. This characteristic could discourage expansion of the use of this innovative study design; however, recent guidance provided by regulatory authorities, i.e. FDA and EMA, about the adaptive trial designs they deem acceptable supports the further embracement of adaptive trial designs [74–76].

11.5.2 Challenges

Nanotheranostics approaches are finding attention in the development of frontier cancer treatment. However, only very few nanocarrier-based cancer therapeutics have effectively entered clinical trials. Thus, it is important to address the challenges nanomedicine produces for clinical use.

In general, the main physicochemical features of nanocarriers are structure, composition, size, surface properties, porosity, charge, and aggregation behavior. Variability within these properties makes it difficult to characterize nanomedicine products before and after administration. Polydispersity (PD) is a measure of heterogeneity of particles with regard to size, shape, or mass. It plays a major role in characterizing nanocarriers because even small variations in PD and physico-chemical characteristics can cause dramatic changes in secondary properties such as biocompatibility, toxicity, and *in vivo* outcomes. Therefore, nanomedicine products should be characterized on a batch-to-batch basis, using multiple methods. The majority of nanomedicine drugs are formulated in aqueous buffers representing the ionic strength of physiological pH. However, nanocarriers may interact with other biological fluids (e.g. blood serum) or biomolecules (e.g. proteins), which may lead to aggregation or agglomeration of particles. Such interactions can significantly alter the function of nanomedicine compounds in biological systems. As discussed earlier, the final form of nanomedicine products should be fully characterized under clinically relevant conditions. Characterization of stability and storage aspects (shelf-life) of nanomedicine products is also challenging. Biodegradable nano-materials such as polymers have been increasingly used in the development of nanomedicine products. After polymer degradation and when reaching biological fluids such as blood serum, nanocarriers can again change their physico-chemical properties such as size, drug loading, and release profile. This may have an impact on their performance *in vivo*. Similarly, storage in aqueous solutions including buffers and even in a lyophilized powder form can alter the properties of nano-carriers. There is a substantial need to improve quality assessment of nanomaterials by developing well-defined and reproducible standards. Moreover, *in vitro* and *in vivo* models accurately representing the clinical setting must be developed [77–79].

Another challenge is to address safety concerns. Widespread use of nanoparticles makes it necessary to address toxicity issues for human health and the environment. The nano-scale dimension of nanomedicine products is similar to that of intra-cellular organelles or biomolecules involved in cell signaling. Several studies have demonstrated that nanoparticles may be associated with detrimental biological interactions. This has led to the emergence of nanotoxicology as an independent field of research [80]. Currently used toxicity assays for nanomaterials are the same as those used for classical drugs. Thus, assessment of nanoparticle toxicity may be inadequate at present and the development of complementary toxicity assays for nanomedicine compounds should be encouraged. Multiple factors modulate the toxicity of nanomaterials. Properties such as size, shape, surface area, surface charge, porosity, or hydrophobicity affect the behavior and performance of nano-medicine drugs at the nano-bio interface. The many variables involved hamper the full toxicological characterization of nanomedicine products. Acute toxicity of nanomedicine compounds usually comprises complement activation, hemolysis, inflammation, oxidative stress, or impaired mitochondrial function. Analyzing chronic toxicity is more demanding, and data are largely missing. In clinical development, risks may be minimized by combining more advanced or even predictive diagnostic tools with novel targeting strategies. Thus, 'safe-responders' may be identified, and individualized cancer therapy can be realized. Theranostic approaches offer a huge potential in this direction. Apart from medical issues, environmental concerns have also been expressed. The biomedical, chemical, and cosmetic industry is increasingly using nanomaterials in large-scale production. This renders the monitoring of environmental or occupational exposure to nanoparticles and its consequences more challenging but also more important [80–83].

Another big challenge is to address regulatory issues. A number of nanomedicine products for cancer therapy have been approved by the FDA and European Medicines Agency (EMA). Despite the fact that they fulfill the current safety requirements set forth by these agencies, these were still not implemented specific guidelines for drug products containing nanomaterials by FDA, EMA, and other regulatory bodies [84].

11.6 Perspective

Several of the nanomedicine formulations in expansion and preclinical and clinical trials are designed, where expansion of the therapeutic window can address issues with permissibility or sub-optimal target exposure that may limit the capability to develop the drug into a viable and effective product. Future prospects for nano-medicines are looking in the direction of delivering the next generation of drugs: molecularly targeted agents, toxin-like agents that induce cell death, DNA and RNA based therapeutics, drug, peptides, combinations, etc. The efficacy of drug or gene delivery to a tumor site is dependent on the physico-chemical properties of the delivery platform. The absence of consistency in preclinical trials of nanoparticle-based delivery systems has prohibited systematic comparison of these studies and has been an impediment to developing design rules for new systems or specific applications. Out of a large number of preclinical and clinical trials, astonishingly few report quantitative data on parameters that would be beneficial in developing design rules for nanomedicines. The poor experimental design and inconsistency of experimental conditions also contribute to slow development of the field and the lack of clinical impact. We highlight some of the problems with preclinical and clinical trials of NPs based delivery systems and propose some solutions to increase the impact of individual studies.

Acknowledgement

The authors would like to acknowledge National Sustainability Programme, project number LO1609 (Czech Ministry of Education, Youth and Sports) and the Seed fund project 940 from European Huntington's Disease Network.

References

[1] Sinn H P and Kreipe H 2013 A brief overview of the WHO classification of breast tumors *Breast Care* **8** 149–54

[2] Weigelt B, Peterse J L and Van't Veer L J 2005 Breast cancer metastasis: markers and models *Nat. Rev. Cancer* **5** 591–602

[3] Park H, Chang S K, Kim J Y, Lee B M and Shin H S 2014 Risk factors for distant metastasis as a primary site of treatment failure in early-stage breast cancer *Chonnam Med. J.* **50** 96–101

[4] Baker D and Fugh-Berman A 2009 Do new drugs increase life expectancy? A critique of a Manhattan institute paper *J. Gen. Intern. Med.* **24** 678–82

[5] Clarke R, Ressom H W, Wang A, Xuan J, Liu M C, Gehan E A and Wang Y 2008 The properties of very high dimensional data spaces: implications for exploring gene and protein expression data *Nat. Rev. Cancer* **8** 37–49

[6] Zhang X-Q, Xu X, Bertrand N, Pridgen E, Swami A and Farokhzad O C 2012 Interactions of nanomaterials and biological systems: implications to personalized nanomedicine *Adv. Drug Deliv. Rev.* **64** 1363–84

[7] Kinders R *et al* 2007 Phase 0 clinical trials in cancer drug development: from FDA guidance to clinical practice *Mol. Interv.* **7** 325–34

[8] Fugh-Berman A and Melnick D 2008 Off-label promotion, on-target sales *PLoS Med.* **5** e210

[9] Olson H *et al* 2000 Concordance of the toxicity of pharmaceuticals in humans and in animals *Regul. Toxicol. Pharmacol.* **32** 56–67

[10] Clark D L, Andrews P A, Smith D D, DeGeorge J J, Justice R L and Beitz J G 1999 Predictive value of preclinical toxicology studies for platinum anticancer drugs *Clin. Cancer Res.* **5** 1161–7

[11] What is Metastasis? *Cancer. Net.* 2 February 2016

[12] Chiang A C and Massagué J 2008 Molecular basis of metastasis *N. Engl. J. Med.* **359** 2814–23

[13] Torre L A, Bray F, Siegel R L, Ferlay J, Lortet-Tieulent J and Jemal A 2015 Global cancer statistics, 2012 *CA Cancer J. Clin.* **65** 87–108

[14] Lu X and Kang Y 2007 Organotropism of breast cancer metastasis *J. Mammary Gland Biol. Neoplasia* **12** 153–62

[15] Godone R L N, Leitão G M, Araújo N B, Castelletti C H M, Lima-Filho J L and Martins D B G 2018 Clinical and molecular aspects of breast cancer: Targets and therapies *Biomed. Pharmacother.* **106** 14–34

[16] Drasin D J, Robin T P and Ford H L 2011 Breast cancer epithelial-to-mesenchymaltransition: examining the functional consequences of plasticity *Breast Cancer Res.* **13** 226

[17] Feng Y, Spezia M, Huang S, Yuan C, Zeng Z, Zhang L, Ji X, Liu W, Huang B and Luo W *et al* 2018 Breast cancer development and progression: Risk factors, cancer stem cells, signaling pathways, genomics, and molecular pathogenesis *Genes Dis.* **5** 77–106

[18] Subramanian A, Manigandan A, Sivashankari P R and Sethuraman S 2015 Development of nanotheranostics against metastatic breast cancer — A focus on the biology & mechanistic approaches *Biotechnol. Adv.* **33** 1897–911

[19] Foroni C, Broggini M, Generali D and Damia G 2012 Epithelial–mesenchymal transition and breast cancer: role, molecular mechanisms and clinical impact *Cancer Treat. Rev.* **38** 689–97

[20] Lammers T, Hennink W E and Storm G 2008 Tumour-targeted nanomedicines: principles and practice *Br. J. Cancer* **99** 392–7

[21] Cruz M E M, Simões S I, Corvo M L, Martins M B and Gaspar M M 2009 Formulation of NPDDS for macromolecules *Drug Delivery Nanoparticles: Formulation and Characterization* ed Y Pathak and D Thassu (New York: Informa Healthcare) 35–49

[22] Bozzuto G and Molinari A 2015 Liposomes as nanomedical devices *Int. J. Nanomed.* **10** 975–99

[23] Deshpande P P, Biswas S and Torchilin V P 2013 Current trends in the use of liposomes for tumor targeting *Nanomedicine* **8** 1509–28

[24] Pattni B S, Chupin V V and Torchilin V P 2015 New developments in liposomal drug delivery *Chem. Rev.* **115** 10938–66

[25] Lamichhane N, Udayakumar T S, D'Souza W D, Simone C B, Raghavan S R, Polf J and Mahmood J 2018 Liposomes: Clinical applications and potential for image-guided drug delivery *Molecules* **23** 288

[26] Petersen A L, Hansen A E, Gabizon A and Andresen T L 2012 Liposome imaging agents in personalized medicine *Adv. Drug Deliv. Rev.* **64** 1417–35

[27] Feng L, Cheng L, Dong Z, Tao D, Barnhart T E, Cai W, Chen M and Liu Z 2017 Theranostic liposomes withhypoxia-activated prodrug to effectively destruct hypoxic tumors post-photodynamic therapy *ACS Nano* **11** 927–37

[28] He Y, Zhang L, Zhu D and Song C 2014 Design of multifunctional magnetic iron oxide nanoparticles/mitoxantrone-loaded liposomes for both magnetic resonance imaging and targeted cancer therapy *Int. J. Nanomed.* **9** 4055–66

[29] Ma M, Lei M, Tan X, Tan F and Li N 2016 Theranostic liposomes containing conjugated polymer dots and doxorubicin for bio-imaging and targeted therapeutic delivery *RSC Adv.* **6** 1945–57

[30] Dai W, Yang F, Ma L, Fan Y, He B, He Q, Wang X, Zhang H and Zhang Q 2014 Combined mTOR inhibitorrapamycin and doxorubicin-loaded cyclic octapeptide modified liposomes for targeting integrin 3 intriple-negative breast cancer *Biomaterials* **35** 5347–58

[31] Bressler E M, Kim J and Shmueli R B *et al* Biomimetic peptide display from a polymeric nanoparticle surface for targeting and antitumoractivity to human triple-negative breast cancer cells *J. Biomed. Mater. Res. A* **106** 1753–64

[32] He L, Gu J and Lim L Y *et al* 2016 Nanomedicine-mediated therapies to target breast cancer stem cells *Front. Pharmacol.* **7** 313

[33] Davis M E 2009 The first targeted delivery of sirna in humans via a self-assembling, cyclodextrin polymer-based nanoparticle: From concept to clinic *Mol. Pharm.* **6** 659–68

[34] Cole A J, Yang V C and David A E 2011 Cancer theranostics: the rise of targeted magnetic nanoparticles *Trends Biotechnol.* **29** 323–32

[35] Akbarzadeh A, Samiei M and Davaran S 2012 Magnetic nanoparticles: preparation, physical properties, and applications in biomedicine *Nanoscale Res. Lett.* **7** 144

[36] Sun C, Lee J S and Zhang M 2008 Magnetic nanoparticles in MR imaging and drug delivery *Adv. Drug Deliv. Rev.* **60** 1252–65

[37] Wicki A, Rochlitz C, Orleth A, Ritschard R, Albrecht I and Herrmann R *et al* 2012 Targeting tumor-associated endothelial cells: anti-VEGFR2 immunoliposomes mediate tumor vessel disruption and inhibit tumor growth *Clin. Cancer Res.* **18** 454–64

[38] Semkina A S, Abakumov M A and Skorikov A S *et al* 2018 Multimodal doxorubicin loaded magnetic nanoparticles for VEGF targeted theranostics of breast cancer *Nanomedicine: Nanotechnol. Biol. Med.* **14** 1733–42

[39] Pizzimenti S, Dianzani C, Zara G P, Ferretti C, Rossi F, Gigliotti C, Daga M, Ciamporcero E, Maina G and Barrera G 2016 Challenges and opportunities of nanoparticle-based theranostics in skin cancer *Nanoscience in Dermatology* ed M R Hamblin, P Avci and T W Prow (Boston, MA: Academic) 177–88

[40] Balakrishnan S, Bhat F and Raja Singh P *et al* 2016 Gold nanoparticle-conjugated quercetin inhibits epithelial-mesenchymal transition, angiogenesis and invasiveness via EGFR/VEGFR-2-mediated pathway in breast cancer *Cell Prolif.* **49** 678–97

[41] Silva C O, Molpeceres J, Batanero B, Fernandes A S, Saraiva N, Costa J G, Rijo P, Figueiredo I V, Faísca P and Reis C P 2016 Functionalized diterpeneparvifloron D-loaded hybrid nanoparticles for targeted delivery in melanoma therapy *Ther. Deliv.* **7** 521–44

[42] Lee S-M, Park H, Choi J-W, Park Y N, Yun C-O and Yoo K-H 2011 Multifunctional nanoparticles for targeted chemophotothermal treatment of cancer cells *Angew. Chem. Int. Ed.* **50** 7581–6

[43] Kojima C, Suehiro T, Watanabe K, Ogawa M, Fukuhara A, Nishisaka E, Harada A, Kono K, Inui T and Magata Y 2013 Doxorubicin-conjugated dendrimer/collagen hybridgels for metastasis-associated drug delivery systems *Acta Biomater.* **9** 5673–80

[44] Xu P, Yin Q, Shen J, Chen L, Yu H, Zhang Z and Li Y 2013 Synergistic inhibition of breast cancer metastasis by silibinin-loaded lipid nanoparticles containing TPGS *Int. J. Pharm.* **454** 21–30

[45] Meng H *et al* 2012 Gadolinium metallofullerenol nanoparticles inhibit cancer metastasis-through matrix metalloproteinase inhibition: imprisoning instead of poisoning cancer cells *Nanomedicine* 8 136–46

[46] Granchi C, Funaioli T, Erler J T, Giaccia A J, Macchia M and Minutolo F 2009 Bioreductively activated lysyl oxidase inhibitors against hypoxic tumours *ChemMedChem* **4** 1590–4

[47] Kanapathipillai M, Mammoto A, Mammoto T, Kang J H, Jiang E, Ghosh K, Korin N, Gibbs A, Mannix R and Ingber D E 2012 Inhibition of mammary tumor growth using lysyl oxidase-targeting nanoparticles to modify extracellular matrix *Nano Lett.* **12** 3213–7

[48] Shen J, Sun H, Xu P, Yin Q, Zhang Z, Wang S, Yu H and Li Y 2013 Simultaneous inhibition of metastasis and growth of breast cancer by co-delivery of twist shRNA and paclitaxel using pluronic P85-PEI/TPGS complex nanoparticles *Biomaterials* **34** 1581–90

[49] Zhang W, Meng J, Ji Y, Li X, Kong H, Wu X and Xu H 2011 Inhibiting metastasis of breast cancer cells *in vitro* using gold nanorod–siRNA delivery system *Nanoscale* **3** 3923–32

[50] Palange A L, DiMascolo D, Carallo C, Gnasso A and Decuzzi P 2014 Lipid–polymer nanoparticles encapsulating curcumin for modulating the vascular deposition of breast cancer cells *Nanomedicine* **10** 991–1002

[51] Xu P, Yin Q, Shen J, Chen L, Yu H, Zhang Z and Li Y 2013 Synergistic inhibition of breast cancer metastasis by silibinin-loaded lipid nanoparticles containing TPGS *Int. J. Pharm.* **454** 21–30

[52] Wang D *et al* 2014 Treatment of metastatic breast cancer by combination ofchemotherapy and photothermal ablation using doxorubicin-loaded DNA wrapped gold nanorods *Biomaterials* **35** 8374–84

[53] Carneiro M L *et al* 2013 Antitumor effect and toxicity of free rhodium(II) citrate and rhodium(II) citrate-loaded maghemite nanoparticles in mice bearing breast cancer *J. Nanobiotechnol.* **11** 4

[54] Pereira E R, Jones D, Jung K and Padera T P 2015 The lymph node microenvironment and its role in the progression of metastatic cancer *Semin. Cell Dev. Biol.* **38** 98–105

[55] Liu R, Gilmore D M, Zubris K A, Xu X, Catalano P J, Padera R F and Grinstaff M W 2013 Prevention of nodal metastases in breast cancer following the lymphatic migration of paclitaxel-loaded expansile nanoparticles *Biomaterials* **34** 1810–9

[56] Zubris K A, Liu R, Colby A, Schulz M D, Colson Y L and Grinstaff M W 2013 *In vitro* activity of paclitaxel-loaded polymeric expansile nanoparticles in breast cancer cells *Biomacromolecules* **14** 2074–82

[57] Murakami M, Ernsting M J, Undzys E, Holwell N, Foltz W D and Li S D 2013 Docetaxel conjugate nanoparticles that target α-smooth muscle actin — expressing stromalcells suppress breast cancer metastasis *Cancer Res.* **73** 4862–71

[58] Li Y, Jin M, Shao S, Huang W, Yang F, Chen W, Zhang S, Xia G and Gao Z 2014 Small-sized polymeric micelles incorporating docetaxel suppress distant metastases in the clinically-relevant 4T1 mouse breast cancer model *BMC Cancer* **14** 329

[59] Frese K K and Tuveson D A 2007 Maximizing mouse cancer models *Nat. Rev. Cancer* **7** 645–58

[60] Steichen S D, Caldorera-Moore M and Peppas N A 2012 A review of current nanoparticle and targeting moieties for the delivery of cancer therapeutics *Eur. J. Pharm. Sci.* **48** 416–27

[61] Couvreur P, Stella B, Reddy L H, Hillaireau H, Dubernet C and Desmaële D *et al* 2006 Squalenoyl nanomedicines as potential therapeutics *Nano Lett.* **6** 2544–8

[62] Cai W, Gao T and Hong H *et al* 2008 Applications of gold nanoparticles in cancer nanotechnology *Nanotechnol. Sci. Appl.* **1** 17

[63] Goss P E and Chambers A F 2010 Does tumour dormancy offer a therapeutic target? *Nat. Rev. Cancer* **10** 871–7

[64] Naumov G N, Townson J L, MacDonald I C, Wilson S M, Bramwell V H, Groom A C and Chambers A F 2003 Ineffectiveness of doxorubicin treatment on solitary dormant mammary carcinoma cells or late-developing metastases *Breast Cancer Res. Treat.* **82** 199–206

[65] Maeda H and Matsumura Y 2011 Epr effect based drug design and clinical outlook for enhanced cancer chemotherapy *Adv. Drug Deliv. Rev.* **63** 129–30

[66] Ailles L E and Weissman I L 2007 Cancer stem cells in solid tumors *Curr. Opin. Biotechnol.* **18** 460–6

[67] Fillmore C M and Kuperwasser C 2008 Human breast cancer cell lines contain stem-like cells that self-renew, give rise to phenotypically diverse progeny and survive chemotherapy *Breast Cancer Res.* **10** R25

[68] Buck E, Eyzaguirre A, Barr S, Thompson S, Sennello R, Young D, Iwata K K, Gibson N W, Cagnoni P and Haley J D 2007 Loss of homotypic cell adhesion by epithelial-mesenchymal transition or mutation limits sensitivity to epidermal growth factor receptor inhibition *Mol. Cancer Ther.* **6** 532–41

[69] Creighton C J, Li X, Landis M, Dixon J M, Neumeister V M, Sjolund A, Rimm D L, Wong H, Rodriguez A and Herschkowitz J I 2009 Residual breast cancers after conventional therapy display mesenchymal as well as tumor-initiating features *Proc. Natl. Acad. Sci. U.S.A.* **106** 13820–5

[70] Desmedt C, Voet T, Sotiriou C and Campbell P J 2012 Next-generation sequencing in breast cancer: first take home messages *Curr. Opin. Oncol.* **24** 597–604

[71] André F, Bachelot T, Commo F, Campone M, Arnedos M and Dieras V *et al* 2014 Comparative genomic hybridisation array and DNA sequencing to direct treatment of metastatic breast cancer: a multicentre, prospective trial (SAFIR01/UNICANCER) *Lancet Oncol.* **15** 267–74

[72] Tsimberidou A M, Iskander N G, Hong D S, Wheler J J, Falchook G S and Fu S *et al* 2012 Personalized medicine in a phase 1 clinical trials program: the MD Anderson Cancer Center initiative *Clin. Cancer Res.* **18** 6373–83

[73] Corcoran R B, Ebi H, Turke A B, Coffee E M, Nishino M and Cogdill A P *et al* 2012 EGFR-mediated re-activation of MAPK signaling contributes to insensitivity of BRAF mutant colorectal cancers to RAF inhibition with vemurafenib *Cancer Discov.* **2** 227–35

[74] Esserman L J, Berry D A, DeMichele A, Carey L, Davis S E and Buxton M *et al* 2012 Pathologic complete response predicts recurrence-free survival more effectively by cancer subset: results from the I-SPY 1 TRIAL-CALGB 150007/150012, ACRIN 6657 *J. Clin. Oncol.* **30** 3242–9

[75] Printz C 2013 I-SPY 2 May change how clinical trials are conducted: researchers aim to accelerate approvals of cancer drugs *Cancer* **119** 1925–7

[76] Ledford H 2010 Clinical drug tests adapted for speed *Nature* **464** 1258

[77] Kettiger H, Schipanski A, Wick P and Huwyler J 2013 Engineered nanomaterial uptake and tissue distribution: from cell to organism *Int. J. Nanomed.* **8** 3255–69

[78] Bertrand N and Leroux J-C 2012 The journey of a drug-carrier in the body: an anatomo-physiological perspective *J. Control. Release* **161** 152–63

[79] Lewinski N, Colvin V and Drezek R 2008 Cytotoxicity of nanoparticles *Small* **4** 26–49

[80] Dobrovolskaia M A and McNeil S E 2007 Immunological properties of engineered nanomaterials *Nat. Nanotechnol.* **2** 469–78

[81] Stone V, Johnston H and Schins R P F 2009 Development of *in vitro* systems for nanotoxicology: methodological considerations *Crit. Rev. Toxicol.* **39** 613–26

[82] Nie S, Xing Y, Kim G J and Simons J W 2007 Nanotechnology applications in cancer *Annu. Rev. Biomed. Eng.* **9** 257–88

[83] Tiede K, Boxall A B A, Tear S P, Lewis J, David H and Hassellov M 2008 Detection and characterization of engineered nanoparticles in food and the environment *Food Addit. Contam. Part A: Chem. Anal. Control Expo Risk Assess.* **25** 795–821

[84] Hamburg M A 2012 Science and regulation. FDA's approach to regulation of products of nanotechnology *Science* **336** 299–300

IOP Publishing

Chapter 12

Biological systems: a challenge for physical stimulation of cancer nanomedicine

Arpita Pandey Tiwari and Sonali Rohiwal

Physical stimulation is a major area of research in cancer theranostics. Physical stimulators are advantageous as they provide a non-invasive alternative over other invasive conventional cancer therapies. In view of this, it becomes necessary to know the common physical stimulators of cancer cells and to understand the underlying mechanism of action on the biological systems. In-depth understanding of all the possible physical stimulators of a biological system, their mechanism of action with the latter, helps to create the proper evaluation of the challenges existing in cancer theranostics. This chapter has extensive coverage of all the physical stimulators of biological cells, their subtypes, their interaction with biological systems and also the challenges against these common physical stimulators of a biological system.

12.1 Introduction

Over the past decade, nanotechnology has emerged as a field of interdisciplinary research. Nanomaterials are showing immense potential in areas of biomedical applications [1–5]. The use of nanomaterials to address the problems of medicine has formed a new branch of science, 'nanomedicine'. Nanomaterials are finding applications in areas like disease imaging and diagnosis, drug delivery and reporters of disease pathogenesis. There are multifunctional nanoparticles which are capable of performing one or more of the above duties and are now in the stages of preclinical and clinical development [6–9].

Cancer is one of the major challenges of the 21st century. Although current advance diagnostic and therapeutic approaches have considerably improved cancer diagnosis and treatment, the current treatment options have several tough issues to be addressed. Successful, non-invasive delivery of biodegradable and non-toxic combination therapies that can target existing cancerous cells and prevent the

doi:10.1088/2053-2563/ab2907ch12

development of new tumors all in one dose is ideal for clinical treatment. These novel technologies that can intelligently and effectively remove the diseased tissue with limited negative effects to the patient, and minimize the chance of cancer recurrence, could make an immediate impact in the clinic. When designing these technologies, it is important to focus on adequate circulation time, specific delivery to cancerous tissue only, evasion of normal tissues and accumulation in the organs, lack of an immune response, and concomitant treatment and non-invasive monitoring in order for successful drug delivery to occur. The use of remotely triggered treatments in combination with imaging such as ultrasound, computed tomography, magnetic resonance imaging, and positron emission tomography could prove to address many of the current issues in cancer treatment, possibly leading to significant clinical outcomes [10–12]. The selectivity and specificity for disease destruction can be enhanced by using externally activatable theranostic agents to produce localized cytotoxicity with little collateral damage. The ability to control drug dosing in terms of quantity, location, and time is a key goal for drug delivery science, as improved control maximizes therapeutic effect while minimizing side effects. Systems responsive to a stimulus such as temperature, pH, applied magnetic or electrical field, ultrasound, light, or enzymatic action have been proposed as triggered delivery systems [13, 14].

This chapter aims to discuss some major advances in the field of remotely triggered nano-theranostics for cancer applications by briefly summarizing the development and clinical potential of various remotely triggered theranostics (photodynamic, photothermal, phototriggered chemotherapeutic release, ultrasound, electro-thermal, magneto-thermal, x-ray, and radiofrequency) and the major challenges in physically triggered cancer therapy are pointed out. Also, the challenges that must be overcome for successful clinical development and implementation of such cancer theranostics are discussed.

12.2 Commonly used physical stimulators in cancer nanomedicine

12.2.1 Photoresponsive

Multifunctional nanoparticles responsive towards light stimulus can be used to locate cancer in a patient using various imaging modalities, including optical imaging. After receiving light stimuli these nanoparticles can then be used for targeted and on-call drug release at the cancer sites. Such photoresponsive theranostics will end in better treatments that eliminate the likelihood of under- or overdosing, and decrease the necessity for several rounds of drug doses. This section will focus on photodynamic therapy, photothermal therapy, and phototriggered chemotherapeutic release. Currently, photodynamic therapy is clinically approved for cancer treatment, and photothermal therapy and phototriggered chemotherapeutic release are still in clinical trials for cancer treatment [15].

Photodynamic therapy in cancer nanomedicine. Photodynamic therapy (PDT) involves the use of a non-toxic agent, known as a photosensitizer (PS), that is irradiated with light to induce the formation of reactive oxygen species (ROS) that stimulate cellular destruction at the region of interest [16]. Targeting of various cancers using this method depends on the ability of the laser. The major challenges are control of the wavelength, fluence, and irradiation time in order to penetrate tissue and reach the PS. For example, an external laser beam can be used to treat head, neck, ocular, and breast cancers while intestinal or pancreatic cancers may be reached using a laser on the end of an endoscope.

Nanoparticle delivery of the PS agents overcomes delivery limitations due to PS hydrophobicity, non-specific targeting, and toxicity. Irradiation of the PMIL via 690 nm near infrared light triggers BPD leading to PDT induced tumor cell apoptosis and microvessel damage and simultaneous release of NP to inhibit anti-apoptotic, vascular endothelial growth factor (VEGF), and the tyrosine kinase for hepatocyte growth factor (MET) receptor signaling. The next steps include specific targeting agents for tumor cells and micrometastases, and optimizing encapsulation of NP to further reduce the small molecule toxicity [17]. There are other promising photo-dynamic therapies for the treatment of cancer. Hsp90 loaded porphyrin based telodendrimers were used to image drug delivery and treat tumors in prostate cancer in mice through the variable release of Hsp90 and localized irradiation, limiting cytotoxicity [18]. Additionally, NaYF4:Yb3+, Er3+ upconversion nanoparticles with the photosensitive molecule Rose Bengal and folic acid targeting agent, which is triggered at 980 nm, show promise for both the imaging and treatment of JAR choriocarcinoma *in vivo* [19]. The same PS was also used with silica nanoparticles to treat breast and oral cancer cell lines *in vitro* [20]. Additionally, folic acid-functionalized carbon nanodots carrying zinc phthalocyanine PS were used to target, image, and treat human cervical cancer HeLa cells [21]. Other nanoparticles used for PDT include upconverted nanoparticles, hyaluronic acid derivatized carbon nanotubes [22], selenium–rubyrin loaded nanoparticles [23], gold vesicles [24], silica coated titanium dioxide [25], cobalt ferrite nanoparticles, lipid calcium phosphate nanoparticles, small molecule quenched activity-based probes, carbon dots, poly-acrylamide nanoparticles, calcium phosphosilicate nanoparticles, mesoporous sili-con nanoparticles, and gold nanoparticles [26–30].

Photothermal theranostics. Photothermal therapy utilizes continuous wave or pulsed lasers to irradiate cancer tissue with an electromagnetic radiation (e.g. visible or near infrared light) to cause a rise in temperature that subsequently leads to cell death. Photo-absorbers convert laser energy to heat, in the range 45–300 °C, which can cause localized diseased tissue destruction through different mechanisms [31]. This therapy, through careful selection of laser parameters and illumination, allows for specific targeting through localized light penetration. This focused targeting can significantly limit systemic effects when using photosensitive agents that are non-cytotoxic unless irradiated by a specific wavelength of light, and can ensure that the diseased tissue will respond differently from the surrounding normal tissue. One particular application of photothermal therapy was developed to target microscopic residual disease (MRD) [32]. MRD leads to tumor resurgence and metastases even after oncosurgery. Use of plasmonic nanobubbles (PNBs), a photomechanical product of photothermal triggering, has recently proved promising in providing intraoperative detection of residual disease *in vivo* in real time. PNBs are generated when clusters of gold nanoparticles (conjugated to panitumumab (60 nm)), are taken up via receptor-mediated endocytosis and subjected to a short laser pulse that induces the formation of transient, photomechanical vapor nanobubbles. In this case, the photothermal application is for forming the photomechanical PNBs that can then be used for MRD detection, rather than using the photothermal effects of the gold nanoclusters for cell destruction. Multimodal imaging was used to guide photothermal therapy in these animals. Some additional nanomaterials used in conjunction with photothermal therapy include silicon nanowires with gold nanoparticles [33]; CuS nanoparticles; multidye mesoporous silica with silane-conjugated heptamethine cyanine dye loaded with silicon 2,3-napthalocyanine dihydroxide dye [34]; single walled carbon nanotubes [35]; porous silicon nanoparticles; NIR resonant silica core, gold shell, PEG nanoparticles; chitosan-coated triangular silver nanoparticles; and palladium nanosheets [36] (figure 12.1).

Phototriggered chemotherapeutic release. The major disadvantages of conventional administration of drugs include an unbalanced drug release profile due to sudden changes in the biological conditions and the irregular dissemination of drugs in the human body leading to severe toxic side effects. Remotely activated delivery methods that can activate release of a drug at the appropriate location (e.g. tumor site) and at a frequency that regulates itself in response to the stage of the cancer are very appealing in oncology. Some light-responsive delivery methods are only good for one-time use. For example, the laser causes an irrevocable physical change in the system that incites a 'burst' release of the payload, while others experience reversible structural changes when the laser is turned off/on and act as multi-controllable systems that release the payload in a pulsatile, on-demand routine [37].

Engineering a non-invasive drug delivery method with increased tumor penetration and limited toxicity is essential for the improved chemotherapeutic treatment of cancer. Tong *et al* constructed a docetaxel encapsulating nanoparticle comprised of spiropyran, polyethylene glycol, and cholesterol (Dtxl/SP NPHCS) that, when triggered by 365 nm light, undergoes a size change from 103 to 49 nm that allows for increased tumor penetration and simultaneous drug release intratumorally via intravenous administration in a subcutaneous implanted fibrosarcoma model [38].

12.2.2 Ultrasound-triggered theranostics

Recently, ultrasound has been explored as a means of exerting external control in drug delivery from biomaterials for pulsatile release. The advantage of ultrasound is that it is non-invasive and may be focused at depth in soft tissue throughout the body [39]. Ultrasound waves transmitted from a transducer are sinusoidal in nature, alternating between high and low pressures. These waves travel through a medium, transfer energy from one element to the next, however, in this process some of the energy is lost in a process which is called dispersion. The combination of both events results in attenuation, which can be summarized as the collective process of

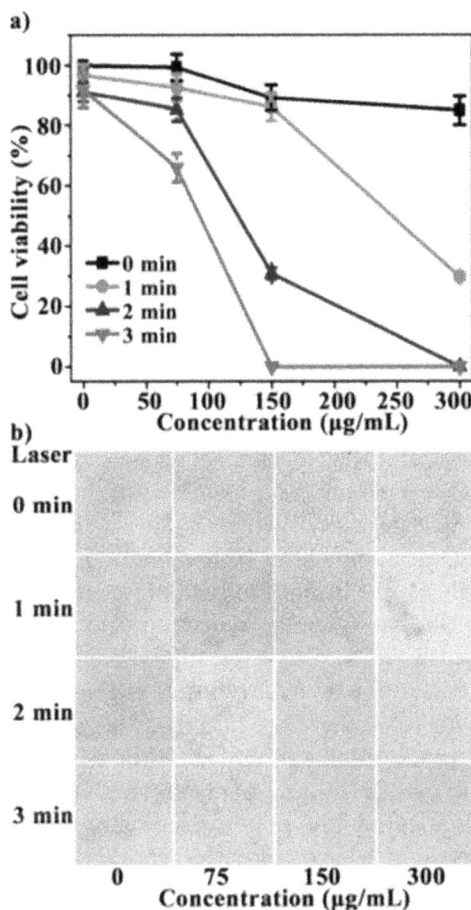

Figure 12.1. Photothermal destruction of KB cancer cells. (a) Viability of KB cells versus NIR laser (808 nm, 2 W cm^{-2}) and time (0, 1, 2, 3 min) as a function of PEG–AuNPs@SiNWs concentrations. (b) Morphology of KB cells incubated with PEG–AuNPs@SiNWs of different concentrations and exposed to different laser irradiation time. Scale bar = 20 μm. Reprinted with permission from [33]. Copyright (2012) American Chemical Society.

absorption and scattering of energy waves. Two important parameters of US waves are the frequency and the power density; frequency of a US wave refers to the number of sinusoidal cycles occurring per second (Hz), whereas the power density is defined as the power per cross-sectional area (W cm^{-2}). In order to achieve targeted delivery and prevent drug accumulation in non-target tissues, the drug carrier needs to travel in the bloodstream until it reaches the target, then extravasate in the space between cells and penetrate the tumor. Using US for triggered drug delivery allows for the time- and location-specific release of anticancer drugs at the desired target cells, thus decreasing the known side effects resulting from chemotherapy spreading to several parts of the body. The effects of US in drug delivery may be either thermal or mechanical. Thermal effects occur due to the absorption of thermal energy by the tissues or cells, which results in an increase in temperature. When hyperthermia occurs, this may result in heating of the drug and carrier, or heating and damaging of the tumor tissues in the absence of chemotherapy. On the other hand, the more prominent effects, which are observed and exploited for triggered drug delivery, are the mechanical effects, which occur due to oscillating bubbles, cavitation, wave pressure, and acoustic streaming. Ultrasound is an attractive mechanism for delivery of proteins, especially insulin, as it offers a reproducible, rapid and reversible method of controlling release with no degradation of proteins. The biological effects of ultrasound can cause formation of cavitation bubbles, localized tissue heating, and radiation force, which can be used for confined drug release from nanosystems, increased extravasation of drugs and/or nanoparticles from blood vessels into tumors, and improved penetration of drugs into tumors [40]. When using ultrasound for temperature-sensitive nanosystems, the encapsulated payload can be released locally through the use of these mechanical forces. Multispectral optoacoustic tomography (MSOT) is a potential, non-invasive alternative to clinically approved sentinel lymph node (SLN) excision, which is an invasive and currently accepted method for diagnosing melanoma, the fifth most common cancer in the United States [41]. MSOT utilizes ultrasound with fluorescent dyes to provide real-time images of melanin in SLN to improve detection and treatment. MSOT *ex vivo* studies showed 100% sensitivity and 62% selectivity. Additionally, this unique use of ultrasound can be taken further by incorporating a chemotherapeutic agent into a nanostructure along with ICG, therefore adding a therapeutic effect to the already successful diagnostic applications of MSOT. Promising cancer therapies using ultrasound include polypyrrole hollow microspheres, microbubbles, temperature-sensitive liposomes, biodegradable poly(methacrylic acid) based nanocapsules, superparamagnetic iron oxide acoustic droplets, crown-ether-coated core/shell nanoparticles, polymer-grafted mesoporous silica nanoparticles, and echogenic glycol chitosan nanoparticles [42–45].

12.2.3 Electro-thermally triggered theranostics

Electrical signals are fairly easy to create and regulate, making them excellent remote triggers for theranostic applications. Electrical stimuli have already been effectively used to activate the release of payloads through the use of conducting polymers or implantable electrical delivery systems [46–48]. One version of an electrically stimulated drug delivery system is the use of another living organism. 'Bacteriobots,' constructed with *S. typhimurium* and conjugated to Cy5.5-coated polystyrene microbeads using biotin and streptavidin, allow for improved targeting because of greater motility and velocity towards tumor cells as compared to 'normal' cells, which can aid in fluorescent imaging. The structure of these bacteriobots is illustrated in figure 12.2. The chosen bacteria strain is easily genetically manipulated, can direct chemotaxis to certain molecular signals through chemotactic receptors, has high-motility flagella that respond to external chemical stimuli, and has been shown to target and proliferate in solid tumors through self-propulsion that aids in penetration. Subcutaneous injection of the bacteriobots in CT-26 tumor-bearing mice demonstrated successful targeting via bioluminescence, as seen in figures

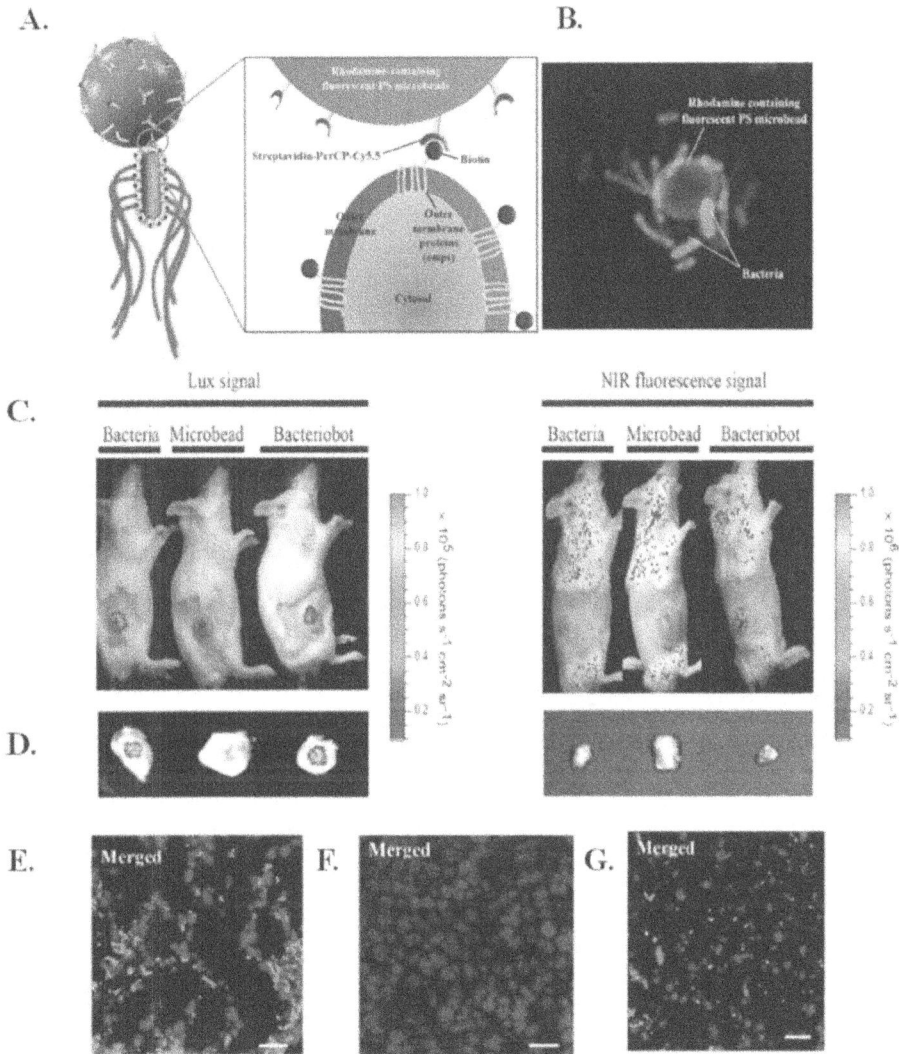

Figure 12.2. Bacteriobots for electro-thermally triggered theranostics. (A) Schematic representation of bacteriobots. Biotin (500 μg) was incubated with omp-expressing *S. typhimurium* (3×10^8 cells mL^{-1}) for 1 h. Rhodamine-containing fluorescent carboxylated PS microbeads (1×10^8 ml^{-1}) were covalently coupled to streptavidin–PerCP–Cy5.5 (500 μg). Biotin-displaying *S. typhimurium* and streptavidin–PerCP–Cy5.5-coated PS microbeads were co-incubated for 30 min at 37 °C. (B) *S. typhimurium*-attached PS microbeads were observed using a confocal laser scanning microscope. (C), (D) Mice (n56) were injected subcutaneously with CT-26 cells (1×10^6). When the tumors reached a volume of approximately 130 mm^3, the tumor-bearing mice were injected with bacteria (3×10^7 CFU/100 μl), microbeads (1×10^7/100 μl) or bacteriobots (bacteria:microbeads ratio = 3:1 per 100 μl). Representative *in vivo* and *ex vivo* bioluminescence and NIR images (Cy5.5 image) were captured 3 days post-injection. (C) *In vivo* bioluminescence and NIR imaging of mouse tumor models. (D) *Ex vivo* bioluminescence and NIR imaging of the dissected tumors. (E), (G) Tumor-bearing mice were injected with (E) bacteria, (F) microbeads, and (G) bacteriobots, the tumor masses were fixed and investigated histologically, and bacteria, microbeads, and bacteriobots were localized by indirect fluorescence. DAPI staining of the same tissue sections and mergence of the DAPI-stained slides (blue); bacteria were detected by indirect immunofluorescence (green); microbeads and bacteriobot were detected by indirect fluorescence (red) in the dissected tumor masses. Reprinted with permission from [48]. Copyright (2013) Springer Nature.

12.2(C), (D). In addition, as seen in figures 12.2(E)–(G), the bacteriobots showed fluorescence in Cy5.5 (red) and there was promising accumulation of bacteriobots in tumors three days after infection. The next steps for this research include using the bacteriobots for payload delivery using appropriate cancer models [49]. To circumvent the limitations of conventional electric stimuli-responsive drug delivery devices, Zare and co-workers recently described a novel dual-stimulus responsive nanosystem for externally triggered payload delivery [50]. This nanosystem was capable of being triggered by temperature changes and use of an electric field. Polypyrrole nanoparticles were loaded with therapeutics and then subcutaneously localized *in vivo* with the help of a temperature-sensitive hydrogel using a triblock polymer, PLGA–PEG–PLGA. Drug delivery from the conductive nanoparticles could be controlled by use of a weak, external electrical trigger. This approach exemplifies an innovative and interactive nanosystem that can be externally activated with decent/ satisfactory/acceptable control on the time, location, and amount of drug released [51]. Other electro-stimulated systems that could prove promising for cancer applications include sensing electrodes in an alginate matrix, electrical impedance sensing system [52], iron–alginate thin-films, electro-active hydrogel based polymer matrix, gold nanoparticles activated by light and an electric field, nanowire substrates, and platinum microelectrodes [53–55].

12.2.4 Magneto-thermally triggered theranostics

Initial experiments with magnetically regulated drug delivery systems used fairly large magnetic beads in the millimeter size range that had to be implanted in an ethylene vinyl acetate milieu. These large beads could be activated through the use of an oscillating magnetic field to open pores for drug release. This approach, however, led to very few real applications in drug delivery due to very slight differences between the 'on' and the 'off' states [56, 57].

Recent progress in nanomedicine has led to experimenting with magnetic nanoparticles, which can be heated by using an alternating current (AC) magnetic field. Use of such magnetic nanoparticles caused a renewed interest in this field of triggered drug delivery. Most of the current work focuses on magneto-thermally-responsive or magnetically-triggered, thermally-sensitive nanomaterials for drug delivery applications. Two fundamental measures are required for magneto-thermal delivery: magnetic heating or magnetic hyperthermia, and a temperature-responsive or thermally-rupturable layer. Once the activation is complete, the advantage of regulating drug release helps the patient by decreasing the net quantity of drug needed to achieve efficacy and by reducing the number of administrations required. Such magnetically-triggered systems could be incorporated into more refined devices that comprise targeting moieties, imaging agents and multimodal therapy. To overcome delivery and toxicity issues associated with pancreatic cancer, Lee *et al* developed urokinase plasminogen activator receptor (uPAR)-targeting iron oxide nanoparticles (IONPs) loaded with gemcitabine (Gem) (ATF-IONP–Gem) to image and treat uPAR-expressing tumor and stromal cells [58].

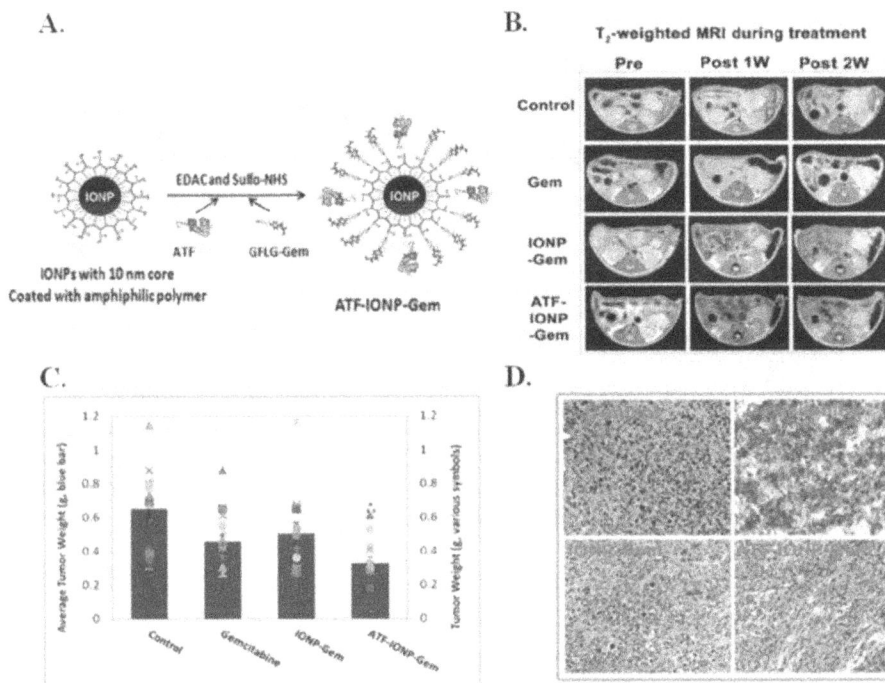

Figure 12.3. IONPs for magneto-thermally triggered theranostics. (A) Diagram of the conjugation of ATF peptides and GFLG–Gem conjugates to IONPs. (B) Coronal T2-weighted MR images and corresponding bright field (BF) images of the tumor-bearing mice after systemic delivery of non-targeted IONP–Gem or ATF-IONP–Gem. Tumor-bearing mice without nanoparticle treatment were used as controls. Yellow dotted circles and arrows indicate the location of primary tumor lesions in the MR and BF images, respectively. (C), (D) Tumor-bearing mice received tail vein injections of 2 mg kg^{-1} of the Gem-equivalent dose of various IONPs five times. At the end of the experimental period, tumors were collected and weighed. (C) The mean tumor weights (navy bar) and individual tumor weight distribution of the tumor-bearing mice in each group are shown as colored symbols. Values represent mean ± S.D. of 16 mice from three repeat studies. *Statistically significant difference versus control, One-Way ANOVA method: $p < 0.0001$; Modified t-test: $p < 0.0002$. **Statistically significant difference. ATF-IONP–Gem versus Gem and IO–Gem groups, One-Way ANOVA method: $p < 0.05$; Modified t-test: $p < 0.05$. (D) Immunohistochemical staining of the cell proliferation marker, Ki-67 in tumor tissue sections. Brown: Ki-67 positive tumor cells. Blue: hematoxylin background staining. Reprinted with permission from [58]. Copyright (2013) American Chemical Society.

As seen in figure 12.3(A), the Gem is conjugated to the IONP using an amino terminal fragment (ATF) so that it is only released after lysosomal or endosomal enzymatic cleavage via receptor-mediated internalization. In addition to delivery of Gem, the IONPs have proven to be an acceptable method for MRI imaging (figure 12.3(B)). The ATF-IONP–Gem nanoparticles (hydrodynamic size 66 nm) yielded 50% tumor growth inhibition in an orthotopic human pancreatic cancer xenograft model, as observed in figure 12.3(C). Post ATF-IOP–Gem treatment (figure 12.3(D)) there was no evident tumor growth or Ki-67 expression, a marker of cell proliferation. The nanoparticle did not damage the liver, spleen, or other organs.

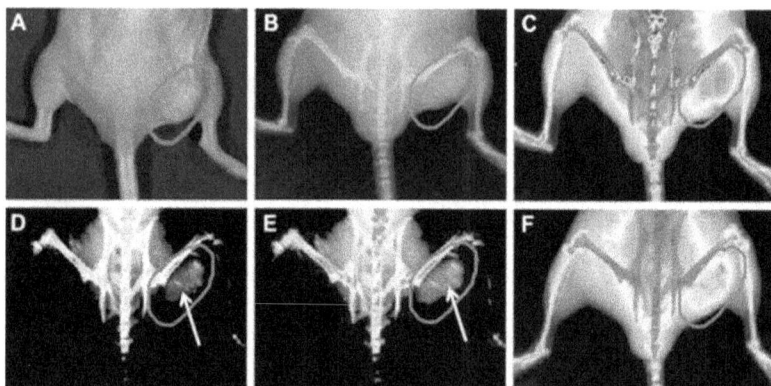

Figure 12.4. Real-time *in vivo* x-ray images after intravenous injection of GNR–SiO$_2$–FA in nude mice at different time points. (A) The photograph of the tumor tissue; (B) the x-ray image at 0 h; (C) the x-ray image at 0 h (in color); (D) the x-ray image at 12 h, (E) the x-ray image at 12 h (in color); (F) the x-ray image at 24 h (in color). (Adapted with permission from [60]. Copyright (2011) Elsevier.)

These results are promising for *in vivo* applications given the effective targeting, and drug conjugation stability [59].

12.2.5 Additional remotely triggered treatments

X-ray irradiation

Another possible remotely triggered treatment includes x-ray induced ionizing radiation. A use of x-rays is to induce the generation of a reactive oxygen species like in photodynamic therapy.

Gold has also shown optimal properties for x-ray imaging and x-ray induced ionizing radiation. A study by Huang *et al* showed that silica-modified folic acid-functionalized gold nanorods could be used for x-ray or CT imaging, as well as for radiation therapy and photothermal therapy, as shown in figure 12.4 [60].

The tumor tissue displays a homogeneous color distribution in figures 12.4(B) and (C). Subsequently, as blood circulation, the tumor tissue displays a strong contrast compared with that of the nontumor tissues, indicating that GNR–SiO$_2$–FA gradually target the tumor tissue during 0 w 12 h (figures 12.4 (D) and (E)). Conversely, there are no obvious changes observed in the x-ray images of the tumor tissues in the control group. After 24 h administration, the tumor tissue displays an inhomogeneity color distribution in figure 12.4(F), indicating the involvement of GNR–SiO$_2$–FA. Another study showed that modified gold nanoparticles exhibited low toxicity in cells and enhanced cancer killing when treated with x-rays compared to cells treated with x-rays without the addition of gold nanoparticles. These gold nanoparticles could then be combined with CT imaging to provide a theranostic platform used for the treatment of cancer [61]. Additionally, there have been studies that focused on the chemotherapeutic release by x-ray irradiation due to increased tissue penetration over light-triggered chemotherapeutic release. One such study focused on doxorubicin conjugated to DNA strands attached to gold nanoparticles, which, upon irradiation, released doxorubicin for chemotherapeutic effects. The properties of gold also allow these particles to be imaged using CT imaging [62]. These are only a few examples of the uses of x-ray irradiation for theranostic treatment methods.

12.2.6 Radiofrequency triggered

Hyperthermia radiofrequency triggered hyperthermia is another popular remotely triggered treatment method. A study by Wang *et al* used gold-coated magneto-liposomes for a variety of purposes including radiofrequency triggered release, chemo-hyperthermia therapy, as well as magnetic resonance and x-ray imaging. These gold-coated magnetoliposomes showed promising results as drug delivery carriers and optimal imaging properties both *in vitro* and *in vivo* [63]. An additional study showed that biodegradable alginate nanoparticles combining radiofrequency triggered hyperthermia and triggered doxorubicin release could be used as an effective combination treatment method. In orthotopic rat liver tumor models, there was enhanced thermal ablation, controlled doxorubicin release, and imaging potential using MRI, as shown in figure 12.5 [64]. X-ray induced ionizing radiation and radiofrequency triggered hyperthermia are only a couple examples of some additional remotely triggered treatments that could be used in theranostic platforms.

12.3 Challenges of current cancer nanomedicine

12.3.1 Toxicity of nanomaterials

The uncertain health hazard potential of nanomaterials is probably the most significant hurdle for regulatory approval and commercialization of nanomedical products [65]. Some nanomaterials such as metal nanoparticles, metal oxide nano-particles, QDs, fullerenes and fibrous nanomaterials were found to induce chromo-somal fragmentation, DNA strand breakages, point mutations, oxidative DNA adducts and alterations in gene expression [66], sometimes even through cellular barriers [67]. In these cases, the safety profile becomes a major concern. Although there have been no reported examples of clinical toxicity due to nanomaterials thus far, early studies indicate that nanomaterials could initiate adverse biological interactions that can lead to toxicological outcomes [68]. Since the mechanisms and severity of nanotoxicity are not fully predictable or testable with current toxicological methods, the toxicity of nanomaterials is rapidly emerging as an important area of tangential study in the nanomedicine research field.

There are many different factors to consider when designing nanomaterials and an understanding of how different parameters affect toxicity can aid in designing safer nanomaterials for medical applications. Some important parameters to consider include size, shape, surface area, charge, state of aggregation, crystallinity, and the potential to generate reactive oxygen species (ROS). Size is a significant factor and can influence the distribution and toxicity of a material. Studies with gold nanoparticles (AuNPs) in four different cell lines demonstrated that both toxicity and the mechanism of cell death were size-dependent [69]. In figure 12.6, 1.4 nm AuNPs were 60-fold more toxic than 15 nm AuNPs and cell death from 1.4 nm

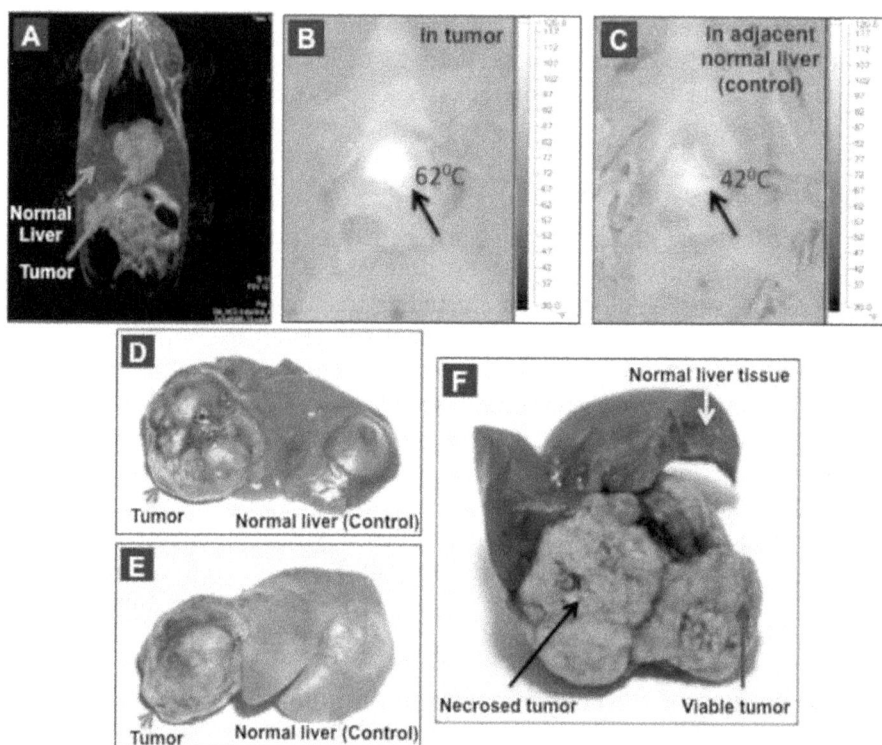

Figure 12.5. (A) MRI imaging confirming the successful development of the N1S1 tumor (red arrow) in liver (green arrow) of Sprague Dawley rats. Infrared imaging performed during the RF ablation procedure showed higher heating (>20 OC difference) at the tumor region (B) than in the adjacent normal liver (C), which was used as the control. This augmented heating was confirmed by gross examination and TTC staining which showed larger areas of ablation at the tumor sites in liver (red arrow) when compared to normal control (green arrow). (F) Cut section of TTC stained liver tumor post RF ablation, which shows ≥80% tumor necrosis (black arrow) even with the short and low energy RF exposure. Blue arrow indicates the viable tumor tissue. Adapted from. Reprinted with permission from [64]. Copyright (2016) American Chemical Society.

AuNPs was due to necrosis while 1.2 nm AuNPs caused apoptosis of the cells. The toxicity of the 1.4 nm AuNPs was due to the ability to intercalate with DNA while AuNPs of larger sizes were unable to intercalate with the DNA. Size can affect both the distribution within the body as well as the distribution within a cell. Studies of QDs in macrophages have shown that QD size influences subcellular trafficking, with the smallest QDs able to target histones in the cell nucleus. Composition is another factor that influences the toxicity of nanomaterials. QDs may create a health hazard due to toxic heavy metal elements such as cadmium that are incorporated into the QDs. It may, however, be possible to reduce the potential toxicity of nanomaterials such as QDs by adding a coating or nanoshell [70].

Carbon nanotubes (CNTs) are a nanomaterial that has great potential in various medical applications. However, concerns have emerged over its toxicity due to its shape, which resembles asbestos fibers [71]. Longer CNTs have been shown to act

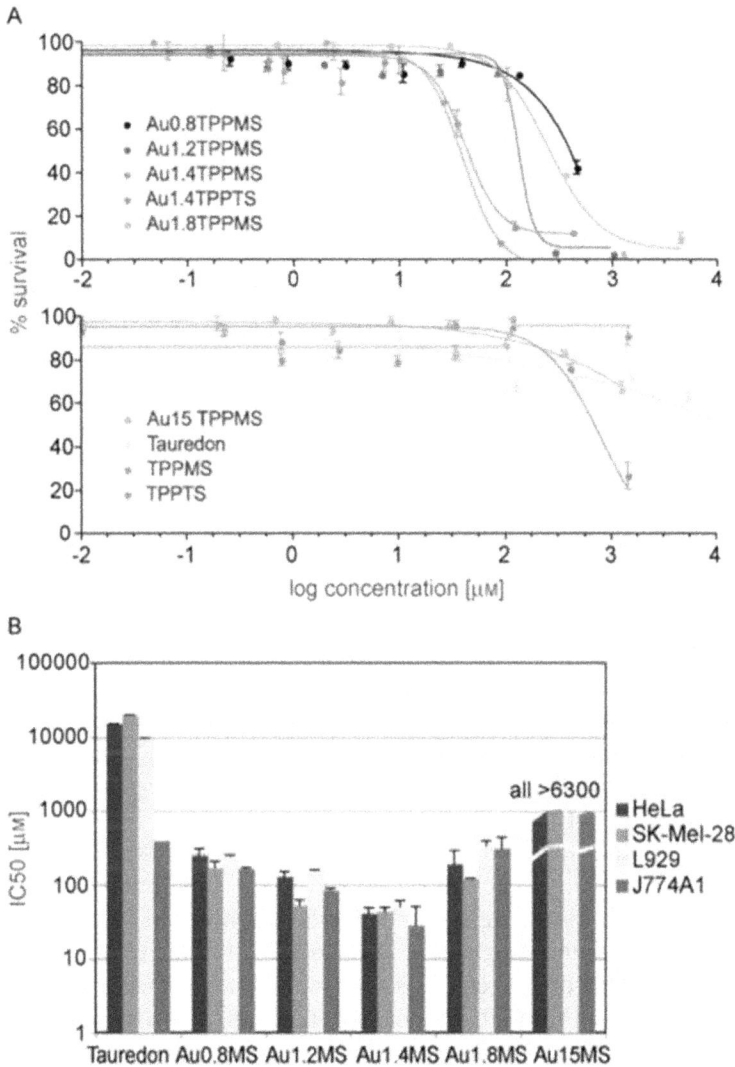

Figure 12.6. Cytotoxicity of Au compounds during the logarithmic growth phase of four cell lines. (A) HeLa cells were seeded at 2000 cells/well and grown for three days into the logarithmic growth phase. Au compounds were added for 48 h and MTT tests were performed as detailed in the experimental section. The logarithmic curve fits of tabulated MTT readings are shown. Each data point represents the mean standard error (SE) of sample triplicates. (B) Note that the IC50 values of Au1.4MS were lowest across all cell lines and that Au compounds of smaller or larger size were progressively less cytotoxic, which suggests a stringent size dependence of cytotoxicity. All concentrations relate to the amount of gold detected by atomic-absorption spectroscope (AAS) in the authentic samples after performing the cytotoxicity test. This procedure ruled out the possibility that cluster synthesis contaminants or dilution errors may have caused erroneous results. Reprinted with permission from [69]. Copyright (2007) WILEY.

like indigestible fibers that lead to frustrated phagocytosis and granuloma formation [72]. Studies in mice have shown that frustrated phagocytosis can lead to massive release of oxygen radicals by immune cells, which can result in chronic granulomatous inflammation and potentially mesothelioma if the CNTs are in the pleural cavity or peritoneum. CNTs can cause mutagenic effects through the generation of inflammation and direct interaction with components of the cell. Exposure of mice to CNTs by inhalation increased the rate of mutation of the K-ras gene locus in the lung, with the mutations occurring during times of maximum inflammation in the tissue. CNTs can also interact directly with the cellular cytoskeleton, including the microtubule system during the formation of the mitotic spindle apparatus, leading to aberrant cell division [73].

Nanomaterials such as titanium dioxide can cause toxicity based on crystalline structure. Cytotoxic studies showed that the anatase form of titanium dioxide was 100 times more toxic than the rutile form, and that the toxicity correlated with the generation of ROS under UV light. Oxidative stress and the generation of ROS is a key injury mechanism that promotes inflammation and atherogenesis, resulting in adverse health events. The surface composition also plays a role in nano-material toxicity. Discontinuous crystal planes and material defects can act as sites for ROS generation. The presence of transition metals or organic chemicals on the surface of nanomaterials can also result in oxygen radical formation and oxidative stress [74].

The degradability of a nanomaterial is another important parameter to consider for toxicity. If nondegradable nanomaterials have no mechanism of clearance from the body, they can accumulate in organs and cells and exert toxic effects. Injectable gold compounds have been used for the treatment of rheumatoid arthritis and the accumulation of gold compounds in the body over time may cause toxic effects in patients. However, biodegradable materials may also cause toxic effects if the degraded components of the material are toxic.

In addition, the nanomaterial charge is a significant contributor to the toxicity of the material. Increased *in vitro* cytotoxicity and *in vivo* pulmonary toxicity has been observed for cationic polystyrene nanospheres when compared with anionic or neutral polystyrene. Interestingly, the mechanism of toxicity for cationic nano-spheres was dependent on the cell type and uptake mechanism. In macrophages, particles entered the cell through phagosomes and caused lysosomal rupture due to the proton sponge effect. Upon entry into the cytosol, the particles caused an increase in Ca^{2+} uptake by mitochondria and oxidative stress, leading to apoptosis. In epithelial cells, cationic particles entered through caveolae. The particles also induced an increase in mitochondrial Ca^{2+} uptake and oxidative stress, but cell death was by necrosis [75].

As new nanomaterials are developed, it is important to consider potential mechanisms of toxicity. Nanomaterials have the increased potential to cross biological barriers and obtain access to tissues and cells as a result of their physicochemical properties. As novel properties are introduced into nanomaterials resulting in new interactions with biological systems, it is possible that new mechanisms of injury and toxicological paradigms might emerge. A further under-standing of how nanomaterials interact with biological systems may provide better methods to engineer nanomaterials to minimize toxicity.

12.3.2 Mass transport

Efficient delivery of nanotherapeutics is another challenge encountered with regards to nanomaterials. The small size of nanoparticles may result in acceleration or delay in their intended action. They may also accumulate non-specifically in certain tissues after administration. Enormous efforts have been expended towards achieving targeted delivery through modification of nanoparticles with antibodies, small molecules, aptamers and/or peptides. However, the biodistribution of nanotherapeutical agents is primarily governed by their ability to negotiate through biological barriers including the mononuclear phagocyte system (MPS), endothelial/epithelial membranes, complex networks of blood vessels, and abnormal blood flow. In addition, drug delivery is further inhibited by barriers such as enzymatic degradation and molecular/ionic efflux pumps that expel drugs from target cells. A full understanding of the interactions between nanomaterials and biological systems will open the door to rational design of nanomedicines and hence improve their biodistribution.

12.3.3 Complexity of nanopharmaceuticals, characterization, stability and storage

To design therapeutics and diagnostics that are functional for personalized use, multiple components will be integrated into a single nanomaterial, requiring multiple steps such as chemical synthesis, formulation and purification. Those procedures will inevitably lower the yield and increase the production cost. In addition, scale-up and manufacturing under current good manufacturing practice (cGMP) will be challenging. In general, multifunctional nanotherapeutics have more variables within their physicochemical properties, which make it more difficult to predict the fate and action after administration. The characterization of nanotherapeutic agents also poses a challenge to manufacturers as well as regulators in terms of chemical, physical, magnetic, optical and biological properties. It would be difficult to monitor a wide range of physicochemical parameters including composition, structure, shape, size, size distribution, concentration, agglomeration, surface functionality, porosity, surface area, surface charge, and surface specification after nanotherapeutic agents are administered.

Stability and storage are also hurdles that must be addressed for clinical practice. For example, biodegradable polymers have been widely used as nanotherapeutic carriers. Depending on their chemical and morphological properties, a polymer will start degrading after nanoparticle formulation in aqueous/organic solvents, which usually results in a change in physicochemical properties (such as agglomeration, particle size, surface charge, drug loading, drug release profile), and can in turn affect the performance *in vivo*. As such, storage conditions may be critical to the shelf life of nanotherapeutics. For example, the measurement result of nanoparticle size, surface charge, polymer degradation rate and drug release profile may be quite different when nanotherapeutics are stored in deionized water, as opposed to phosphate buffered saline (PBS) or human blood serum.

The incorporation of nanomaterials and nanotechnology into personalized medicine also brings up ethical issues. Nanodiagnostics and genetic testing offer the opportunity to collect more personal data on patients than ever before [76]. In particular, the use of point-of-care nanodevices that may bypass a health care professional will have a large impact on mass collection of personal data. This large volume of molecular-level data collected by such nanodevices will challenge the health care system in terms of storage and handling as well as privacy issues, and may raise questions for patients who will receive a torrent of medical information that will inevitably contain false positives and other misleading data.

The advances in nanomaterials and nanobiotechnology will play an important role in the development of cutting-edge diagnostic and therapeutic tools, which are an essential component of personalized medicine. While nanomedicine products face safety, scientific, regulatory and ethical issues, personalized medicine also encounters challenges and obstacles. A major obstacle with personalized approaches such as genetic testing is heterogeneity. A recent study demonstrated that a tumor's genetic makeup can vary significantly within a single tumor [77]. The study showed that, within a single tumor, about 2/3 of the mutations found in a single biopsy were not uniformly detected throughout all the sampled regions of the same patient's tumors. These results elucidated that a single biopsy cannot be considered representative of the landscape of genetic abnormalities in a tumor and that current practices may miss important genetic mutations that could affect the treatment of the disease [78]. Moreover, there were significant differences between mutations in the original tumor and the site of metastasis. The tumor discovered at diagnosis may be very different from the tumor that is growing or exposed to different treatments. However, getting additional biopsies from patients at different stages could be costly and inconvenient for patients. These findings represent a significant challenge for personalized medicine, as the use of genetic testing to direct therapy may be more complex than currently thought.

12.3.4 Economic considerations

The economical conundrums behind the advance of personalized nanomedicine are intricate. On the one hand, given the important resources devoted to the development of complex nanomaterial systems, the choice to focus only on the treatment of a subset of the population (i.e. HER-2 positive breast cancer patients) might be a difficult one to make. The aforementioned risks and challenges associated with the design of nanomaterial remain similar whether it is to treat all patients suffering from cancer or just a cohort showing a specific mutation. Therefore, the financial gain-to-risk ratio strongly leans towards applications which benefit larger populations. On the other hand, the proof of efficacy needed to obtain regulatory approval might be easier to obtain with a system rationally designed for a specific sub-population where the prognosis with standard treatment is particularly grim. The evaluation of therapeutic candidates in patients that are more likely to benefit from it might speed up clinical trials and facilitate regulatory approval of the nanomaterial.

In this context, what makes nanomaterials remarkably appealing is their versatility and the ability to transfer the efforts dedicated to the development of one platform to other applications, liposomes similar to the commercially-available doxorubicin liposomal formulations were recently proposed to act as scavenging nanomaterials for drug detoxification. Similarly, 2-hydroxypropyl-β-cyclodextrin, an excipient which forms nanosized complexes with multiple drugs, was shown to overcome cholesterol metabolism dysfunction in Niemann-Pick Type C. It was approved in 2011 for the intravenous and intrathecal treatment of this very rare LSD [79, 80].

Finally, the development of treatments for orphan or 'niche' diseases might provide attractive entryways to the clinic for nanomaterials. The favorable benefit-to-risk ratio expressly encountered in disorders for which no current treatment exist can prove an efficient way of showing the feasibility of an approach as well as the tolerability and safety of a novel material. In this perspective, scientists at the Children's Hospital of Philadelphia have invested tremendous efforts in developing an adenovirus-based treatment for Leber Congenital Amaurosis (LCA), a very rare degenerative disease which irremediably leads to blindness. This gene delivery vector, which is now in phase II/III for LCA, was developed in parallel with an analogous formulation containing encoding DNA for the human coagulation factor IX, for the treatment of hemophilia B. These examples, showcasing the versatility of drug delivery systems, offer strong support to the future contribution of nano-medicine to personalized medicine [81]. In order to promote the development of nanomedicines into clinically feasible therapies, there is an urgent need for complete characterization of nanomaterial interactions with biological milieus that drive possible toxicological responses. Medical products must be demonstrated to not only be effective but also safe before they are approved for patient use. Some experimental studies have indicated that engineered nanomaterials could exhibit unique toxicological properties in cell culture and in animal models that may not be predicted from the toxicological assessment of the bulk version of the same materials. To establish a database and appropriated standardized protocols for toxicity assessment, the mechanism of nanomaterial-induced toxicity must be fully explored and nanomaterials must be investigated *in vitro* and *in vivo* (e.g. absorption, distribution, metabolism, excretion and toxicological studies) on a particle-by-particle basis.

In parallel, the concept of personalized medicine is also particularly appealing from the perspective of optimizing treatments for an individual patient. Nevertheless, this is a nascent field that has yet to reach its full potential. A potential error may be to succumb to over-enthusiasm and adopt personalized therapeutic practices without strong evidence that personalized treatment is superior to conventional approaches. Even in the field of antibody-based targeted anticancer treatments, which benefited from a head-start in individualized therapies, each clinical or genomic study brings new understanding of the intricate phenomena involved in treating the disease [82]. The understanding of all genomic components of complex diseases like cancer is still unraveling. One should therefore be careful before jumping to conclusions in identifying a particular biomarker as the new ubiquitous target that will eradicate the disease once and for all.

Although significant challenges exist, including regulatory issues and scientific challenges associated with manufacturing nanomedical products, the development and deployment of personalized nanomedicines holds enormous promise for the future treatment of complex diseases. Some nanomedicine products are already in clinical trials, and many others are in various phases of preclinical development. Critical and rational assessment of clinical needs coupled with an improved understanding of physicochemical parameters of nanomaterials that define their effects on the biological system will foster the development of efficient and safe nanomedicine. It is therefore practical to envision a future translation of personalized nanomedicine to the bedside.

12.4 Conclusions and future directions

Advances in cancer nanomedicine can go ahead as an accepted cancer treatment, if the challenges associated with the nanomaterial's interaction with biological systems can be addressed. For all 'smart' nanotheranostic systems, there are several biological impediments that challenge the efficacy of nanoparticle delivery. An ideal nanotheranostic would have to be nonimmunogenic, allow targeted and rapid buildup in tumor tissues, report relevant characteristics of the tissues of interest, deliver effective therapy on-demand, monitor response, initiate secondary treatments, and be biocompatible and biodegradable with innocuous derivatives. An important challenge in the translation of remotely triggered theranostics is that some of the current externally activated systems have very low efficacies with significant toxicities. For example, phototriggered systems have low photothermal conversion, or reactive oxygen species generation efficiency of photosensitizers, with low tumor-to-normal cell ratios during biodistribution studies. An additional issue is the limited tissue dissemination of the radiation that is required for remote initiation that impedes the treatment of deep tissue cancers. Longer wavelength radiation sources can penetrate deeper into the tissue, but the selection of the source is governed by the choice of the photosensitizers. Toxicity is also a major challenge for the clinical translation of inorganic nanoparticle-based remotely triggered nanosystems that may present acute and chronic toxicities. These challenges must be addressed prior to designing a remotely activated theranostic nanosystem.

References

[1] Shahbazi R, Ozpolat B and Ulubayram K 2016 Oligonucleotide-based theranostic nanoparticles in cancertherapy *Nanomedicine* **11** 1287–308
[2] Lim W Q, Phua Z F, Xu S, Sreejith H V and Zhao Y 2016 Recent advances in multifunctional silica based hybrid nanocarriers for bioimaging and cancer therapy *Nanoscale* **8** 12510–9
[3] Park H, Yang J, Seo S, Kim K, Suh J and Kim D *et al* 2008 Multifunctional nanoparticles for photo thermally controlled drug delivery and magnetic resonance imaging enhancement *Small* **4** 192–6

[4] Ali M R K, Panikkanvalappil S R and Sayed M A E 2014 Enhancing the efficiency of gold nanoparticles treatment of cancer by increasing their rate of endocytosis and cell accumulation using rifampicin *J. Am. Chem. Soc.* **136** 4464–7

[5] Kumar S, Harrison N, Kortum R R and Sokolov K 2007 Plasmonic nanosensors for imaging intracellular biomarkers in live cells *Nano Lett.* **7** 1338–43

[6] Buxton D B 2009 Current status of nanotechnology approaches for cardiovascular disease: a personal perspective *Wiley Interdiscip. Rev. Nanomed. Nanobiotechnol.* **1** 149–55

[7] Morris K 2009 Nanotechnology crucial in fighting infectious disease *Lancet Infect. Dis.* **9** 215

[8] Nazem A and Mansoori G A 2008 Nanotechnology solutions for Alzheimer's disease: advances in research tools, diagnostic methods and therapeutic agents *J. Alzheimers Dis.* **13** 199–223

[9] Couvreur P and Vauthier C 2006 Nanotechnology: intelligent design to treat complex disease *Pharm. Res.* **23** 1417–50

[10] Ikeda T and Mitsuyama S 2007 New challenges in the field of breast cancer therapy-do we need surgery for the patients with breast cancer? *Breast Cancer* **14** 37–8

[11] Meurisse M, Defechereux T, Meurisse N, Bataille Y and Hamoir E 2002 New challenges in the treatment of early breast cancer or surgery for early breast cancer … can less be more? *Acta Chir. Belg.* **102** 97–109

[12] Giacomantonio C A and Temple W J 2000 Quality of cancer surgery: challenges and controversies *Surg. Oncol. Clin. N. Am.* **9** 51–60

[13] Funkhouser J 2002 Reinventing pharma: The theranostic revolution *Curr. Drug Discov.* **2** 17–9

[14] Lukianova-Hleb E, Hanna E Y, Hafner J H and Lapotko D O 2010 Tunable plasmonic nanobubbles for cell theranostics *Nanotechnology* **21** 85102

[15] Rai P *et al* 2010 Development and applications of photo-triggered theranostic agents *Adv. Drug Deliv. Rev.* **62** 1094–124

[16] Dolmans D E, Fukumura D and Jain R K 2003 Photodynamic therapy for cancer *Nat. Rev. Cancer* **3** 380–7

[17] Spring B Q *et al* 2016 A photoactivable multi-inhibitor nanoliposome for tumour control and simultaneous inhibition of treatment escape pathways *Nat. Nanotechnol.* **11** 378–87

[18] Lin T *et al* 2016 HSP90 inhibitor encapsulated photo-theranostic nanoparticles for synergistic combination cancer therapy *Theranostics* **6** 1324–35

[19] Liu K *et al* 2012 Covalently assembled NIR nanoplatform for simultaneous fluorescence imaging and photodynamic therapy of cancer cells *ACS Nano* **6** 4054–62

[20] Uppal A, Jain B, Gupta P K and Das K 2011 Photodynamic action of Rose Bengal silica nanoparticle complex on breast and oral cancer cell lines *Photochem. Photobiol.* **87** 1146–51

[21] Tian G *et al* 2013 Red-emitting upconverting nanoparticles for photodynamic therapy in cancer cells under near-infrared excitation *Small* **9** 1929–38

[22] Shi J *et al* 2013 The application of hyaluronic acid-derivatized carbon nanotubes in hematoporphyrin monomethyl ether-based photodynamic therapy for *in vivo* and *in vitro* cancer treatment *Int. J. Nanomed.* **8** 2361–73

[23] Tian J *et al* 2013 Cell-specific and pH-activatable rubyrin-loaded nanoparticles for highly selective near-infrared photodynamic therapy against cancer *JACS* **135** 18850–8

[24] Lin J *et al* 2013 Photosensitizer-loaded gold vesicles with strong plasmonic coupling effect for imaging-guided photothermal/photodynamic therapy *ACS Nano* **7** 5320–9

[25] Feng X *et al* 2015 A novel folic acid-conjugated TiO$_2$–SiO$_2$ photosensitizer for cancer targeting in photodynamic therapy *Colloids Surf.* B **125** 197–205

[26] Chen R, Zhang L and Gao J *et al* 2011 Chemiluminescent nanomicelles for imaging hydrogen peroxide and self-therapy in photodynamic therapy *J. Biomed. Biotechnol.* **2011** 679492

[27] Barth B M *et al* 2011 Targeted indocyanine-green-loaded calcium phosphosilicate nanoparticles for *in vivo* photodynamic therapy of leukemia *ACS Nano* **5** 5325–37

[28] Gary-Bobo M, Hocine O and Brevet D *et al* 2012 Cancer therapy improvement with mesoporous silica nanoparticles combining targeting, drug delivery and PDT *Int. J. Pharm.* **423** 509–15

[29] Gary-Bobo M, Mir Y and Rouxel C *et al* 2012 Multifunctionalized mesoporous silica nanoparticles for the *in vitro* treatment of retinoblastoma: Drug delivery, one and two-photon photodynamic therapy *Int. J. Pharm.* **432** 99–104

[30] Khaing Oo M K, Yang Y and Hu Y *et al* 2012 Gold nanoparticle-enhanced and size-dependent generation of reactive oxygen species from protoporphyrin IX *ACS Nano* **6** 1939–47

[31] Lindner U, Weersink R A and Haider M A *et al* 2009 Image guided photothermal focal therapy for localized prostate cancer: phase I trial *J. Urol.* **182** 57

[32] Lukianova-Hleb E Y, Kim Y S and Belatsarkouski I *et al* 2016 Intraoperative diagnostics and elimination of residual microtumours with plasmonic nanobubbles *Nat. Nanotechnol.* **11** 525–32

[33] Su Y, Wei X and Peng F *et al* 2012 Gold nanoparticles-decorated silicon nanowires as highly efficient near-infrared hyperthermia agents for cancer cells destruction *Nano Lett.* **12** 1845–50

[34] Hong C, Lee J and Zheng H *et al* 2011 Porous silicon nanoparticles for cancer photo-thermotherapy *Nanoscale Res. Lett.* **6** 321

[35] Gobin A M, Lee M H and Halas N J *et al* 2007 Near-infrared resonant nanoshells for combined optical imaging and photothermal cancer therapy *Nano Lett.* **7** 1929–34

[36] Tang S *et al* 2014 Sub-10-nm Pd nanosheets with renal clearance for efficient near-infrared photothermal cancer therapy *Small* **10** 3139–44

[37] Alvarez-Lorenzo C, Bromberg L and Concheiro A 2009 Light-sensitive intelligent drug delivery systems *Photochem. Photobiol.* **85** 848–60

[38] Luo D *et al* 2016 Doxorubicin encapsulated in stealth liposomes conferred with light-triggered drug release *Biomaterials* **75** 193–202

[39] Zhong Y *et al* 2013 Gold nanorod-cored biodegradable micelles as a robust and remotely controllable doxorubicin release system for potent inhibition of drug-sensitive and -resistant cancer cells *Biomacromolecules* **14** 2411–9

[40] Yudina A *et al* 2011 Ultrasound-mediated intracellular drug delivery using microbubbles and temperature-sensitive liposomes *J. Control. Release* **155** 442–8

[41] Ranjan A, Jacobs G C and Woods D L *et al* 2012 Image-guided drug delivery with magnetic resonance guided high intensity focused ultrasound and temperature sensitive liposomes in a rabbit Vx2 tumor model *J. Control. Release* **158** 487–94

[42] Wang C H *et al* 2013 Superparamagnetic iron oxide and drug complex-embedded acoustic droplets for ultrasound targeted theranosis *Biomaterials* **34** 1852–61

[43] Lee S F *et al* 2013 Ultrasound, pH, and magnetically responsive crown-ether-coated core/shell nanoparticles as drug encapsulation and release systems *ACS Appl. Mater. Interfaces* **5** 1566–74

[44] Paris J L *et al* 2013 Polymer-grafted mesoporous silica nanoparticles as ultrasound-responsive drug carriers *ACS Nano* **9** 11023–33

[45] Min H S *et al* 2015 Echogenic glycol chitosan nanoparticles for ultra-sound-triggered cancer theranostics *Theranostics* **5** 1402–18

[46] Moscicka-Studzinska A, Czarnecka K and Ciach T 2008 Electrically enhanced and controlled drug delivery through buccal mucosa *Acta Pol. Pharm.* **65** 767–9

[47] Svirskis D, Travas-Sejdic J, Rodgers A and Garg S 2010 Electrochemically controlled drug delivery based on intrinsically conducting polymers *J. Control. Release* **146** 6–15

[48] Park S J *et al* 2013 New paradigm for tumor theranostic methodology using bacteria-based microrobot *Sci. Rep.* **3** 3394

[49] J, Neofytou E and Cahill T J *et al* 2012 Drug release from electric-field-responsive nanoparticles *ACS Nano* **6** 227–33

[50] Katz E *et al* 2015 Substance release triggered by biomolecular signals in bioelectronic systems *J. Phys. Chem. Lett.* **6** 1340–7

[51] Ge J, Neofytou E and Cahill T J *et al* 2012 Drug release from electric-field-responsive nanoparticles *ACS Nano* **6** 227–33

[52] Hondroulis E *et al* 2014 Immuno nanoparticles integrated electrical control of targeted cancer cell development using whole cell bioelectronic device *Theranostics* **4** 919–30

[53] Jin Z *et al* 2012 Electrochemically controlled drug-mimicking protein release from iron-alginate thin-films associated with an electrode *ACS Appl. Mater. Interfaces* **4** 466–75

[54] Timko B P and Kohane D S 2012 External electric fields trigger drug release from new hydrogel formulation *Nanomedicine* **7** 316

[55] Venkatanarayanan A, Keyes T E and Forster R J 2013 Label-free impedance detection of cancer cells *Anal. Chem.* **85** 2216–22

[56] Edelman E R, Brown L, Taylor J and Langer R 1987 *In vitro* and *in vivo* kinetics of regulated drug release from polymer matrices by oscillating magnetic fields *J. Biomed. Mater. Res.* **21** 339–53

[57] Edelman E R and Langer R 1993 Optimization of release from magnetically controlled polymeric drug release devices *Biomaterials* **14** 621–6

[58] Lee G Y *et al* 2013 Theranostic nanoparticles with controlled release of gemcitabine for targeted therapy and MRI of pancreatic cancer *ACS Nano* **7** 2078–89

[59] Brazel C S 2009 Magnetothermally-responsive nanomaterials: combining magnetic nano-structures and thermally-sensitive polymers for triggered drug release *Pharm. Res.* **26** 644–56

[60] Huang P *et al* 2011 Folic acid-conjugated Silica-modified gold nanorods for X-ray/CT imaging-guided dual-mode radiation and photo-thermal therapy *Biomaterials* **32** 9796–809

[61] Kong T *et al* 2008 Enhancement of radiation cytotoxicity in breast-cancer cells by localized attachment of gold nanoparticles *Small* **4** 1537–43

[62] Starkewolf B, Miyachi L, Wong J and Guo T 2013 X-ray triggered release of doxorubicin from nanoparticle drug carriers for cancer therapy *Chem. Commun.* **49** 2545–7

[63] Wang L *et al* 2015 Radiofrequency-triggered tumor-targeting delivery system for theranos-tics application *ACS Appl. Mater. Interfaces* **7** 5736–47

[64] Somasundaram V H *et al* 2016 Biodegradable radiofrequency responsive nanoparticles for augmented thermal ablation combined with triggered drug release in liver tumors *ACS Biomater. Sci. Eng.* **2** 768–79

[65] De Jong W H and Borm P J 2008 Drug delivery and nanoparticles: applications and hazards *Int. J. Nanomed.* **3** 133–49

[66] Singh N, Manshian B, Jenkins G J, Griffiths S M, Williams P M, Maffeis T G, Wright C J and Doak S H 2009 Nanogenotoxicology: the DNA damaging potential of engineered nanomaterials *Biomaterials* **30** 3891–914

[67] Bhabra G *et al* 2009 Nanoparticles can cause DNA damage across a cellular barrier *Nat. Nanotechnol.* **4** 876–83

[68] Xia T, Li N and Nel A E 2009 Potential health impact of nanoparticles *Annu. Rev. Publ. Health* **30** 137–50

[69] Pan Y, Neuss S, Leifert A, Fischler M, Wen F, Simon U, Schmid G, Brandau W and Jahnen-Dechent W 2007 Size-dependent cytotoxicity of gold nanoparticles *Small* **3** 1941–9

[70] Sandhiya S, Dkhar S A and Surendiran A 2009 Emerging trends of nanomedicine — an overview *Fundam. Clin. Pharm.* **23** 263–9

[71] Poland C A, Duffin R, Kinloch I, Maynard A, Wallace W A, Seaton A, Stone V, Brown S, Macnee W and Donaldson K 2008 Carbon nanotubes introduced into the abdominal cavity of mice show asbestos-like pathogenicity in a pilot study *Nat. Nanotechnol.* **3** 423–8

[72] Dostert C, Petrilli V, Van Bruggen R, Steele C, Mossman B T and Tschopp J 2008 Innate immune activation through Nalp3 inflammasome sensing of asbestos and silica *Science* **320** 674–7

[73] Sargent L M *et al* 2009 Induction of aneuploidy by single-walled carbon nanotubes *Environ. Mol. Mutagen.* **50** 708–17

[74] Nel A E, Diaz-Sanchez D and Li N 2001 The role of particulate pollutants in pulmonary inflammation and asthma: evidence for the involvement of organic chemicals and oxidative stress *Curr. Opin. Pulm. Med.* **7** 20–6

[75] Fischer H C and Chan W C 2007 Nanotoxicity: the growing need for *in vivo* study *Curr. Opin. Biotechnol.* **18** 565–71

[76] Kircher M F, Mahmood U, King R S, Weissleder R and Josephson L 2003 A multimodal nanoparticle for preoperative magnetic resonance imaging and intraoperative optical brain tumor delineation *Cancer Res.* **63** 8122–5

[77] Chen J, Tung C H, Mahmood U, Ntziachristos V, Gyurko R, Fishman M C, Huang P L and Weissleder R 2002 *In vivo* imaging of proteolytic activity in atherosclerosis *Circulation* **105** 2766–71

[78] Weissleder R, Tung C H, Mahmood U and Bogdanov A Jr 1999 *In vivo* imaging of tumors with protease-activated near-infrared fluorescent probes *Nat. Biotechnol.* **17** 375–8

[79] Abi-Mosleh L, Infante R E, Radhakrishnan A, Goldstein J L and Brown M S 2009 Cyclodextrin overcomes deficient lysosome-to-endoplasmic reticulum transport of cholesterol in Niemann–Pick type C cells *Proc. Natl. Acad. Sci. U. S. A.* **106** 19316–21

[80] Liu B, Turley S D, Burns D K, Miller A M, Repa J J and Dietschy J M 2009 Reversal of defective lysosomal transport in NPC disease ameliorates liver dysfunction and neuro-degeneration in the npc1−/− mouse *Proc. Natl. Acad. Sci. U. S. A.* **106** 2377–82

[81] Maguire A M *et al* 2009 Age-dependent effects of RPE65 gene therapy for Leber's congenital amaurosis: a phase 1 dose-escalation trial *Lancet* **374** 1597–605

[82] Guarneri V, Barbieri E and Conte P 2011 Biomarkers predicting clinical benefit: fact or fiction? *J. Natl. Cancer Inst.* **43** 63–6

www.ingramcontent.com/pod-product-compliance
Lightning Source LLC
Chambersburg PA
CBHW082135210326
41599CB00031B/5992